FOURTH CZECHOSLOVAKIAN SYMPOSIUM ON COMBINATORICS, GRAPHS AND COMPLEXITY

ANNALS OF DISCRETE MATHEMATICS 51

General Editor: Peter L. HAMMER
Rutgers University, New Brunswick, NJ, USA

Advisory Editors:

C. BERGE, Université de Paris, France
R.L. GRAHAM, AT&T Bell Laboratories, NJ, USA
M.A. HARRISON, University of California, Berkeley, CA, USA
V. KLEE, University of Washington, Seattle, WA, USA
J.H. VAN LINT, California Institute of Technology, Pasadena, CA, USA
G.C. ROTA, Massachusetts Institute of Technology, Cambridge, MA, USA
T. TROTTER, Arizona State University, Tempe, AZ, USA

FOURTH CZECHOSLOVAKIAN SYMPOSIUM ON COMBINATORICS, GRAPHS AND COMPLEXITY

Edited by

Jaroslav NEŠETŘIL
Department of Applied Mathematics
Charles University
Prague, Czechoslovakia

Miroslav FIEDLER
Mathematics Institute
Czechoslovak Academy of Sciences
Prague, Czechoslovakia

1992

NORTH-HOLLAND – AMSTERDAM • LONDON • NEW YORK • TOKYO

ELSEVIER SCIENCE PUBLISHERS B.V.
Sara Burgerhartstraat 25
P.O. Box 211, 1000 AE Amsterdam, The Netherlands

Library of Congress Cataloging-in-Publication Data

```
Czechoslovakian Symposium on Combinatorics, Graphs, and Complexity
  (4th : 1990 : Prachatice, Czechoslovakia)
    Fourth Czechoslovakian symposium on combinatorics, graphs and
  complexity / J. Nešetřil, M. Fiedler, editors.
        p.    cm. -- (Annals of discrete mathematics ; 51)
    Includes bibliographical references and index.
    ISBN 0-444-89543-4
    1. Combinatorial analysis--Congresses.  2. Graph theory-
  -Congresses.  3. Computational complexity--Congresses.
  I. Nešetřil, J. (Jaroslav), 1946-   .  II. Fiedler, Miroslav.
  III. Title.  IV. Series.
  QA164.C94  1990
  511'.6--dc20                                              92-14160
                                                                CIP
```

ISBN: 0 444 89543 4

© 1992 Elsevier Science Publishers B.V. All rights reserved.

No part of this publication may be reproduced, stored in a retrieval system or transmitted, in any form or by any means, electronic, mechanical, photocopying, recording or otherwise, without the prior written permission of the publisher, Elsevier Science Publishers B.V., Copyright & Permissions Department, P.O. Box 521, 1000 AM Amsterdam, The Netherlands.

Special regulations for readers in the U.S.A. – This publication has been registered with the Copyright Clearance Center Inc. (CCC), Salem, Massachusetts. Information can be obtained from the CCC about conditions under which photocopies of parts of this publication may be made in the U.S.A. All other copyright questions, including photocopying outside of the U.S.A, should be referred to the copyright owner, Elsevier Science Publishers B.V., unless otherwise specified.

No responsibility is assumed by the publisher for any injury and/or damage to persons or property as a matter of products liability, negligence or otherwise, or from any use or operation of any methods, products, instructions or ideas contained in the material herein.

p.p. 59-62, 141-144, 315-320: Copyright not transferred

This book is printed on acid-free paper.

Printed in The Netherlands

Preface

This book contains the Proceedings of the Fourth Czechoslovak Symposium on Combinatorics held in Prachatice, a medieval town in South Bohemia. The conference was attended by 170 mathematicians and computer scientists from Czechoslovakia and abroad. The list of participants is included in this volume.

The Prachatice-meeting was the fourth in the series which began with the Smolenice meeting held in 1963. This was perhaps the first truly international meeting on Graph Theory. In particular for the first time specialists from East and West were able to come together. Its Proceedings "Theory of Graphs and its Applications", were published in 1964 jointly by Academia, Prague and Academic Press, New York. The second symposium was held in 1974 in Prague under the title "Recent Advances in Graph Theory". This was a much larger meeting than the smaller Smolenice meeting.

The third symposium was held in 1982 in Prague and its Proceedings were published in 1983 by Teubner, Leipzig under title "Graphs and other Combinatorial Topics". A total of 119 participants took part in this meeting.

The organizing committee wishes to thank all the participants for contributing to the success of the 1990 Symposium and especially the authors for their contributions. We hope that the participants as well as other researchers will find the present volume useful.

These proceedings are being published by North Holland in the Annals of Discrete Mathematics series. Each participant in the Symposium received a prepublication copy of the volume.

We give thanks to dr. P. Liebl (Prague) and dr. A. Sevenster (North Holland Amsterdam) for their efforts and efficiency which have made this volume possible.

Prague, September 12th, 1991

M. Fiedler

J. Nešetřil

Prachatice — sv. Jakub

Contents

Preface v

J. ABRHAM, A. KOTZIG †: Two Sequences of 2-Regular Graceful Graphs Consisting of 4-gons 1

D. ARCHDEACON: A Survey of Self-Dual Polyhedra 5

M. BAČA: On Magic Labellings of Convex Polytopes 13

V. BÁLINT: A Packing Problem and Geometrical Series 17

R. BODENDIEK AND K. WAGNER: On the Bananas Surface B_2 23

O. V. BORODIN: Structural Properties and Colorings of Plane Graphs ... 31

M. BOROWIECKI: The Binding Number of Graphs 39

P. BUGATA: Note on Algorithmic Solvability of Trahtenbrot-Zykov Problem ... 45

G. BUROSCH, P. V. CECCHERINI: Cartesian Dimensions of a Graph 51

D. CIESLIK: The Steiner Minimal Tree Problem in L_p^2 59

T. DVOŘÁK: On k-Connected Subgraphs of the Hypercube 63

P. ERDŐS: On Some of My Favourite Problems in Various Branches of Combinatorics 69

D. FRONČEK: Realizability of Some Starlike Trees 81

H. GROPP: The Construction of All Configurations $(12_4, 16_3)$ 85

M. HORŇÁK: (p,q)-realizability of Integer Sequences with Respect to Möbius Strip 93

O. HUDEC: Vertex Location Problems 103

A. J. C. HURKENS, C. A. J. HURKENS, R. W. WHITTY: On Generation of a Class of Flowgraphs 107

J. IVANČO: The Weight of a Graph 113

F. JAEGER: On the Kauffman Polynomial of Planar Matroids 117

S. JENDROĽ: On Symmetry Groups of Selfdual Convex Polyhedra 129

V. JURÁK: A Remark on 2-(v,k,λ) Designs 137

M. KOEBE: On a New Class of Intersection Graphs 141

URSZULA KONIECZNA: Asymptotic Normality of Isolated Edges in Random Subgraphs of the n-Cube 145

A. V. Kostochka, L. S. Mel'nikov: On Bounds of the Bisection Width of Cubic Graphs 151

A. V. Kostochka, A. A. Sapozhenko, K. Weber: On Random Cubical Graphs 155

J. Kratochvíl, J. Nešetřil, O. Zýka: On the Computational Complexity of Seidel's Switching 161

A. Kundrík: The Harmonious Chromatic Number of a Graph 167

A. Kurek: Arboricity and Star Arboricity of Graphs 171

C. Lin, W. D. Wallis, Zhu Lie: Extended 4-Profiles of Hadamard Matrices 175

M. Loebl, S. Poljak: Good Family Packing 181

Z. Lonc: Solution of an Extremal Problem Concerning Edge-Partitions of Graphs 187

T. Luczak, A. Ruciński: Balanced Extensions of Spare Graphs 191

A. Malnič and B. Mohar: Two Results on Antisocial Families of Balls 205

D. Marušič: Hamiltonicity of Vertex-transitive pq-Graphs 209

A. Meir and J. W. Moon: On Nodes of Given Out-Degree in Random Trees 213

E. Mendelsohn, N. Shalaby, Shen Hao: All Leaves and Excesses Are Realizable for $k = 3$ and All λ 223

Danuta Michalak: The Binding Number of k-Trees 229

P. Mihók: An Extension of Brook's Theorem 235

L. Nebeský: On Sectors in a Connected Graph 237

V. Nýdl: Irreconstructability of Finite Undirected Graphs from Large Subgraphs 241

C. H. Papadimitriou: On Inefficient Proofs of Existence and Complexity Classes 245

C. H. Papadimitriou and Martha Sideri: Optimal Coteries on a Network 251

J. Plesník: On Some Heuristics for the Steiner Problem in Graphs 255

A. Raspaud: Cycle Covers of Graphs with a Nowhere-Zero 4-Flow (Abstract) 259

A. Recski: Minimax Results and Polynomial Algorithms in VLSI Routing 261

D. G. Rogers: Critical Perfect Systems of Difference Sets 275

Anna Rycerz: Some Operations (Not) Preserving the Integer Rounding Property 281

Petra Scheffler: Optimal Embedding of a Tree into an Interval Graph in Linear Time 287

W. SCHÖNE: Construction of Polytopal Graphs	293
J. J. SEIDEL: More About Two-Graphs	297
J. SHEEHAN AND C. R. J. CLAPHAM: These are the Two-free Trees	309
S. K. SIMIĆ: A Note on Reconstructing the Characteristic Polynomial of a Graph	315
Z. SKUPIEŃ: Exponential Constructions of Some Nonhamiltonian Minima	321
M. SONNTAG: Hamiltonicity of Products of Hypergraphs	329
M. TKÁČ: Non-Hamiltonian Simple 3-Polytopal Graphs with Edges of Only Two Types	333
S. V. TSARANOV: On Spectra of Trees and Related Two-Graphs	337
V. VETCHÝ: Metrically Regular Square of Metrically Regular Bigraphs	341
M. WOŹNIAK: Embedding of Graphs in the Complements of Their Squares	345
A. Z. ZELIKOVSKY: An $\frac{11}{6}$-Approximation Algorithm for the Steiner Problem on Graphs	351
B. ZELINKA: Distances Between Graphs (Extended Abstract)	355
B. ZELINKA: Domatic Number of a Graph and its Variants (Extended Abstract)	363
A. A. ZYKOV: The Space of Graphs and its Factorizations	371
Problems Proposed at the Problem Session of the Prachatice Conference on Graph Theory	375
List of Participants	385
Name Index	389
Subject Index	397

This is a prepublication copy of *Combinatorics, Graphs, Complexity*.
The final version of this book is published by North Holland
as a volume of *Annals of Discrete Mathematics*.
North Holland is the sole distributor of this book.

Typeset by TeX. All pictures were created by METAFONT.

Two Sequences of 2-Regular Graceful Graphs Consisting of 4-gons

Jaromir Abrham, Anton Kotzig†

A *graceful numbering* (*valuation*) of a graph G with m vertices and n edges is a one-to-one mapping ψ of $V(G)$ into the set $\{0, 1, 2, \ldots, n\}$ with the following property: If we define, for any $e \in E(G)$ with the end vertices u, v, the value $\bar\psi(e)$ of the edge e by $\bar\psi(e) = |\psi(u) - \psi(v)|$ then $\bar\psi$ is a one-to-one mapping of $E(G)$ onto the set $\{1, 2, \ldots, n\}$. A graph is called *graceful* if it has a graceful valuation.

An *α-valuation* ψ of a graph G is a graceful valuation of G which has the following additional property: There exists a number γ ($0 \leqslant \gamma < |E(G)|$) such that for any edge e with the end vertices u, v, it is

$$\min[\psi(u), \psi(v)] \leqslant \gamma < \max[\psi(u), \psi(v)].$$

The term "graceful valuation" was introduced by S. W. Golomb [4]; however, the concepts of a graceful valuation (under the name β-valuation) and of an α-valuation were introduced by A. Rosa [8]. In the same paper, Rosa proved the following theorem: *If an eulerian graph G is graceful then $|E(G)| \equiv 0$ or $3 \pmod 4$.* This implies that if an eulerian graph G has an α-valuation it is $|E(G)| \equiv 0 \pmod 4$. In this theorem, an eulerian graph is any graph in which the degree of each vertex is positive and even; it does not have to be connected.

In particular, the above theorem holds for 2-regular graphs. It is well known that the condition of the above theorem is also sufficient for 2-regular connected graphs (cycles) (see Kotzig [5], Rosa [8]) and for 2-regular graphs with two isomorphic components (Kotzig [6] proved that a 2-regular graph consisting of two s-cycles (s even) has an α-valuation). A partial extension to the case of three components can be found in the same paper; in this case, the condition of the above theorem is not always sufficient. Some of the results concerning graceful graphs can also be found in Bosák's book [3]. A relation between graceful valuations of 2-regular graphs and Skolem sequences was studied in [1]. More recently, it has been proved in [2] that the number of α-valuations of $4k$-cycles and the number of graceful valuations of $(4k + 3)$-cycles grow at least exponentially with k. Some additional results on graceful graphs can be found in [7].

For the special case of 2-regular graphs consisting of 4-gons, it is known that a graph consisting of k 4-gons has an α-valuation for $1 \leqslant k \leqslant 10$, $k \neq 3$; for $k = 3$, this graph is graceful but it does not have an α-valuation (see [6]).

In the remaining part of this paper, the symbol Q_k will denote the 2-regular graph consisting of k 4-gons. The main result of this paper is the following.

Theorem. *The graphs Q_{n^2} and Q_{n^2+n} have α-valuations for every positive integer n.*

Proof. **1. The graph Q_{n^2}**

We will decompose Q_{n^2} into subgraphs in the following way:
$$Q_{n^2} = [Q_1^1 Q_2^1 Q_3^1 \ldots Q_{n-1}^1 Q_n Q_{n-1}^2 Q_{n-2}^2 \ldots Q_3^2 Q_2^2 Q_1^2];$$
for $1 \leqslant k \leqslant n-1$, Q_k^1, Q_k^2 denote the first and the second occurence of Q_k, respectively. Clearly, it is sufficient to give the values of the vertices (in cyclic order) for each 4-gon in each subgraph Q_k^1, Q_k^2, and in Q_n.

The values of the vertices end edges of the 4-gons in Q_k^1:

4-gon #1: The values of the vertices (in cyclic order) are:
$(k-1)^2$, $4n^2 - k^2 + 1$, $k^2 + 1$, $4n^2 - k^2 - k + 1$;
the resulting values of the edges are:
$4n^2 - 2k^2 + 2k$, $4n^2 - 2k^2$, $4n^2 - 2k^2 - k$, $4n^2 - 2k^2 + k$.

4-gon #2: Vertices:
$(k-1)^2 + 2$, $4n^2 - k^2 + 2$, $k^2 + 3$, $4n^2 - k^2 - k + 2$.
Edges:
$4n^2 - 2k^2 + 2k - 1$, $4n^2 - 2k^2 - 1$, $4n^2 - 2k^2 - k - 1$,
$4n^2 - 2k^2 + k - 1$.

$\vdots$

4-gon #k: Vertices:
$(k-1)^2 + 2(k-1)$, $4n^2 - k^2 + k$, $k^2 + 2k - 1$, $4n^2 - k^2$.
Edges:
$4n^2 - 2k^2 + k + 1$, $4n^2 - 2k^2 - k + 1$, $4n^2 - 2k^2 - 2k + 1$,
$4n^2 - 2k^2 + 1$.

For $k = n$, we obtain the values of the vertices and edges of the 4-gons in Q_n. The values of the 4-gons in each Q_k^2 can be obtained from those in the corresponding Q_k^1 according to the following rule: If a, b, c, d are the values of the vertices in a 4-gon in Q_k^1 such that $a > \gamma$, $d > \gamma$, $b \leqslant \gamma$, $c \leqslant \gamma$ (where γ is the number from the definition of an α-valuation) then the values of the vertices of the corresponding 4-gon in Q_k^2 will be $a - 2n^2 + 2k^2$, $b + 2n^2 - 2k^2$, $c + 2n^2 - 2k^2$, $d - 2n^2 + 2k^2$.

Example 1. We will give the values of the vertices in Q_{16}. Q_{16} will be expressed in the form $[Q_1^1\ Q_2^1\ Q_3^1\ Q_4\ Q_3^2\ Q_2^2\ Q_1^2]$. We have:

Q_1^1: $0, 64, 2, 63$

Q_2^1: $1, 61, 5, 59;\ 3, 62, 7, 60$

Q_3^1: $4, 56, 10, 53;\ 6, 57, 12, 54;\ 8, 58, 14, 55$

Q_4: $9, 49, 17, 45;\ 11, 50, 19, 46;\ 13, 51, 21, 47;\ 15, 52, 23, 48$

Q_3^2: $18, 42, 24, 39;\ 20, 43, 26, 40;\ 22, 44, 28, 41$

Q_2^2: $25, 37, 29, 35;\ 27, 38, 31, 36$

Q_1^2: $30, 34, 32, 33$.

2. The graph Q_{n^2+n}

Using the same notation as in the previous case we can decompose Q_{n^2+n} into subgraphs in the following way:

$$Q_{n^2+n} = [Q_1^1 Q_2^1 Q_3^1 \ldots Q_{n-1}^1 Q_n^1 Q_n^2 Q_{n-1}^2 \ldots Q_3^2 Q_2^2 Q_1^2]$$

Let us put $\delta_{nk} = 4(n^2+n) - (k-1)^2$. Then each subgraph Q_k^1 ($1 \leqslant k \leqslant n$) will be numbered in the following way (we give the values of the vertices in a cyclic order):

4-gon #1: $\delta_{nk} - 2k + 2,\ k^2,\ \delta_{nk} - 4k + 2,\ k^2 - k$
4-gon #2: $\delta_{nk} - 2k + 4,\ k^2 + 1,\ \delta_{nk} - 4k + 4,\ k^2 - k + 1$
$\vdots$
4-gon #k: $\delta_{nk},\ k^2 + k - 1,\ \delta_{nk} - 2k,\ k^2 - 1$

The subgraphs Q_k^2 ($1 \leqslant k \leqslant n$) will be numbered as follows:

4-gon #1: $2n^2 + k^2 + 2n + 2,\ 2n^2 - k^2 + 2n + 1,\ 2n^2 + k^2 + 2n - 2k + 2,\ 2n^2 - k^2 + 2n - k + 1$
4-gon #2: $2n^2 + k^2 + 2n + 4,\ 2n^2 - k^2 + 2n + 2,\ 2n^2 + k^2 + 2n - 2k + 4,\ 2n^2 - k^2 + 2n - k + 2$
$\vdots$
4-gon #k: $2n^2 + k^2 + 2n + 2k,\ 2n^2 - k^2 + 2n + k,\ 2n^2 + k^2 + 2n,\ 2n^2 - k^2 + 2n.$

Example 2. We will give the values of the vertices in Q_{20}. Q_{20} will be represented in the form $[Q_1^1\ Q_2^1\ Q_3^1\ Q_4^1\ Q_4^2\ Q_3^2\ Q_2^2\ Q_1^2]$.

Q_1^1: $0, 80, 1, 78$
Q_2^1: $2, 77, 4, 73;\ 3, 79, 5, 75$
Q_3^1: $6, 72, 9, 66;\ 7, 74, 10, 68;\ 8, 76, 11, 70$
Q_4^1: $12, 65, 16, 57;\ 13, 67, 17, 59;\ 14, 69, 18, 61;\ 15, 71, 19, 63$
Q_4^2: $21, 58, 25, 50;\ 22, 60, 26, 52;\ 23, 62, 27, 54;\ 24, 64, 28, 56$
Q_3^2: $29, 51, 32, 45;\ 30, 53, 33, 47;\ 31, 55, 34, 49$
Q_2^2: $35, 46, 37, 42;\ 36, 48, 38, 44$
Q_1^2: $39, 43, 40, 41.$

□

Acknowledgements. The research was sponsored by NSERC grants No. A7329 and A9232.

References

[1] J. Abrham, *Graceful 2-regular graphs and Skolem sequences*, Working paper 87-8, March 1987, Department of Industrial Engineering, University of Toronto. To appear in Annals of Discrete Mathematics, Volume on Advances in Graph Labelings.

[2] J. Abrham and A. Kotzig, *Estimates of the number of graceful valuations of cycles*, Working paper 89-08, December 1989, Department of Industrial Engineering, University of Toronto. Submitted to Congressus Numerantium.

[3] J. Bosák, *Rozklady grafov (Decompositions of Graphs)*, Veda, Bratislava, 1986.

[4] S. W. Golomb, *How to number a graph*, Graph Theory and Computing (C. R. Read, ed.), Academic Press, N.Y., 1972, pp. 23–37.

[5] A. Kotzig, *On decomposition of a complete graph into 4k-angles*; in Russian, Matematický časopis **15** (1965), 229–233.

[6] ———, *β-valuations of quadratic graphs with isomorphic components*, Utilitas Mathematica **7** (1975), 263–279.

[7] ———, *Recent results and open problems in graceful graphs*, Congressus Numerantium **44** (1984), 197–219.

[8] A. Rosa, *On certain valuations of the vertices of a graph*, Théorie des graphes — Journées internationales d'étude Rome, juillet 1966, Dunod, Paris, 1967, pp. 349–355.

Jaromir Abrham
Dept. of Industrial Engineering,
University of Toronto,
Toronto, Ontario, Canada M5S 1A4

Anton Kotzig †
deceased in Spring 1991

Fourth Czechoslovakian Symposium on
Combinatorics, Graphs and Complexity
J. Nešetřil and M. Fiedler (Editors)
© 1992 Elsevier Science Publishers B.V. All rights reserved.

A Survey of Self-Dual Polyhedra

DAN ARCHDEACON

This paper surveys recent results on self-dual polyhedra. These results include a complete set of constructions for spherical self-dual polyhedra. We discuss similar constructions for projective-planar polyhedra. We next give a construction of involutory polyhedra for general surfaces. We describe an additive construction and a recursive construction. We give some specific results concerning embeddings where both the primal and dual graphs are either complete graphs or complete bipartite graphs. Finally, we give some directions for future research.

§1 Introduction

Let G be a graph cellularly embedded in some surface. Then G has a natural geometric dual G^*. Each face f of the embedded G corresponds to a vertex v_f of G^*. Each edge e of G incident with say faces f and g of G corresponds to an edge $e^* = v_f v_g$ in G^*.

Suppose that G is a simple graph of minimum degree at least 3. Furthermore suppose that G is embedded so that no two faces of G have a multiply connected union. Then we call this embedded graph a *polyhedron*. The definition is analogous to that of a *spherical polyhedron* (a 3-connected plane graph under the equivalence due to Steinitz [Ste]), except that here we allow an arbitrary surface. Note that the graph of any polyhedron is 3-connected [B]. Also note that the dual of any polyhedron is itself a polyhedron [B].

Our primary interest is:

The Main Problem. *Construct and classify all self-dual polyhedra.*

In this paper we describe recent progress towards the main problem. Most results are stated without proof. For details we refer the reader to the indicated citations.

The paper is organized as follows. In Section 2 we give some necessary background material and rephrase the main problem. Section 3 focuses on self-dual spherical polyhedra; Section 4 on self-dual projective polyhedra. In Section 5 we describe a general quotient construction. In Section 6 we give an addition construction, and in Section we give 7 a recursive construction. In Section 8 we describe some duality relations between complete and complete bipartite graphs. Finally, in Section 9 we offer some concluding remarks.

§2 Background Material

In this section we introduce the requisite background material. The reader is referred to [AR] where most of the material is developed in more detail. Also [GT] discusses the general theory of embedded graphs and coverings thereof.

Let G_1 and G_2 be graphs cellularly embedded in some surface. A *map isomorphism* is a graph isomorphism φ from G_1 to G_2 which preserves facial walks. Such a map isomorphism extends to an isomorphism between the surfaces. In particular, the two surfaces are homeomorphic. Note that it is not sufficient that the graphs alone be isomorphic, even when the two ambient surfaces are identical.

We now define the *radial graph*, $R(G)$, of an embedded graph G. The vertices of R are the vertices and faces of G. An edge of the radial graph joins two vertices of R which represent incident elements of G. So the radial graph is naturally bipartite with the vertex bipartition induced by vertices versus faces of G. Observe that $R(G) = R(G^*)$, where G^* denotes the dual of the embedded G. Note also that R embeds in a natural manner in the same surface as does G. In particular, the faces of this embedding are quadrilaterals and are in one-to-one correspondence with the edges of G. Conversely, any bipartite quadrangulation is the radial graph of some dual pair G, G^*.

Let φ be a map isomorphism between an embedded graph G and its dual G^*. Then φ induces an automorphism of the radial graph R, also denoted φ, which swaps the vertex parts. Conversely, any part-reversing face-preserving automorphism of R induces a self-duality map on the underlying G. Call such an automorphism φ a *radial self-duality map*.

If G is a polyhedron, then R is a *clean quadrangulation*; that is, the set of faces is exactly the set of quadrilaterals. In this case any part-reversing automorphism of R is a map automorphism, and hence is a radial self-duality map.

Our main problem may be reformulated as:

The main problem reformulated. *Construct and classify all bipartite quadrangulations with a part-reversing map automorphism.*

§3 Spherical Polyhedra

In this section we consider self-dual polyhedra whose underlying surface is the sphere (see also [SC, Sh, McK]). We first offer a typical construction.

Planar Construction

We begin with a nonbipartite quadrangulation H of the projective plane; for example, K_4 embeds in the projective plane with 3 quadrilateral faces. It is well known that the projective plane has a 2-fold covering by the sphere. In fact, the quotient map which identifies antipodal points on the sphere is such a covering. Consider the embedded graph R on the sphere formed by lifting H using this covering. Then R has every face a quadrilateral, and since the face boundaries generate the $\mathbb{Z}_2$-cycle space of the sphere, R is bipartite. In particular R is the radial map of some polyhedron. Since H was nonbipartite, the map which swaps the two vertices in a fiber (the antipodal points) is part-reversing. It follows that this deck-transformation is a radial self-duality map. From this R it is easy to recover the self-dual polyhedron. For example, the lifting of the aforementioned embedding of K_4 is the 3-dimensional cube, Q_3. In turn, Q_3 is the radial graph of the spherical self-dual embedding of the tetrahedron. (It is only a coincidence that the graph involved is again K_4.) We note that any radial self-duality map thus constructed is of order 2.

In [AR] the authors gave a set of six possible constructions for spherical self-dual polyhedra. The other five constructions are similar in nature to the above. They then proved the following.

Spherical Classification Theorem. *Any self-dual spherical polyhedron comes from one of the six constructions.*

We note that each of the six constructions are necessary in the statement of the Classification Theorem. The authors have examples for each of the constructions which yield polyhedra that cannot arise from any of the other constructions.

Grünbaum and Shephard [GS] noted that it had been incorrectly stated that the square of a self-duality map φ was always the identity. But even though "the dual of the dual is the original," φ^2 need not be the identity. In [GS] they give an example of a self-duality map φ whose square is not the identity, although their example did admit a second self-duality map which was of order 2. They defined the *rank* of a self-dual polyhedron as the minimum order of all its duality maps, and they asked if every polyhedron was of rank 2. This question was negatively answered by Jendrol' [J] who gave a polyhedron of rank 4. The problem was completed by McCanna [McC], who found a polyhedron of rank 2^n for every n, and showed that these were the only possible ranks. The problem was independently solved by Archdeacon and Richter [AR]. The classification theorem above also yields some insight into the nature of large rank polyhedra; for example, if the rank exceeds 4 then the polyhedron is chiral.

§4 Projective Polyhedra

It is natural to look for a classification theorem for self-dual polyhedra on other surfaces, analogous to the spherical classification theorem in the previous section. The next "simplest" surface after the sphere is the real projective plane. Archdeacon and Negami [AN] have found two constructions for self-dual projective polyhedra and have shown that any such polyhedron must arise from one of these constructions. We do not describe the constructions here, but we do note the following corollary.

Projective Order Corollary. *Any self-duality map of a projective polyhedron is of order 2 or 4.*

We thus have specific sets of constructions for self-dual polyhedra on the sphere and the projective plane. A natural question is the generalization of our results to other surfaces. For example, could one find all self-dual 3-connected graphs on the torus? On Klein's bottle? The author suspects that there are some additional difficulties with surfaces of nonpositive Euler characteristic.

§5 A General Quotient Construction

We now sketch a construction for self-dual polyhedra on an arbitrary surface. The construction here is a generalization of that given in Section 3. It is presented in full detail in [A2].

General Construction

We begin with an embedded graph H together with a distinguished set of faces $\mathcal{F}$ such that **(a)** no vertex has more than one incidence with faces in $\mathcal{F}$, and **(b)** every face not in $\mathcal{F}$ is a quadrilateral. We define a graph $G = G(H)$ as follows. The vertex set of G is the vertex set of H. Two vertices of G are joined if and only if either **(a)** they are on opposite corners of a quadrilateral face not in $\mathcal{F}$, or **(b)** they are joined by an edge in the boundary cycle of a face in $\mathcal{F}$.

We next describe a self-dual embedding of G. Let $f_1, \ldots, f_k$ be the faces of the embedded H incident with a vertex v of G. Let F_v be the set of edges of G of two types: **(a)** the diagonal of f_i not incident with v, or **(b)** the edges of $f \in \mathcal{F}$ incident with v. Then F_v induces a simple cycle in G. Moreover, the set $\{F_v : v \in V(G)\}$ forms the set of faces for an embedding of G. (It is straightforward to check that every edge of G is in exactly two such F_v, a little harder to check that these faces give a cyclic rotation at each vertex.) This embedding of G is always self-dual, the involutory self-duality map swaps the vertex v with the face F_v for each $v \in V(G)$.

In general the constructed G need not be a polyhedron—it need not even be connected. Necessary and sufficient conditions for yielding a polyhedron are given in [A2]. The planar construction given in §3 is (non-obviously) the special case where H is embedded in the real projective plane and $\mathcal{F}$ is the empty set. A more general form of the construction given in [A2], together with variations of the upcoming addition construction, can be shown to give all possible involutory self-dual polyhedra.

§6 The Addition Construction

In this section we give a way to "add" two self-dual embeddings and obtain a third self-dual embedding. The resulting graph is a union of the two initial graphs. Specifically, let G_1 be a graph with vertices $v_1, \ldots, v_k$ and let G_2 be a graph with vertices $u_1, \ldots, u_k$. Define the *amalgamation* $G = G_1 \underset{k}{\cup} G_2$ along these vertices as the graph formed from the disjoint union of G_1 and G_2 by pairwise identifying $v_1 = u_1, v_2 = u_2, \ldots, v_k = u_k$.

We need the following definition. Let G be an embedded graph and let $\varphi: G \to G^*$ be a graph isomorphism. A vertex $v \in V(G)$ is called *reflexive* if it is incident with its dual face $\varphi(v)$.

The Addition Construction. *Let G_1 be a self-dual embedded graph with reflexive vertices $v_1, \ldots, v_k$ and let G_2 be a self-dual embedded graph with reflexive vertices $u_1, \ldots, u_k$. Then there exists a self-dual embedding of $G = G_1 \underset{k}{\cup} G_2$. Moreover, the vertices $v_1 = u_1, \ldots, v_k = u_k$ are all reflexive in this embedding.*

As an example we begin with an orientable self-dual embedding of $K_{4,4}$ on the torus such that each of the four vertices in the first part is reflexive (such an embedding exists). Using the addition construction we can add two such embeddings and obtain a self-dual embedding of $K_{4,8}$. In fact, we can add n such embeddings to get a self-dual embedding of $K_{4,4n}$.

§7 A Recursive Construction

In this section we describe a recursive method for constructing involutory self-dual polyhedra. It will be slightly easier to describe our process as a reduction algorithm, one which starts with a self-dual map and modifies it to obtain a smaller self-dual map. (By smaller we mean one with fewer vertices.) By a sequence of such reductions we will eventually arrive at a cellular map with but a single vertex and a single face. The promised recursive construction will arise by reversing the reduction steps.

Let e be an edge of a self-dual G with a map automorphism $\varphi: G \to G^*$. Define its *mate*, $\bar{e}$, as the edge of G dual to $\varphi(e)$. So that if e is incident with vertices u and v, then $\bar{e}$ is incident with faces $\varphi(u)$ and $\varphi(v)$. But by assumption any map automorphism is of order 2. Thus if $\varphi(\bar{e})$ is incident with vertices a and b, then e is incident with faces $\varphi(a)$ and $\varphi(b)$. In other words, mating is a symmetric relation. Note that an edge may be mated with itself, call such an edge *self-mated*.

We now describe the first of our reduction operations. Suppose that one of the two edges, say e, is a link (i.e., $u \neq v$) and further suppose that $e \neq \bar{e}$. Define the *squash* of G by $G' = G \backslash \bar{e}/e$, that is, as the graph formed by deleting $\bar{e}$ and contracting e. If G' is a squash of G then we say that we obtain G by *cleaving* G'. So G is formed by adding in an edge across a face and splitting the corresponding mated vertex. Note that G is self-dual if and only if G' is self-dual, as deleting $\bar{e}$ removes an adjacency between a and b and merges $\varphi(u)$ and $\varphi(v)$ into one face, while contracting e removes an adjacency between $\varphi(a)$ and $\varphi(b)$ and merges u and v into one vertex.

The second of our reduction operations is *trimming*. Suppose that v is a degree two vertex of a self-dual G which is incident with its dual face $\varphi(v)$. Then v is incident with a single other face, which implies that $\varphi(v)$ is a digon. Let u be the other vertex adjacent to v, so that $\varphi(u)$ is the other incident face. The *trim* G' is formed by deleting v and its two incident edges, while merging the face $\varphi(v)$ into $\varphi(u)$. The reverse of trimming is called *dangling*, where we can only dangle from u in the face $\varphi(u)$. As before, it is easy to check that G is self-dual if and only if G' is.

The following theorem is proved in [A2].

Recursive Theorem. *The set of involutory self-duality maps are exactly those formed from a map with one vertex by some sequence of cleavings and/or danglings.*

Servatius [Se] first introduced this operation to generate *all* self-dual plane graphs together with their self-duality maps.

§8 Duality Relations Between Complete and Complete Bipartite Graphs

We begin by examining self-dual embeddings of the complete graph. In 1898 Heffter [H] constructed orientable self-dual embeddings of K_n for n prime and congruent to 1 modulo 4. Then in 1973 White [W] constructed orientable self-dual embeddings of Cayley graphs, including K_n for every n congruent to 1 modulo 4. Pengelley [P] completed the solution for the orientable K_n in 1975 by constructing self-dual embeddings for n divisible by 4. An Euler characteristic calculation will reveal that these are the only two congruence classes for which these embeddings are possible, so the question is solved in general.

We next examine self-dual embeddings of complete bipartite graphs. The size of a face in an embedded graph is equal to the degree of the corresponding vertex in the dual. It follows that if a graph is bipartite, then every vertex in the dual is of even degree. Also note that if a graph has a vertex of degree 2, then the dual graph has parallel edges. So that a self-dual embedding of $K_{n,m}$ can exist only if both n and m are even integers exceeding 2. The following theorem from [AH] asserts that these necessary conditions are sufficient, with one exception.

Theorem. $K_{n,m}$ *has both an orientable and a nonorientable self-dual embedding for all even integers n and m exceeding 2, except that there is no orientable self-dual embedding of $K_{6,6}$.*

The preceding Theorem is proven using a form of the Addition Construction together with orientable self-dual embeddings of $K_{4,4}$, $K_{4,6}$, $K_{6,10}$, and nonorientable self-dual embeddings of $K_{4,4}$, $K_{4,6}$, and $K_{6,6}$.

Starting with these self-dual embeddings and using covering constructions we can prove [A1] something much stronger.

Theorem. *Let p, q, r, and s be even integers at least 4 with $pq = rs$. Then there exists an orientable and a nonorientable embedding of $K_{p,q}$ with dual $K_{r,s}$, except that there is no orientable self-dual embedding of $K_{6,6}$.*

We close by asking what dualities exist among complete and complete bipartite graphs. Let n, p, q be integers with n odd and $2pq = n(n-1)$. When does there exist an embedding of K_n with dual $K_{p,q}$? For example, the 6 hexagons (783156), (795432), (891264), (473682), (493581), (752961), together with the 6 hexagons (781245), (892356), (794613), (476958), (143625), and (172839) give an embedding of K_9 in the sphere with 17 crosscaps having dual $K_{6,6}$. Aside from several other small examples, and an infinite class for n prime based on a difference construction, little is known here.

Conclusion

We have made considerable progress towards our main problem of constructing and/or characterizing self-dual polyhedra. But much remains to be done.

Of particular interest are self-dual embeddings of Cayley graphs. Particularly nice are the works of Stahl [Sta], and of Hartsfield, Marušič, and Ringel [HMR]. Hartsfield asks:

Query. *Does every group of order at least 4 have some Cayley graph with a self-dual embedding?*

The author conjectures that the answer is *yes*.

We ask what is the general form of the recursive construction for noninvolutory self-duality maps.

Finally, we ask what duality relationships exist among complete multipartite graphs (see also [Sta]). In [A1] some constructions are given which (among other things) lift self-dual embeddings of complete graphs K_n to self-dual embeddings of regular complete multipartite graphs $K_{n(m)}$ for infinitely many values of m. How can these embeddings be combined with the Addition Theorem?

Acknowledgment

The author thanks Richard Wilson for the embedding of K_9 dual to $K_{6,6}$. He also thanks Nora Hartsfield, Bruce Richter, Rick Vitray, and Andy Woldar for helpful discussions.

References

[A1] D. Archdeacon, *The medial graph and voltage-current duality*, preprint (1990).

[A2] _____, *The construction of involutory self-dual polyhedra*, in preparation.

[AH] D. Archdeacon and N. Hartsfield, *Self-dual embeddings of complete multipartite graphs*, preprint (1989).

[AN] D. Archdeacon and S. Negami, *The construction and classification of self-dual projective polyhedra*, in preparation.

[AR] D. Archdeacon and B. Richter, *The construction and classification of self-dual polyhedra*, preprint (1989).

[B] D. W. Barnette, *Graph theorems for manifolds*, Israel J. Math. **16** (1973), 62–72.

[GS] B. Grünbaum and G. C. Shephard, *Is selfduality involutory?*, Amer. Math. Monthly **95** (1988), 729–733.

[GT] J. L. Gross and T. W. Tucker, *Topological Graph Theory*, John Wiley & Sons, New York, 1987.

[HMR] N. Hartsfield, D. Marušič, and G. Ringel, *Nonorientable self-dual embeddings of Cayley graphs*, preprint (1988).

[H] L. Heffter, *Ueber metacyklische Gruppen und Nachbarconfigurationen*, Math. Ann. **50** (1898), 261–268.

[J] S. Jendrol', *A non-involutory selfduality*, Discrete Mathematics **74** (1989), 325–326.

[McC] J. McCanna, *Selfduality is not involutory*, preprint (1989).

[McK] T. McKee, *Neighborhood and self-dual multigraphs*, preprint (1988).

[P] D. Pengelley, *Self-dual orientable embeddings of K_n*, J. of Comb. Th. Ser. B **18** (1975), 46–52.

[Sh] H. Shank, *personal communication*.

[SC] B. Servatius and P. R. Christopher, *Construction of self-dual graphs*, American Mathematical Monthly (to appear).

[Se] B. Servatius, *NSF-Report*.

[Sta] S. Stahl, *Self-dual embeddings of Cayley graphs*, J. of Comb. Th. Ser. B **27** (1979), 92–107.

[Ste] E. Steinitz, *Polyeder und Raumeinteilungen*, Enzykl. math. Wiss. Vol. 3 (Geometrie), Part 3AB12 (1922), 1–139.

[W] A. White, *Orientable embeddings of Cayley graphs*, Duke Math. J. **41** (1974), 353–371.

Dan Archdeacon
Department of Mathematics and Statistics,
University of Vermont,
Burlington, VT, USA 05405
Email: "archdeac@uvm.edu"

Fourth Czechoslovakian Symposium on
Combinatorics, Graphs and Complexity
J. Nešetřil and M. Fiedler (Editors)
© 1992 Elsevier Science Publishers B.V. All rights reserved.

On Magic Labellings of Convex Polytopes

MARTIN BAČA

The paper describes magic labellings of type $(1,1,0)$ for graphs of two classes of convex polytopes which are obtained by a combination of vertex and edge labellings.

1. Introduction

The notion of magic labelling of plane graphs was defined by Ko-Wei Lih [4]. However, the subject can be traced back to the 13th century when similar notions were investigated by Yang Hui (1275) and later by Chang Chhao (1670), Pao Chhi-Shou (1880) and Li Nien (1935), (see [4]).

Magic labellings of type $(1,1,0)$ for wheels, friendship graphs, prisms and some of the Platonic polyhedra are given in [4]. Magic labellings for antiprisms and m-antiprisms are described in [1] and [2].

2. Notation

Let $G = (V, E)$ be a finite connected plane graph which contains no loops and no parallel edges.

A *labelling of type* $(1,1,0)$ is a bijection from the set $\{1, 2, \ldots, |V(G)|+|E(G)|\}$ onto the vertices and edges of the plane graph G. If we label only vertices or only edges, we call such a labelling a *vertex labelling* or an *edge labelling*, respectively. The *weight of a face* under a labelling is the sum of the labels of vertices and edges surrounding that face.

A labelling is said to be *magic* if, for every integer s, all s-sided faces have the same weight. We allow different weights for different values of s.

This notion of magicality is different from the definition given by Sedláček in [5]. However, a magic edge labelling of a plane graph, in our sense, is equal to a supermagic labelling of the plane dual graph G^* of G as defined, for instance, in [6].

3. Statement of the main results

For $n \geq 5$ by R_n we denote the graph of the convex polytope which is obtained as a combination of the graph of a prism and the graph of an antiprism (Fig. 1). The prism and the antiprism are Archimedean convex polytopes defined e.g. in [3].

We make the convention that $a_{n+1} = a_1$, $b_{n+1} = b_1$, $c_{n+1} = c_1$ and $d_{n+1} = d_1$ to simplify the notation.

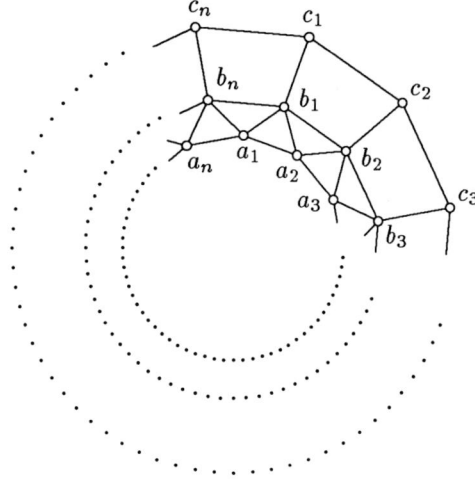

FIGURE 1

Theorem 1. *For $n \geqslant 5$ the graph of the convex polytope R_n has a magic labelling of type $(1, 1, 0)$.*

Proof. We construct an edge labelling f_1 and a vertex labelling f_2 of R_n in the following way.

$$\begin{aligned}
f_1(b_i a_{i+1}) &= \tfrac{4}{3}|V(R_n)| - i & &\text{if} & 1 \leqslant i \leqslant n-1 \\
&= \tfrac{4}{3}|V(R_n)| & &\text{if} & i = n \\
f_1(b_i c_i) &= \tfrac{5}{3}|V(R_n)| + 1 & &\text{if} & i = 1 \\
&= 2|V(R_n)| - i + 2 & &\text{if} & 2 \leqslant i \leqslant n \\
f_1(c_i c_{i+1}) &= i - 1 & &\text{if} & 2 \leqslant i \leqslant n \\
&= \tfrac{1}{3}|V(R_n)| & &\text{if} & i = 1 \\
f_1(a_i a_{i+1}) &= |V(R_n)| - i + 1 & f_2(a_i) &= |E(R_n)| + i \\
f_1(a_i b_i) &= \tfrac{4}{3}|V(R_n)| + i & f_2(b_i) &= \tfrac{2}{3}|V(R_n)| + |E(R_n)| - i + 1 \\
f_1(b_i b_{i+1}) &= \tfrac{1}{3}|V(R_n)| + i & f_2(c_i) &= \tfrac{2}{3}|V(R_n)| + |E(R_n)| + i
\end{aligned}$$

for $i = 1, 2, \ldots, n$.

It is simple to verify that the values of f_1 and f_2 are $1, 2, \ldots, |V(R_n)| + |E(R_n)|$ and further that the common weight of all 3-sided faces is $\tfrac{1}{3}|V(R_n)| + 5|E(R_n)| + 3$ and the weight of all 4-sided faces is $|V(R_n)| + 7|E(R_n)| + 4$.

For the weights of the n-sided faces we have

$$\sum_{i=1}^{n}\bigl(f_1(a_i a_{i+1}) + f_2(a_i)\bigr) = \sum_{i=1}^{n}\bigl(f_1(c_i c_{i+1}) + f_2(c_i)\bigr) = $$
$$= n\bigl(|V(R_n)| + |E(R_n)| + 1\bigr).$$

Clearly, such a labelling of the graph of the convex polytope R_n is a magic labelling of type $(1,1,0)$. □

In the sequel we shall investigate labellings of the graph of the convex polytope Q_n consisting of 3-sided faces, 4-sided faces, 5-sided faces and n-sided faces. (Fig.2).

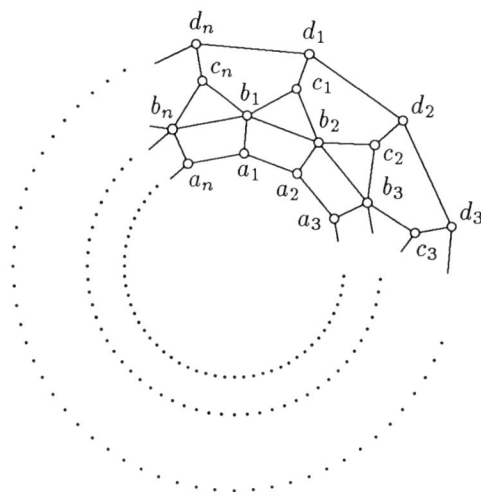

FIGURE 2

Theorem 2. For $n \geq 6$ the graph of the convex polytope Q_n has a magic labelling of type $(1,1,0)$.

Proof. Define the vertex labelling f_3 as follows.

$$f_3(a_i) = |E(Q_n)| + i \qquad f_3(c_i) = \tfrac{3}{4}|V(Q_n)| + |E(Q_n)| + i$$
$$f_3(b_i) = \tfrac{3}{4}|V(Q_n)| + |E(Q_n)| - i + 1 \qquad f_3(d_i) = \tfrac{1}{2}|V(Q_n)| + |E(Q_n)| - i + 1$$

for $i = 1, 2, \ldots, n$ and further define the edge labelling f_4:

$$\begin{aligned}
f_4(a_i a_{i+1}) &= \tfrac{1}{4}|V(Q_n)| + i + 1 & \text{if} &\quad 1 \leq i \leq n-1 \\
&= \tfrac{1}{4}|V(Q_n)| + 1 & \text{if} &\quad i = n \\
f_4(c_i b_{i+1}) &= |V(Q_n)| + i + 1 & \text{if} &\quad 1 \leq i \leq n-1 \\
&= |V(Q_n)| + 1 & \text{if} &\quad i = n \\
f_4(a_i b_i) &= \tfrac{3}{4}|V(Q_n)| - i + 1 & f_4(b_i b_{i+1}) &= \tfrac{3}{4}|V(Q_n)| + i \\
f_4(b_i c_i) &= \tfrac{3}{2}|V(Q_n)| - i + 1 & f_4(d_i d_{i+1}) &= \tfrac{1}{4}|V(Q_n)| - i + 1 \\
f_4(c_i d_i) &= \tfrac{3}{2}|V(Q_n)| + i & \text{for } i = 1, 2, \ldots, n.
\end{aligned}$$

It is easy to see that the edge labelling f_4 uses each integer $1, 2, \ldots, |E(Q_n)|$ and the vertex labelling f_3 uses each integer $|E(Q_n)|+1, |E(Q_n)|+2, \ldots, |E(Q_n)|+|V(Q_n)|$ exactly once.

It is not difficult to check that the common weight of all 3-sided faces is $9|V(Q_n)| + |E(Q_n)| + 3$, for all 4-sided faces it is $4|V(Q_n)| + 4|E(Q_n)| + 4$ and for all 5-sided faces it is $9|V(Q_n)| + 5|E(Q_n)| + 5$ and further that the weight of the external n-sided face is equal to the weight of the internal n-sided face and is $n\big(4|V(Q_n)| - |E(Q_n)| + 1\big)$. This proves that the graph of the convex polytope Q_n has a magic labelling of type $(1, 1, 0)$. □

References

[1] M. Bača, *Labellings of Two Classes of Convex Polytopes*, Utilitas Math. **34** (1988), 24–31.
[2] _____, *Labellings of m-Antiprisms*, Ars Combinat. **28** (1989), 242–245.
[3] E. Jucovič, *Convex Polyhedra*, Veda, Bratislava, 1981. (Slovak)
[4] Lih Ko-Wei, *On Magic and Consecutive Labellings of Plane*, Graphs. Utilitas Math. **24** (1983), 165–197.
[5] J. Sedláček, *Problem 27*, Theory of Graphs and Its Applications. Proc. Symposium Smolenice, June 1963, Praha, 1964, pp. 163–167.
[6] B. M. Stewart, *Supermagic Complete Graphs.*, Canad. J. Math. **19** (1967), 427–438.

Martin Bača
Department of Mathematics,
Technical University, Košice,
Czechoslovakia

Fourth Czechoslovakian Symposium on
Combinatorics, Graphs and Complexity
J. Nešetřil and M. Fiedler (Editors)
© 1992 Elsevier Science Publishers B.V. All rights reserved.

A Packing Problem and Geometrical Series

Vojtech Bálint

In the paper [1] the following problem is posed: What is the area R of the smallest rectangle in which can be packed the set of rectangles of total area 1 and sides of lengths $\frac{1}{n}$ and $\frac{1}{n+1}$ for $n = 1, 2, \ldots$? It can be shown that $R \leqslant \frac{113}{96}$. Is $R > 1$? In this paper we are going to show $R < 1.0024$.

The estimate $R \leqslant \frac{113}{96}$ is strengthened to $R \leqslant \left(\frac{31}{30}\right)^2$ in [2], where the packing into a square is considered, and a further improvement is required: "It would be desirable to find the smallest value of $\varepsilon \geqslant 0$ such that all these rectangles can be packed into a square of side $1 + \varepsilon$. Whether $\varepsilon > 0$ or $\varepsilon = 0$ is an open question." The problem is repeated in [3], too.

We denote by P_k the rectangle with sides $\frac{1}{k}$ and $\frac{1}{k+1}$ for $k = 1, 2, 3, \ldots$.

Lemma 1. *Let r be a natural number. The system $\{P_k \colon k = r, r+1, \ldots, 2r-1\}$ of r rectangles can be packed into the rectangle of sides 1 and $\frac{1}{r}$.*

Proof. The rectangle of sides 1 and $\frac{1}{r}$ we divide into r equal squares with sides $\frac{1}{r}$. Into each of these squares we can pack one of the rectangles. □

Theorem 1. *If the $r - 1$ largest rectangles $P_1, P_2, \ldots, P_{r-1}$ can be packed into the unit square, then the set of rectangles*

$$\frac{1}{n} \times \frac{1}{n+1}, \quad n = 1, 2, 3, \ldots \tag{1}$$

can be packed into a rectangle with sides of lengths 1 and R, where $R \leqslant 1 + \frac{2}{r}$.

Proof. Denote $r_i = r \cdot 2^i$ for $i = 0, 1, 2, \ldots$. The rectangles $P_{r_i}, P_{r_i+1}, \ldots, P_{2r_i-1}$ we pack into the rectangle $1 \times \frac{1}{r_i}$ according to Lemma 1. Adding the shorter sides of the strips, we obtain the asked result $\sum_{i=0}^{\infty} \frac{1}{r_i} = \frac{2}{r}$. □

If the problem is to pack the system (1) into a square, then we pack $P_1, P_2, \ldots, P_{r-1}$ into a unit square, the rectangles $P_r, P_{r+1}, \ldots, P_{2r-1}$ into the strip $1 \times \frac{1}{r}$ (see Fig. 1), and the union of all further strips $1 \times \frac{1}{r_i}$,

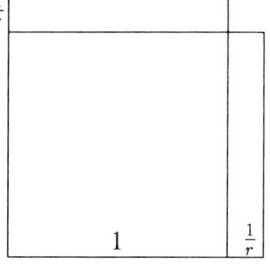

Figure 1

17

$i = 1, 2, \ldots$ gives $1 \times \frac{1}{r}$, too, i.e. the system (1) can be packed into a square $(1 + \frac{1}{r})^2$. Therefore it remains to answer the question, how large is r? It is easily seen that $r \geqslant 100$ and with some patience $r \geqslant 500$, and this strengthens the estimate $R \leqslant (\frac{31}{30})^2 \doteq 1.0677$ to $R \leqslant (\frac{501}{500})^2 = 1.004004$. (Of course, r can be still increased). For the first mentioned question on packing into a rectangle we have $R < 1 + \frac{2}{500} = 1.004$, but we are still able to strengthen this estimate.

Lemma 2. *Let $k \geqslant 4$ be an arbitrary natural number. Let $r = 4k$. The set $\{P_i : i = r, r+1, \ldots, 6r - 1\}$ of $5r$ rectangles can be packed into the rectangle $1 \times \frac{1}{r}$.*

Proof. We divide the rectangle $1 \times \frac{1}{r}$ into r equal squares. Into these squares we pack the rectangles $P_{4k}, P_{4k+1}, \ldots, P_{24k-1}$ so that their longer side is parallel with the longer side of the rectangle $1 \times \frac{1}{r}$. For the sake of brevity we show the packing algorithm only by a figure and write the list of packed rectangles. Due to technical reasons the rectangle P_m is denoted by m only.

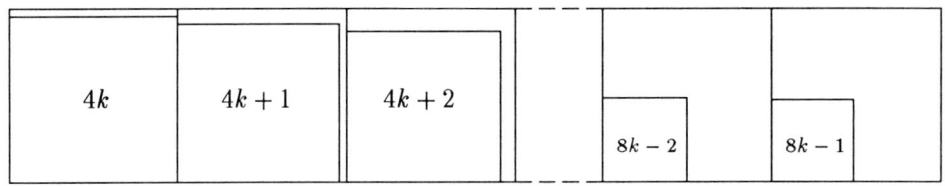

FIGURE 2

The $P_{4k}, P_{4k+1}, \ldots, P_{8k-1}$ are packed.

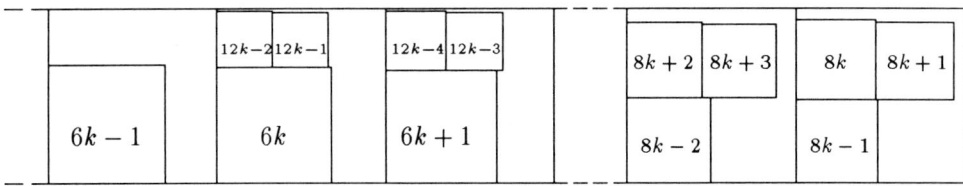

FIGURE 3

The $P_{8k}, P_{8k+1}, \ldots, P_{12k-1}$ are packed.

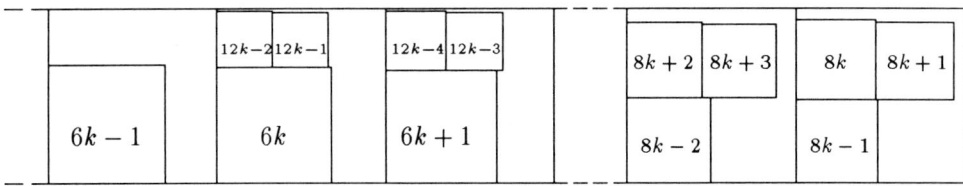

FIGURE 4

The $P_{12k}, P_{12k+1}, P_{12k+2}, \ldots, P_{16k}, P_{16k+1}$ are packed. (Here, $z = 16k$.)

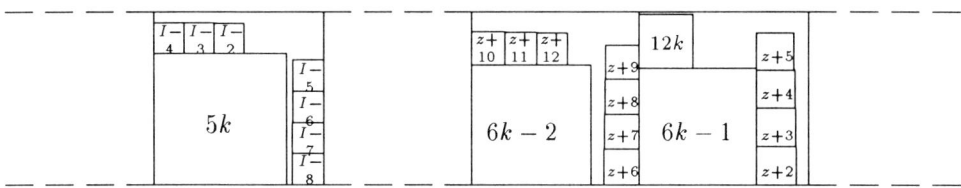

FIGURE 5

The $P_{16k+2}, \ldots, P_{16k+5}$ and also $P_{16k+6}, P_{16k+7}, \ldots, P_{23k-2}$ are packed. (Here, $I = 23k$.)

We omit the proofs of possibility of these packings, because they lead to inequalities of at most second order.

The rectangle P_{23k-1} we place above the rectangle P_{5k-1}.

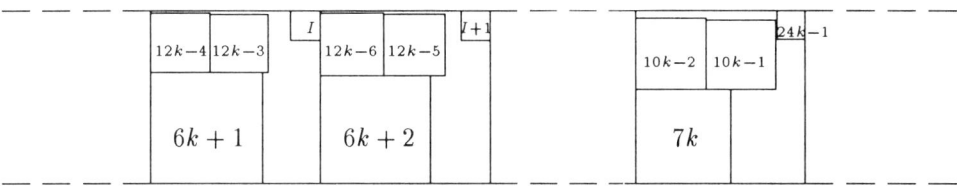

FIGURE 6

The $P_{23k}, P_{23k+1}, \ldots, P_{24k-1}$ are packed. It is necessary to prove that

$$\frac{1}{12k-2j-2} + \frac{1}{12k-2j-1} + \frac{1}{23k+j-1} \leq \frac{1}{4k} \quad \text{for } j = 1, 2, \ldots, k, \quad (2)$$

which is equivalent to

$$0 \leq 2j^3 + (22k+1)j^2 - (248k^2 - 61k + 2)j + 264k^3 - 228k^2 + 31k - 1. \quad (3)$$

The right hand side of (3) we denote by $f(j)$ and find the minimum of function $f(j)$ for $j \in \langle 1, k \rangle$. It holds $\dfrac{df(j)}{dj} = 6j^2 + 2(22k+1)j - (248k^2 - 61k + 2)$. The equation of stationary points $\dfrac{df(j)}{dj} = 0$ is quadratic with roots

$$j_{1,2} = \frac{-22k - 1 \pm \sqrt{1\,972k^2 - 322k + 13}}{6} \quad (4)$$

where $D = 1\,972k^2 - 322k + 13 > 0$ for $k \geq 4$. Trivially $j_2 = \dfrac{-22k - 1 - \sqrt{D}}{6} < 0$ holds. On the other side $198k^2 - 63k + 2 > 0$ holds for $k \geq 4$. From this we get

$$1\,188k^2 - 378k + 12 > 0$$
$$1\,972k^2 - 322k + 13 > 784k^2 + 56k + 1$$
$$\sqrt{D} > 28k + 1$$
$$-22k - 1 + \sqrt{D} > 6k$$
$$j_1 = \frac{-22k - 1 + \sqrt{D}}{6} > k.$$

So $\frac{df(j)}{dj} < 0$ for $j \in \langle j_2, j_1 \rangle$ and, of course, also for $j \in \langle 1, k \rangle$. The function f decreases on the interval $\langle 1, k \rangle$ and so it has the minimum $f_{\min}(k) = 2k^3 + (22k + 1)k^2 - (248k^2 - 61k + 2)k + 264k^3 - 228k^2 + 31k - 1 = 40k^3 - 166k^2 + 29k - 1 = g(k)$.

Further $\frac{dg(k)}{dk} = 120k^2 - 322k + 29$ and

$$\frac{dg(k)}{dk} = 0 \quad \Leftrightarrow \quad k_{1,2} = \frac{166 \pm \sqrt{24\,076}}{120} \doteq \begin{cases} 2.67 \\ 0.09 \end{cases}$$

For $k \geqslant 4$, $\frac{dg(k)}{dk} > 0$ holds, g is increasing and $g_{\min}(4) = 19$. Hence $g(k) > 0$ for $k \geqslant 4$, from that $f_{\min}(k) > 0$ and so $f(j) > 0$ for $k \geqslant 4$ and for every $j \in \{1, 2, \ldots, k\}$. $\square$

Theorem 2. *Let $k \geqslant 4$ be an arbitrary natural number. Let $r = 4k$. If the $r - 1$ largest rectangles $P_1, P_2, \ldots, P_{r-1}$ can be packed in the unit square, then $R \leqslant 1 + \frac{6}{5r}$.*

Proof. Similarly, as in the proof of Theorem 1, we obtain $\frac{1}{r} + \frac{1}{r_1} + \frac{1}{r_2} + \ldots = \frac{1}{r} + \frac{1}{6r} + \frac{1}{36r} + \ldots = \frac{6}{5r}$. $\square$

Remark 1. *Theorem 2 and $r \geqslant 500$ give $R < 1.0024$.*

Remark 2. *Coxeter in (4) asked the question: "Find the radius of the smallest circle inside which discs of radius $1/n$ ($n = 1, 2, 3, \ldots$) can all be packed."*

This problem was solved in (5). Another solution we can easily obtain from the above mentioned statements. By Lemma 1, the circles of radii $\frac{1}{k}, \frac{1}{k+1}, \ldots, \frac{1}{2k-1}$ can be packed into the rectangle of sides 2 and $\frac{2}{k}$. By repeated application of this we obtain that all discs of radii $\frac{1}{k}, \frac{1}{k+1}, \ldots$ can be packed into a rectangle of sides 2 and $\frac{4}{k}$. So the discs of radii $\frac{1}{20}, \frac{1}{21}, \ldots$ we pack into a rectangle of sides 2 and $\frac{1}{5}$. This rectangle and the largest 19 discs can be comfortably placed into a circle of radius $\varrho = \frac{3}{2}$. The inequality $\varrho \geqslant \frac{3}{2}$ is trivial. Therefore $\varrho = \frac{3}{2}$.

References

[1] J. W. Moon, L. Moser, *Some packing and covering theorems*, Colloquim mathematicum **17** (1979), 103–110.

[2] A. Meir, L. Moser, *On packing of squares and cubes*, Journal of Combinatorial Theory **5** (1968), 126–134.

[3] W. Moser, *Research Problems of Discrete Geometry*, 1986.

[4] H. S. M. Coxeter, *Problem P-276*, Canad. Math. Bull. **22** (1979), 248.

[5] D. V. Boyd, *Solution to problem 276*, Canad. Math. Bull. **23** (1980), 251–252.

Vojtech Bálint
Katedra matematiky VŠDS,
01026 Žilina,
Czechoslovakia

Fourth Czechoslovakian Symposium on
Combinatorics, Graphs and Complexity
J. Nešetřil and M. Fiedler (Editors)
© 1992 Elsevier Science Publishers B.V. All rights reserved.

On the Bananas Surface B_2

R. BODENDIEK AND K. WAGNER

1. Introduction

Let Γ denote the set of all finite undirected graphs without loops and multiple edges. Let B_2 be the bananas surface arising from the torus by contracting two different meridians of the torus to a simple point, each. These two points are called the singular points of B_2 and denoted by s_1 and s_2. There is a further definition of B as a pseudosurface in the way that B_2 is the 2-amalgamation of two spheres sharing the two singular points s_1 and s_2. Finally, let $\Gamma(B_2)$ be the set of all graphs in Γ which do not embed in B_2. According to [4] we define five elementary relations R_i, $i = 0, 1, 2, 3, 4$ in the following way:

Let $G, H \in \Gamma$ (or $\Gamma(B_2)$). Then it holds:

(0) $(G, H) \in R_0$ or $R_0(G) = H$ means that H arises from G by deleting an edge or an isolated vertex of G.

(1) $(G, H) \in R_1$ iff H arises from G by contracting an edge $e = \{v_1, v_2\} \in E(G)$ with the property that at least one of the two endpoints v_1, v_2 of e has degree 1 or 2, and that v_1, v_2 do not have a common neighbour in G.

(2) In a similar way, R_2 means the contraction of an edge $e = \{v_1, v_2\}$ in G, where v_1 and v_2 have degree at least 3 and do not have a common neighbour in G.

(3) $R_3(G) = H$ means that H arises from G by substituting for the trihedral $\{v\} * \{v_1, v_2, v_3\}$ the triangle $\{v_1, v_2, v_3, v_1\}$ provided that v_1, v_2, v_3 are independent in G.

(4) $R_4(G) = H$ iff H arises from G by substituting for the double trihedral $\{v_1, v_2\} * \{v\} * \{u\} * \{u_1, u_2\}$ the double triangle $K_2 * \{u = v\} * K_2'$, where K_2, K_2' are two copies of the complete graph of order 2 with $V(K_2) = \{v_1, v_2\}$ and $V(K_2') = \{u_1, u_2\}$ provided that the two edges $\{v_1, v_2\}$, $\{u_1, u_2\}$ do not belong to $E(G)$.

By means of these five elementary relations we define the following five partial ordering relations $>_i \subseteq \Gamma \times \Gamma$ (or $\subseteq \Gamma(B_2) \times \Gamma(B_2)$) in the following way.

For $G, H \in \Gamma$ (or $\Gamma(B_2)$) we define: $(G, H) \in >_i \subseteq \Gamma \times \Gamma$ (or $\Gamma(B_2) \times \Gamma(B_2)$) or $G >_i H$, $i = 0, 1, 2, 3, 4$, iff either $G = H$ or there is a sequence of graphs $G_1, G_2, \ldots, G_n \in \Gamma$, $n \geq 2$, such that $G_1 = G$, $G_n = H$ and $(G_\nu, G_{\nu+1}) \in R_{j_\nu}$, $j_\nu \in \{0, \ldots, i\}$ for $\nu = 1, 2, \ldots, n - 1$.

It is obvious that $G >_0 H$ means $G \supseteq H$, $>_1$ is the well-known subdivision relation and that $>_2$ represents the Wagner homomorphism.

For each $i \in \{0, \ldots, 4\}$ we are able to define the so-called minimal basis $M_i(B_2) = \{G \in \Gamma(B_2) \mid G$ is $>_i$-minimal$\}$. According to [4] it is easy to see that the following inclusion chains are true:

$$>_0 \subseteq >_1 \subseteq >_2 \subseteq >_3 \subseteq >_4 \tag{a}$$

and

$$M_4(B_2) \subseteq M_3(B_2) \subseteq M_2(B_2) \subseteq M_1(B_2) \subseteq M_0(B_2). \tag{b}$$

2. A Kuratowski-type theorem for B_2

In November 1988, we asked J. Širáň from Bratislava to help us to decide the question whether there exists a Kuratowski-type theorem for B_2 in the sense that there is a finite set $M_i(B_2)$ for some $i = 1, 2, 3, 4$ such that the following assertion is true:

$$\bigwedge_{G \in \Gamma} (G \in \Gamma(B_2) \leftrightarrow \bigvee_{H \in M_i(B_2)} G >_i H). \tag{c}$$

It is obvious that this Kuratowski-type theorem can only be true if $>_i$, $i = 1, 2, 3, 4$, preserves the embedding of graphs in B_2 in the sense that the following proposition is true:

$$\bigwedge_{G, H \in \Gamma} (G \notin \Gamma(B_2) \wedge G >_i H \to H \notin \Gamma(B_2)). \tag{d}$$

K. Wagner, R. Bodendiek and J. Širáň could independently prove that $>_1$ satisfies (d) whereas $>_i$, $i = 2, 3, 4$, does not. Instead of (d) it holds:

$$\bigvee_{G, H \in \Gamma} (G \notin \Gamma(B_2) \wedge G >_i H \wedge H \in \Gamma(B_2)). \tag{e}$$

This proof of existence is not very difficult for we only need to find two graphs $G, H \in \Gamma$ satisfying (e). In [2] it is verified that the two graphs $G = R_0(G_{85}) =$

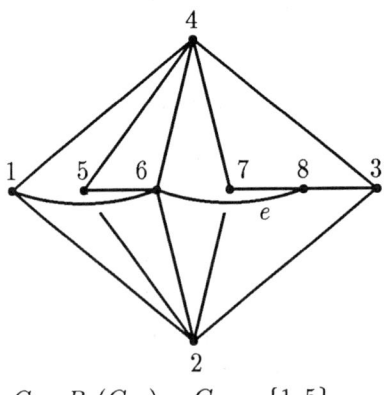

$G = R_0(G_{85}) = G_{85} - \{1, 5\}$

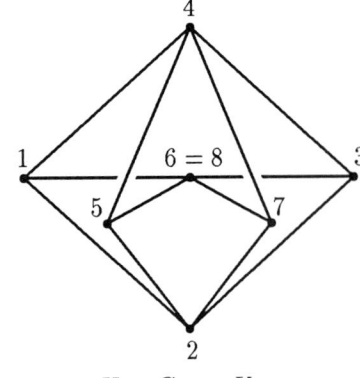

$H = G_{86} = K_{3,4}$

FIGURE 1

$G_{85} - \{1,5\}$ and $H = G_{86} = K_{3,4}$ depicted in Figure 1 make (e) true because of the following facts: At first $H = G_{86} = K_{3,4}$ belongs to $M_2(B_2) \subseteq M_1(B_2)$. Secondly, $G_{85} = G - \{1,5\}$ is an element of $M_1(B_2) - M_2(B_2)$ with the property that $G = G_{85} - \{1,5\}$ embeds in B_2. Finally, $G >_2 H$ because H arises from G by removing the two edges $\{2,6\}$ and $\{4,6\}$ (i.e. applying R_0 twice) and contracting the edge $e = \{6,8\}$ (i.e. applying R_2).

For the sake of completeness we mention and show that the elementary relations R_3 and R_4 also do not preserve the embedding of graphs in B_2. In order to show that R_3 and R_4 satisfy the statement

$$\bigvee_{G,H \in \Gamma} (G \notin \Gamma(B_2) \wedge H = R_i(G) \wedge H \in \Gamma(B_2)), \qquad (e')$$

where $i = 3, 4$, we consider the graph $G = U(K_{3,3}) \bigcup S_3$ depicted in Figure 2, where

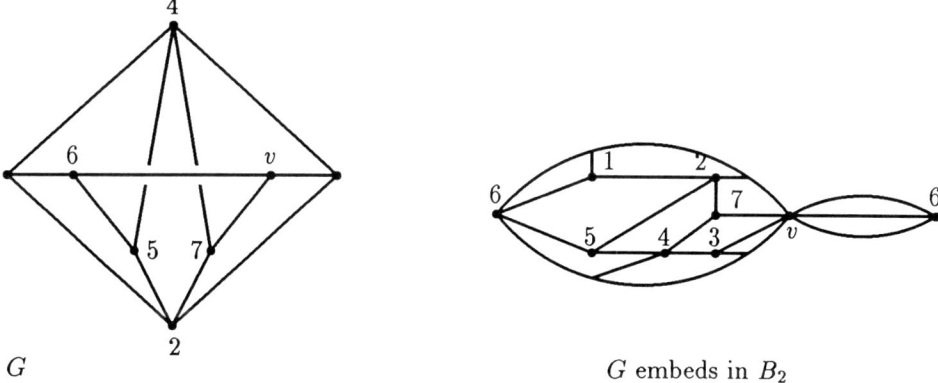

G   G embeds in B_2

FIGURE 2

$U(K_{3,3})$ is the subdivision graph of order 7 arising from the complete bipartite graph $K_{3,3} = \{1,3,5\} * \{2,4,6\}$ with the property that the edge $e = \{3,6\}$ with the endpoints 3 and 6 is divided by the inserted vertex v and where S_3 is the star $S_3 = \{7\} * \{2,4,v\}$. In Figure 2, we have got an embedding of G in B_2 such that G satisfies the two first conditions of (e'). If we apply R_3 to G by replacing the trihedral $\{v\} * \{3,6,7\}$ by the triangle $(3,6,7,3)$, we obviously obtain $R_3(G) = H = K_{3,4} \bigcup e$ with $e = \{3,7\}$ and $K_{3,4} = \{1,3,5,7\} * \{2,4,6\}$ which does not embed in B_2 because of $K_{3,4} \in M_1(B_2)$. If we apply R_4 to G by replacing the double trihedral $\{1,5\} * \{6\} * \{v\} * \{3,7\}$ by the double triangle $K_2 * \{6 = v\} * K_2'$ with $V(K_2) = \{1,5\}$ and $V(K_2') = \{3,7\}$, we obviously obtain $R_4(G) = H' = H \bigcup e'$ where $H = R_3(G)$ and $e' = \{1,5\}$.

The consequence of (e') for $i = 2, 3, 4$ is that we cannot use the minimal bases $M_2(B_2) - M_4(B_2)$ for a Kuratowski-type theorem for B_2 and that we have to restrict ourselves to $M_1(B_2)$. Considering $M_1(B_2)$ there immediately arises the question whether it is finite or not. At first, we were optimistic that it could be finite for we succeeded in finding exactly 84 $>_1$-minimal at most 2-connected graphs in $M_1(B_2)$

depicted in [3]. Investigating the subset of all 3-connected graphs in $M_1(B_2)$ we found an infinite set $M = \{X_n \in \Gamma(B_2) \mid n \in \mathbb{N} - \{1\}\}$ of graphs $X_2, X_3, \ldots$ in $\Gamma(B_2)$ which are mutually $>_1$-independent and a graph $X \in \Gamma(B_2)$ with the property that $X_n >_1 X$ for every $n \in \mathbb{N} - \{1\}$.

In order to describe this matter we define M and X in the following way: Let $C_{2n} = (v_1, \ldots, v_{2n})$, $n \geq 2$, denote a circuit of length $2n$, where $E(C_{2n}) = \{e_1, \ldots, e_{2n}\}$ with $e_i = \{v_i, v_{i+1}\}$, $i = 1, \ldots, 2n$ and $v_{2n+1} = v_1$. Let $H_1, \ldots, H_{2n}$ be the set of $2n$ mutually disjoint copies of K_5 without one edge such that $H_i = K_5 - e_i = v_i * K_3 * v_{i+1}$, $i = 1, \ldots, 2n$, $v_{2n+1} = v_1$ and H_i, H_{i+1} share the vertex v_i with $H_{2n+1} = H_1$. Then X_n, $n \geq 2$, arises from C_{2n} by substituting the edge $e_i = \{v_i, v_{i+1}\}$ by H_i and by adding the diagonal $e' = \{v_1, v_{n+1}\}$. X is defined by Figure 3. Figure 4 shows X_3.

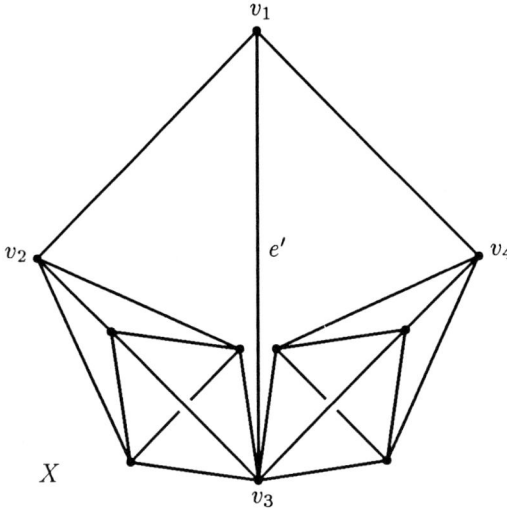

FIGURE 3

If we denote $M' = M \bigcup \{X\}$, then we can prove the following

Theorem 1. X and X_n, $n \geq 2$, do not embed in B_2 so that $M' \subseteq \Gamma(B_2)$. (1)

$$\bigwedge_{\substack{X_m, X_n \in M \\ m \neq n}} (X_m \not>_1 X_n \wedge X_n \not>_1 X_m) \quad (2)$$

$$\bigwedge_{X_n \in M} X_n >_1 X \quad (3)$$

For the sake of shortness we have to omit details of the proof. But it is necessary to make some remarks concerning the proof of Theorem 1. At first, the fact that X and X_n, $n \geq 2$, do not embed in B_2, follows from an embedding criterion given in [2] and [3]. Assertion (2) is true because the elementary relation R_1 cannot be

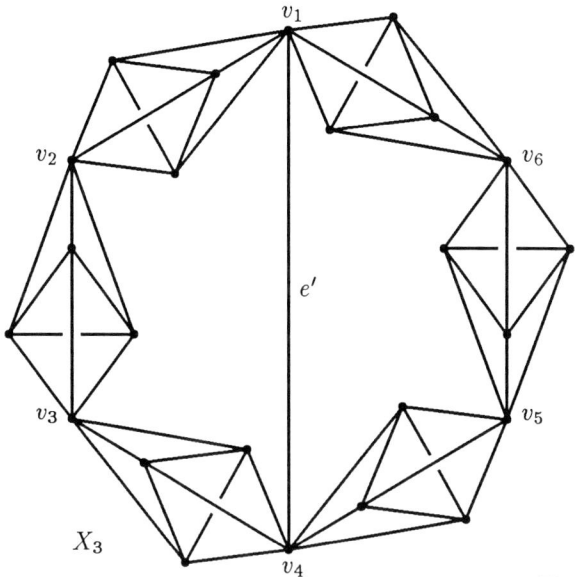

FIGURE 4

applied to $X_n \in M$, $n \geqslant 2$. By repeated application of R_0 and R_1 to X_n, $n \geqslant 2$, we obtain $X_n >_1 X$.

Because of Theorem 1 it became more probable that $M_1(B_2)$ is infinite. Indeed, at the Fourth Czechoslovak Graph Theory Conference at Prachatice in 1990, J. Širáň introduced an infinite set of graphs which are not embeddable in B_2 and belong to $M_1(B_2)$. The whole proof can be found in [5]. The Širáň proof implies that the Kuratowski theorem cannot be extended to B_2. In other words, there does not exist a Kuratowski-type theorem for B_2.

3. The minimal basis $M_2(B_2)$

Although there is not a Kuratowski theorem for B_2 as mentioned above it remains an open interesting question if $M_2(B_2)$ is finite or not because we do not know whether (e') influences the validity of the Bodendiek/Wagner proof and the Robertson/Seymour proof given in [1] and [6]. If we assume that $M_2(B_2)$ is infinite then we would obtain a contradiction to Wagner's theorem proved by Robertson and Seymour. Although we were not able to check the question up to this time whether both proofs mentioned above can be extended to B_2 we want to determine as many 3-connected $>_2$-minimal graphs in $M_2(B_2)$ as possible. Theorem 2 gives a first incomplete answer in the following way:

Theorem 2. *(1) In $M_2(B_2)$ there exist exactly three disconnected $>_2$-minimal graphs, namely $G_1 = K_5 \bigcup K_5'$, $G_2 = K_5 \bigcup K_{3,3}$ and $G_3 = K_{3,3} \bigcup K_{3,3}'$, where K_5, K_5', $K_{3,3}$, $K_{3,3}'$ are mutually disjoint Kuratowski graphs.*
(2) In $M_2(B_2)$ there are precisely three 1-connected $>_2$-minimal graphs G_4, G_5, G_6, where $G_4 = K_5 \overset{v}{\bigcup} K_5' = K_4 * \{v\} * K_4'$ (i.e., G_4 is the 1-amalgamation of K_5

and K_5' at a main vertex v), $G_5 = K_5 \overset{v}{\bigcup} K_{3,3} = K_4 * \{v\} * 3 * 2$ (i.e., G_5 is the 1-amalgamation of K_5 and $K_{3,3}$ at a main vertex v), and $G_6 = K_{3,3} \overset{v}{\bigcup} K_{3,3}' = 2 * 3 * \{v\} * 3 * 2$ (i.e., G_6 is the 1-amalgamation of $K_{3,3}$ and $K_{3,3}'$ at a main vertex).

(3) The graphs $G_{14} = K_{3,3} \overset{u,v}{\bigcup} K_{3,3}'$, where $K_{3,3}$ and $K_{3,3}'$ are two different copies of the complete bipartite graph $3 * 3$ with $K_{3,3} = \{1,3,5\} * \{2,u,v\}$ and $K_{3,3}' = \{u,v,4\} * \{7,9,11\}$ sharing exactly u and v, $G_{15} = K_5 \overset{e}{\bigcup} K_5'$ (arising from two different copies of the complete graph of order 5 sharing precisely the edge e), $G_{22} = K_5 \overset{e}{\bigcup} K_{3,3}$ (arising from K_5 and $K_{3,3}$ sharing the edge e), $G_{32} = K_{3,3} \overset{e}{\bigcup} K_{3,3}'$ (arising from two different copies of the complete bipartite graph $3 * 3$ sharing one and only one edge e), $G_{39} = K_{3,3} \overset{u,v}{\bigcup} K_{3,3}$, where $K_{3,3}$ and $K_{3,3}'$ are two copies of the complete bipartite graph $3 * 3$ with $K_{3,3} = \{1,3,u\} * \{2,4,v\}$ and $K_{3,3}' = \{u,v,6\} * \{5,7,9\}$ sharing exactly the two vertices u, v, $G_{52} = (K_5 - e) \overset{u,v}{\bigcup} K_{3,3}$ where e is the edge of K_5 with the endpoints u, v and where K_5 and $K_{3,3} = \{u,v,2\} * \{1,3,5\}$ are two Kuratowski graphs with the property that $K_5 - e$ and $K_{3,3}$ share exactly the two vertices u, v, are the only 2-connected graphs in $M_2(B_2)$.

(4) The graphs $G_{86} = K_{3,4}$ (3-connected) and $G_{87} = K_6 - e = 1 * K_4 * 1$ (4-connected) belong to $M_2(B_2)$.

(5) The graph G_{85} (Figure 1) and the graphs $G_{88}, G_{89}, G_{90}, G_{91}, G_{92}$ and G_{93} depicted in Figure 5 belong to $M_1(B_2)$.

For the sake of shortness we have to omit details of the proof. But we want to make some concluding remarks.

(a) The proof of (1) – (3) is given in [3]. Hence it is quite clear that we choose the denotations given in [3].

(b) The graphs $G_{86} = K_{3,4}$ and $G_{87} = K_6 - e$ are the most beautiful graphs in $M_2(B_2)$. The proof of $G_{85}, G_{87} \in M_2(B_2)$ is a matter of routine checking.

(c) It is really simple to verify the belonging of $G_{88}, \ldots, G_{93}$ to $M_1(B_2)$. It is remarkable that G_{92} and G_{93} arise from G_{88} by splitting up the vertex v_1 or the vertices v_1 and v_2, respectively. While it is immediately clear that $G_{86} = K_{3,4}$ and $G_{87} = K_6 - e$ also belong to $M_3(B_2)$ and $M_4(B_2)$ it remains to investigate which of the graphs $G_{88}, \ldots, G_{99}$ are elements of $M_i(B_2)$, $i = 2, 3, 4$.

(d) Using $>_4$ and $M_4(B_2)$ instead of $>_2$ and $M_2(B_2)$ we can reduce the number of minimal graphs. Instead of $G_1, \ldots, G_6 \in M_2(B_2)$ we only need to consider $G_1, G_4 \in M_4(B_2)$.

On the Banana Surface B_2

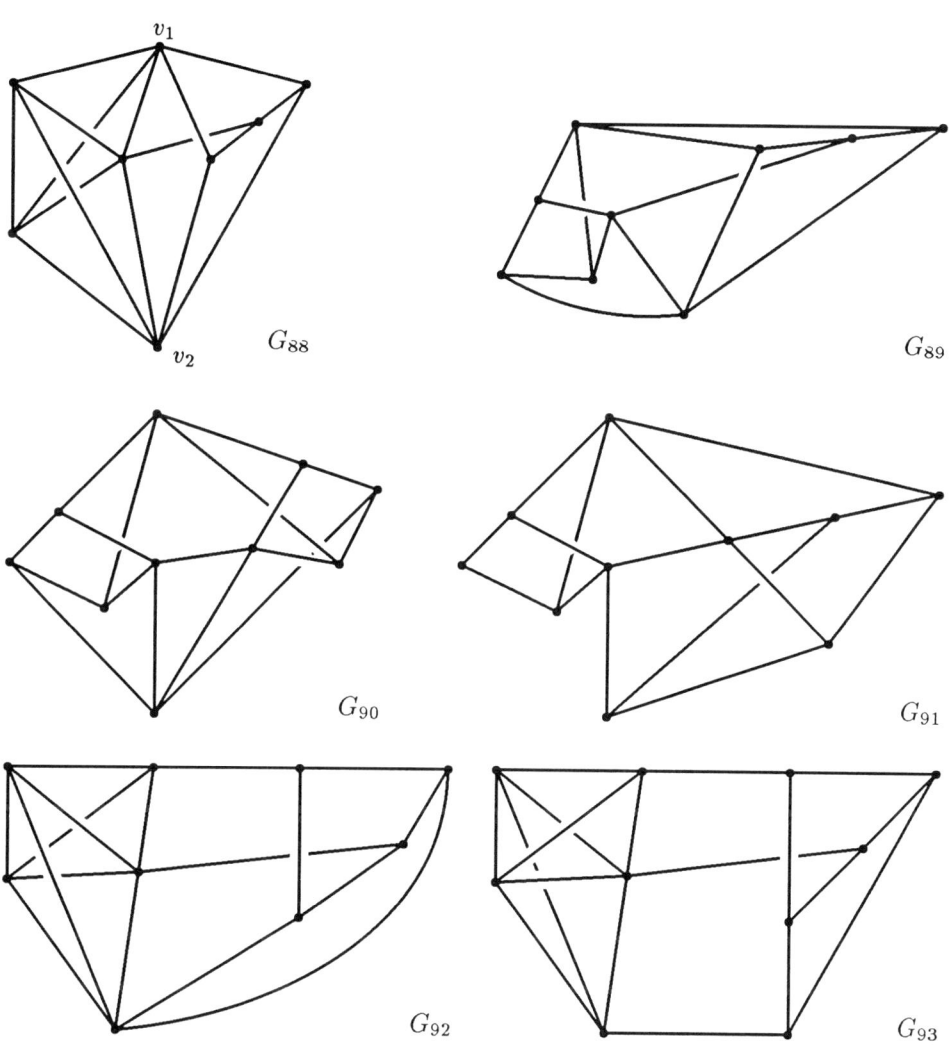

FIGURE 5

References

[1] R. Bodendiek, K. Wagner, *Solution to König's Graph Embedding Problem*, Math. Nachr. **140** (1989), 251–271.

[2] ———, *Solution to König's embedding problem bananas surfaces*, submitted.

[3] R. Bodendiek, P. Gvozdjak, J. Širáň, *On Minimal Graphs for Bananas Surface B_2*, R. Bodendiek: Contemporary Methods in Graph Theory (in honour to K. Wagner), Mannheim, 1990.

[4] R. Bodendiek, H. Schumacher, K. Wagner, *Über Relationen auf Graphenmengen*, Abh. Math. Sem. Universität Hamburg **51** (1981), 232–243.

[5] P. Gvozdjak, J. Širáň, *Kuratowski Theorem does not extend to Pseudosurfaces, Proceedings*, to appear in J. Combinatorial Theory.

[6] N. Robertson, P. D. Seymour, *Graph Minors: VIII A Kuratowski theorem for general surfaces*, submitted.

R. Bodendiek and K. Wagner
Institut für Mathematik und ihre Didaktik,
Pädagogische Hochschule Kiel,
D-2300 Kiel, FRG

Fourth Czechoslovakian Symposium on
Combinatorics, Graphs and Complexity
J. Nešetřil and M. Fiedler (Editors)
© 1992 Elsevier Science Publishers B.V. All rights reserved.

Structural Properties and Colorings of Plane Graphs

OLEG V. BORODIN

The purpose of this paper is to present some relationships between colorings (cyclic, simultaneous, and assigned) and structural properties of plane graphs including the concept of a weight (of an edge or a face) introduced in 1955 by Anton Kotzig.

1. Six color theorem

A graph is 1-*planar* if it can be represented in the plane so that every its edge intersects with at most one other edge (in inner points). In 1965 Ringel proved that each 1-planar graph is 7-colorable and conjectured that in fact it can be 6-colorable [36]. He also pointed out the other two forms of this conjecture. One of them is the simultaneous 6-colorability of the vertices and faces of each plane graph so that every two neighbour elements are colored differently [36]. The other form consists in such 6-colorability of the vertices of each plane graph having 3- and 4-faces only that the boundary vertices of each face receive pairwise different colors [36].

In [6] I confirmed Ringel's conjecture on 6 colors; for more information on the problem see [5]. The two abovementioned forms of this conjecture gave rise to the two separate directions in coloring theory: simultaneous and cyclic coloring, respectively. These topics are briefly discussed in Sections 4 and 2.

2. Cyclic coloring of plane graphs

In 1969, Ore and Plummer [32] defined a plane graph to be k-*cyclically* colored if any two vertices which are the boundary vertices of a face of the size at most k are colored with different colors. The minimal number of colors needed to k-cyclically color the graph in question is denoted by x_k. Obviously, the Four Color Problem solved by Appel and Haken [1] may be written as $x_3 \leqslant 4$, and our theorem on six colors [6] as $x_4 \leqslant 6$.

For arbitrary k, Ore and Plummer [32] proved the bound $x_k \leqslant 2k$. In [6,7] I announced some strengthenings of this result, and in [8] proved the following: $x_5 \leqslant 9$, $x_6 \leqslant 10$, $x_7 \leqslant 12$, and $x_k \leqslant 2k - 3$ for all $k \geqslant 8$.

On the other hand, one may easily derive from a 3-prism an example of a plane graph having $x_k = \lfloor \frac{3}{2}k \rfloor$ (each of the three vertical edges is replaced by a simple chain of appropriate length) [6, 34].

For a special case of 3-polytopes (i.e., 3-connected planar graphs), the upper bound for x_k was significantly improved by Plummer and Toft [34]: $x_k \leqslant k+9$. In fact, they obtained even tighter results when k is large [34]: $x_k \leqslant 22$ if $k \leqslant 15$; $x_k \leqslant k+7$ if $k \geqslant 16$; $x_k \leqslant k+6$ if $k \geqslant 18$; $x_k \leqslant k+5$ if $k \geqslant 24$; and finally, $x_k \leqslant k+4$ if $k \geqslant 42$. They also conjectured that, moreover, $x_k \leqslant k+2$ for each $k \geqslant 3$. In [9] I proved, in particular, the following result intermediate between Plummer and Toft's bound and their conjecture: $x_k \leqslant k+3$ if $k \geqslant 24$.

3. Cyclic degree of a vertex in planar graphs

The main tool for proving coloring results in [34, 9] were upper bounds for the *cyclic degree*, $\delta(v)$ of a vertex v in a 3-polytope defined as the number of vertices lying in common faces with v. Namely, Plummer and Toft used [34] a theorem of Lebesgue [31, p. 42] to prove in particular, that if $k \geqslant 42$ and the size of each face of a 3-polytope is at most k, then there exists a vertex v with $\deg v \leqslant 5$ and $\delta(v) \leqslant k+3$. I proved a similar stronger result in which $k \geqslant 24$ and $\delta(v) \leqslant k+2$ [9]; an example of k-prism shows that this bound whenever true is the best possible for all $k \geqslant 3$: each vertex is incident with a k-face and two 4-faces, i.e., it lies in common faces precisely with $k - 2 + 4 - 2 + 4 - 2 = k+2$ vertices.

4. Simultaneous and assigned colorings

The basic elements of a plane graph are its vertices, edges, and faces denoted by V, E, and F, respectively. The idea of simultaneously coloring heterogeneous elements of a graph belongs to Ringel [36]. It is generally assumed that every two neighbour elements to color should be colored with different colors. For instance, one of the forms of Ringel's Six Color Conjecture [36] was $x_{vf} \leqslant 6$, where x_{vf} is the minimal number of colors needed to color the graph's vertices and faces so that every two adjacent or incident of them receive different colors. Note that $x_{vf} \leqslant 6$ was first proved for triangulations [36] by Ringel, afterwards for the opposite special case of triangle-free plane graphs by Archdeacon [3], and at last in [6] in general form.

The total graph conjecture [4, 37] states that $x_{ve} \leqslant \Delta + 2$ for arbitrary (not necessary planar) graph, where Δ is its maximal degree. It is confirmed nowadays for $\Delta \leqslant 5$ only [26]. For the planar graphs, however, we have [10] more information: $x_{ve} \leqslant \Delta + 2$ if $\Delta \geqslant 9$, and $x_{ve} = \Delta + 1$ if $\Delta \geqslant 14$.

In 1973, Kronk and Mitchem conjectured [30] that $x_{vef} \leqslant \Delta + 4$ for all planar graphs with $\Delta \geqslant 3$ and proved this conjecture for $\Delta = 3$. In [11–13] I proved that $x_{vef} \leqslant \Delta + 4$ for all $\Delta \geqslant 8$ and $x_{vef} \leqslant \Delta + 2$ for all $\Delta \geqslant 14$. The bound $x_{vef} \leqslant \Delta + 2$ whenever true is the best possible for each $\Delta \geqslant 3$ as shown by an example of a plane graph $K_{1,\Delta}$ in which the internal vertex, all Δ edges and the infinite face are $\Delta + 2$ mutually adjacent or incident elements, so should be colored with different colors.

For the numbers x_{ef}, similar conjecture was proposed by Mel'nikov in [35, Problem Section, p. 543], and similar results obtained by the present author in [14, 15].

Well-known Vizing's Theorem [38] on the edge chromatic number, $x_e \leqslant \Delta + 1$, was sharpened [39] for the planar graphs as follows: if $\Delta \geqslant 8$, then $x_e = \Delta + 1$. As contrast to the usual coloring, in *assigned coloring* introduced independently in [40] and [21], the set of admissible colors varies from one element to another. The problem consists of choosing one color from those assigned to each element so that each two adjacent elements receive different colors. A usual coloring is clearly a special case of assigned one in which all the elements are assigned the same set of colors. A temptative extension $x_e^* \leqslant \Delta + 1$ of Vizing's Theorem remains unproved for all $\Delta > 3$. For the case of planar graphs we again have the following almost final results: $x_e^* \leqslant \Delta + 1$ if $\Delta \geqslant 9$; $x_e^* = \Delta$ if $\Delta \geqslant 14$ [10].

5. Minimal weight of edges in some classes of planar graphs

The *weight* $w(e)$ of an edge e was defined by Kotzig [27] as the degree sum of the end vertices of e. Kotzig proved [27] that each 3-connected planar graph contains an edge of the weight at most 13, the bound being sharp. In [10] I proved the same bound in a broader class of planar graphs, namely of those with the minimal degree, δ, at least 3 (this result was conjectured by Erdős, see [22]).

A number of extensions and analogues [10–15] of Kotzig's Theorem [27] appeared to become a tool for proving results on simultaneous colorings described in Section 4. Note that Kotzig's Theorem itself did not provide coloring applications, since the class of 3-connected graphs is not closed with respect to removing edges. We want to formulate here a few such theorems.

A *k-bunch* is defined to be an induced subgraph on $k+2$ vertices in which certain two vertices are joined by k chains of the length 2, with the intermediate vertices having degree 2. Each planar graph with $\delta \geqslant 2$ and without k-bunches, where $k \geqslant 2$, appears to contain an edge e such that $w(e) \leqslant 5k + 7$, the bound being sharp [11].

A cycle $v_1 v_2 \cdots v_{2k}$ is k-alternating if $\deg v_1 = \deg v_3 = \cdots = \deg v_{2k-1} = k$. It is clear that a 2-bunch is a special case of a 2-alternating cycle. If $\delta \geqslant 2$ and there are no 2-alternating cycles in a planar graph, then there exists an edge with the weight at most 15, the bound is also sharp [10]. Similarly, if $\delta \geqslant 3$ and there are no 3-alternating 4-cycles, then the minimal weight of an edge is at most 11 [10].

Lebesgue proved [31] that in each plane graph with $\delta \geqslant 3$ there exists either a 5-vertex incident with four triangles, or a 4-vertex incident with a triangle, or a 3-vertex incident with a face of the size at most 5. The following strengthening [13] of the Lebesgue's result also has coloring applications: there exists an edge incident either with two triangles, 5-vertex, and $\leqslant$ 7-vertex, or with a triangle, 4-vertex, and $\leqslant$ 7-vertex, or else with a 3-vertex and $\leqslant$ 4-face, or finally with two 3-vertices and $\leqslant$ 5-face.

6. Separability (weight) of faces

Kotzig [28] proved that in each plane triangulation without vertices of degree 3 and 4, it can be found a triangle with the degree sum (*weight*) of boundary vertices at most 18 and conjectured that the sharp bound here is 17. The double n-pyramid provides an example of a triangulation in which the minimal weight of a triangle is arbitrarily large.

On the other hand, Grünbaum conjectured [23] that the *cyclic connectivity* of the 5-connected planar graphs is at most 11, i.e., one can remove ≤ 11 edges from a graph so that to obtain two connected components each of which containing a cycle. Previously, Plummer proved [33] the cyclic 13-connectivity of 5-connected plane graphs.

Both conjectures were confirmed by the following statement [16]: in each plane graph with the minimal degree 5 there exists a triangle of the weight at most 17. Thus, at most 11 edges may be removed so that to separate, more specifically, some triangle from the rest of a graph. Observe that in dual terms, this result may be restated as existence of a 3-vertex with the cyclic degree at most 11 in any plane graph without 3- and 4-faces, having $\delta \geq 3$. If 3-faces are allowed and 4-faces are still forbidden, then the minimal cyclic degree is sharply restricted by 23 provided that the graph is cubic (each vertex has degree 3), and may be arbitrarily large otherwise [17].

Returning to original terms, each triangulation without 4-vertices contains a triangle of the weight at most 29 (the bound is sharp) [17], which is a solution to a problem raised by Kotzig [29] who proved that under the same assumptions the weight does not exceed 39. This result cannot be extended to all plane graphs without 4-vertices: the minimal weight of a triangle may then increase to infinity [17].

We conclude this section by emphasizing a close relation between the concepts of the cyclic connectivity and cyclic coloring which seem to appear quite independently of each other; a natural link between them is of course the concept of the cyclic degree.

7. Number of light edges and faces

Let $e_{i,j}$ be the number of edges joining i-vertices with j-vertices in the graph at hand and, similarly, let $f_{i,j,k}$ be the number of triangles formed by an i-vertex, a j-vertex, and a k-vertex.

A typical situation is: If in this or that class of plane graphs there exists an upper bound, say U for the minimal weight (of an edge or of a face), then the number of elements whose weight does not exceed U is great enough.

Thus, in 1904 Wernicke proved [41] for planar triangulations with $\delta = 5$ the bound $e_{5,5} + e_{5,6} > 0$. Grünbaum showed [24] that in fact $4e_{5,5} + e_{5,6} \geq 60$, and I strengthened [18] this to $2\frac{4}{7}e_{5,5} + e_{5,6} \geq 60$. Kotzig's Theorem $\sum_{i+j \leq 13} e_{i,j} > 0$ [27] was generalized by Jucovič as follows [25]:

$$20e_{3,3} + 25e_{3,4} + 16e_{3,5} + 10e_{3,6} + 6\frac{2}{3}e_{3,7} + 5e_{3,8} + 2\frac{1}{2}e_{3,9} + 2e_{3,10} +$$
$$+ 20e_{4,4} + 11e_{4,5} + 5e_{4,6} + 5e_{4,7} + 5e_{4,8} + 3e_{4,9} +$$
$$+ 8e_{5,5} + 2e_{5,6} + 2e_{5,7} + 2e_{5,8} \geq 120.$$

For all planar graphs with $\delta \geqslant 3$, I proved the following relation in which every coefficient is the best possible as shown by specific constructions [19]:

$$40e_{3,3} + 25e_{3,4} + 16e_{3,5} + 10e_{3,6} + 6\frac{2}{3}e_{3,7} + 5e_{3,8} + 2\frac{1}{2}e_{3,9} + 2e_{3,10}+$$
$$+ 16\frac{2}{3}e_{4,4} + 11e_{4,5} + 5e_{4,6} + 1\frac{2}{3}e_{4,7} + 5\frac{1}{3}e_{5,5} + 2e_{5,6} \geqslant 120.$$

For an intermediate result see [20]. To obtain similar final relation for triangulations, one should only replace 40 by 20 in the first term above [19]. Complete answers to some other questions of Jucovič [25] also follow easily from the inequality above [19].

For the plane triangulations with $\delta = 5$, the following results were obtained: $e_{5,5} + f_{5,6,6} > 0$ [2], $f_{5,5,5} + \frac{2}{3}f_{5,5,6} + +\frac{3}{7}f_{5,5,7} + \frac{1}{4}f_{5,5,8} + \frac{1}{9}f_{5,5,9} + \frac{1}{3}f_{5,6,6} + \frac{2}{21}f_{5,6,7} \geqslant 20$ [33], and $f_{5,5,5} + f_{5,5,6} + f_{5,5,7} + f_{5,5,8} + f_{5,6,6} + f_{5,6,7} + f_{6,6,6} > 0$ [29].

My result already mentioned in Section 6, may be written as: $f_{5,5,5} + f_{5,5,6} + f_{5,5,7} + f_{5,6,6} > 0$ for each plane graph with $\delta = 5$ [16]. Moreover, as shown by constructions [18], no term of this inequality may be removed without upsetting it. In fact, the following bound is valid for the plane graphs with $\delta = 5$ [18]: $18f_{5,5,5} + 9f_{5,5,6} + 5f_{5,5,7} + 4f_{5,6,6} \geqslant 144$ in which every coefficient, except probably the third, is the best possible, as shown by constructions. This relation may be regarded as a non-superfluous version of the Lebesgue's estimate cited above [31].

8. Conclusion

First we have seen a close relationship between the cyclic coloring and the vertex-face chromatic number x_{vf} on the one hand, and the cyclic degree, cyclic connectivity, and the weight of a triangle in a plane graph on the other.

Secondly, those simultaneous colorings comprising edges, i.e., the numbers x_{ve}, x_{ef}, and x_{vef}, as well as the assigned coloring of edges (x_e^*), may be successfully treated, using structural properties of plane graphs including the concept of the weight of an edge.

References

[1] K. Appel, W. Haken, *The existence of unavoidable sets of geografically good configurations*, Illinois J. Math. **20** (1976), 218–297.

[2] ———, *The solution of the four-color-map problem*, Scientific American **237**, No. 4 (1977), 108–121.

[3] D. Archdeacon, *Coupled colorings of planar maps*, Congressus Numerantium **39** (1983), 89–99.

[4] M. Behzad, G. Chartrand, J. Cooper, *The coloring numbers of complete graphs*, J. London Math. Soc. **42** (1967), 226–229.

[5] R. Bodendiek, H. Schumacher, K. Wagner, *Bemerkungen zu einem Sechsfarbenproblem von G. Ringel*, Abh. Math. Sem. Univ. Hamburg **53** (1983), 41–52.

[6] O. V. Borodin, *Solution of Ringel's problems on vertex-face coloring of plane graphs and coloring of 1-planar graphs*, Met. diskret. anal. Novosibirsk **41** (1984), 12–26.
[7] _____, *On cyclic coloring the vertices of plane graphs*, Abstracts I World Congress of Bernoulli Soc., Tashkent, 1986, pp. 499.
[8] _____, *Cyclic coloring of plane graphs*, (submitted).
[9] _____, *Structure, contraction of edges, and cyclic colorings of 3-polytopes*, (submitted).
[10] _____, *On the total coloring of planar graphs*, J. reine angew. Math. **394** (1989), 180–185.
[11] _____, *Consistent colorings of graphs on the plane*, Met. diskret. anal. Novosibirsk **45** (1987), 21–27.
[12] _____, *Simultaneous coloring of vertices, edges, and faces of plane graphs*, Met. diskret. anal. Novosibirsk **47** (1988), 27–37.
[13] _____, *A structural theorem on plane graphs and its application to coloring*, (submitted).
[14] _____, *Simultaneous coloring of edges and faces of plane graphs (submitted)*.
[15] _____, *A structural property of planar graphs and the simultaneous colouring of their edges and faces*, Mathematica Slovaca 40 **2** (1990), 113–116.
[16] _____, *Solution of Kotzig's and Grünbaum's problems on separability of a cycle in plane graphs*, Matem. zametki 46 **5** (1989), 9–12.
[17] _____, *Faces with restricted degree sum of boundary vertices in plane triangulations*, (submitted).
[18] _____, *Structural properties of planar maps with the minimal degree 5*, (submitted).
[19] _____, *Precise lower bound for the number of edges of minor weight in planar maps*, (submitted).
[20] _____, *Computing light edges in planar graphs*, G. Ringel's 70th birthday Festschrift, Physica-Verlag, Heidelberg, to appear.
[21] P. Erdős, A. L. Rubin, H. Taylor, *Choosability in graphs*, Proc. West Coast Conf. Combin., Humboldt State Univ., 1979, pp. 125–157.
[22] B. Grünbaum, *New views on some old questions of combinatorial geometry*, Proc. Int. Colloq. Rome 1973, Accad. nac. dei lincei Rome, 1976, pp. 451–468.
[23] _____, *Polytopal graphs*, Math. Assoc. of Amer. Studies in Math. **12** (1975), 201–224.
[24] B. Grünbaum, G. C. Shephard, *Analogues for tillings of Kotzig's theorem on minimal weight of edges*, Ann. discr. math **12** (1981), 129–140.
[25] E. Jucovič, *Strengthening of a theorem about 3-polytopes*, Goem. Dedic **13** (1974), 20–34.
[26] A. V. Kostochka, *Precise upper bound for the total chromatic number of multigraphs*, 24th Int. Wiss. Koll., TH Ilmenau, 1979, pp. 33–36.

[27] A. Kotzig, *Contribution to the theory of eulerian polyhedra*, Mat. čas. **5** (1955), 101–103.

[28] _____, *From the theory of Euler's polyhedrons*, Mat. čas. **13** (1963), 20–34.

[29] _____, *Extremal polyhedral graphs*, Proc. Sec. Int. Conf. Combin. Math., New York, 1978, pp. 569–570.

[30] H. Kronk, J. Mitchem, *A seven-color theorem on the sphere*, Discrete Math. **5** (1973), 253–260.

[31] H. Lebesgue, *Quelques conséquences simple de la formula d'Euler*, J. de Math. **9** (1940), 27–43.

[32] O. Ore, M. D. Plummer, *Cyclic coloration of planar graphs*, Recent progr. in combin., New York-London, 1969, pp. 287–293.

[33] M. D. Plummer, *On the cyclic connectivity of planar graphs*, Graph theory and applications, Berlin, 1972, pp. 235–242.

[34] M. D. Plummer, B. Toft, *Cyclic coloration of 3-polytopes*, J. Graph Theory **11** (1987), 507–516.

[35] *Recent advances in graph theory*, Proc. Int. Symp. Prague, 1974, vol. 1975, Academia Praha, p. 544.

[36] G. Ringel, *Ein Sechsfarbenproblem auf der Kugel*, Abh. Math. Sem. Univ. Hamburg **29** (1965), 107–117.

[37] V. G. Vizing, *Some unsolved problems in graph theory*, Uspehi mat. nauk **23** (1968), 117–134.

[38] _____, *On the bound for the chromatic number of p-graphs*, Diskret. anal. Novosibirsk **3** (1964), 25–30.

[39] _____, *Critical graphs with given chromatic class*, Diskret. anal. Novosibirsk **5** (1965), 9–13.

[40] _____, *Coloring the vertices of a graph with assigned colors*, Met. diskret. anal. Novosibirsk **29** (1976), 3–10.

[41] P. Wernicke, *Über den Kartographischen Vierfarbensatz*, Math. Ann. **58** (1904), 413–426.

Oleg V. Borodin
Institute of Mathematics,
Siberian Branch,
Academy of Sciences of the USSR,
630090 Novosibirsk,
USSR

The Binding Number of Graphs

MIECZYSŁAW BOROWIECKI

1. Introduction

All graphs considered in this paper are finite and simple. We follow the terminology of Harary [9]. The *order* of a graph G is $|V(G)|$ and its *size* $|E(G)|$. If $X \subseteq V(G)$ we denote by $\Gamma(X)$ the set of vertices adjacent to vertices in X and by $d(v) = |\Gamma(v)|$ the *vertex degree* of v, and by $\delta(G)$ the *minimal vertex degree* of a vertex of G.

For a given graph G let $F_G = \{X \subseteq V(G): X \neq \emptyset \text{ and } \Gamma(X) \neq V(G)\}$. According to [21], the *binding number* of a graph G, denoted by $\text{bind}(G)$, is defined by
$$\text{bind}(G) = \min\{|\Gamma(X)|/|X|: X \in F_G\}.$$
A set $X \in F_G$ is the *realizing set* for $\text{bind}(G)$ if
$$\text{bind}(G) = |\Gamma(X)|/|X|.$$

This paper is intended to survey and to complete the collection of main results, problems and papers for the binding number. We describe here also the structure of the realizing sets of the binding number.

2. $\text{bind}(G) \leqslant 1$

An important result on realizing sets for the binding number in this case is given in [13], namely:

(2.1) *Let G be a graph with n vertices.*
  (a) *If $\text{bind}(G) < 1$, then every realizing set X for $\text{bind}(G)$ is independent and $|\Gamma(X)| + |X| < n - 2$ or $= n$.*
  (b) *If $\text{bind}(G) = 1$ and G is connected, then G has an independent realizing set for $\text{bind}(G)$.*

Combining (2.1) with results of Faragó [7] we have immediately

(2.2) *There exists a polynomial-time algorithm which computes $\text{bind}(G)$ for the class of graphs with $\text{bind}(G) \leqslant 1$, i.e. for graphs with $|\Gamma(X)| \leqslant |X|$ for some $X \in F_G$.*

A linear-time algorithm for computing the binding number of trees was found by D. Michalak [17].

3. bind(G) > 1

Problems concerning the binding number which were investigated can be formulated as follows:
(i) Let G be a graph with a property P. Find the exact value or some bounds for bind(G).
(Part 4 of this paper contains some details).
(ii) Characterize graphs which are binding minimal, i.e., such that bind($G - e$) < bind(G) for any $e \in E(G)$.
Some results are contained in [15].
(iii) Suppose that bind(G) $\geq c$, where c is a given constant. What properties has the graph G?
Now we review the main results of kind (iii).

If $c \geq \frac{4}{3}$, then G has a perfect matching ([1]).
If $c \geq \frac{3}{2}$, then
[21]: G is Hamiltonian;
[18]: G has a triangle;
[19]: G is pancyclic.
[14]: If $c > (2k-1)(n-1)/[k(n-2)+3]$, $k \geq 2$, $V(G) = n \geq 4k - 6$ and kn is even, then G has a k-factor.

Some other results of this kind can be found in [11].

Let $X \in F_G$ be a realizing set for bind(G) > 1 and let $I = X \setminus \Gamma(X)$, $Z = \Gamma(X) \cap X$, $C = \Gamma(X) \setminus X$, $Y = V \setminus (\Gamma(X) \cup X)$. Then

(3.1) $I = \emptyset$ or I is an independent set of G.

(3.2) $I \neq \emptyset$ if and only if $Y = \emptyset$.

Proof. $\Rightarrow$: Let $x \in I$ and suppose that there is $y \in Y$. Let $X' = X \cup Y$. Since $x \notin \Gamma(X')$, then $\Gamma(X') \subseteq \Gamma(X) \cup Y \neq V$. Hence,

$$\text{bind}(G) = (|\Gamma(X)|/|X|) \leq (|\Gamma(X')|/|X'|) \leq (|\Gamma(X)| + |Y|)/(|X| + |Y|).$$

This implies $|Y|(|\Gamma(X)| - |X|) \leq 0$. Since $|Y| > 0$, then $|\Gamma(X)| - |X| \leq 0$, a contradiction. Thus, $Y = \emptyset$.

$\Leftarrow$: If $I = \emptyset$, then $X \subseteq \Gamma(X) = V$, which is not admissible.

Since either I or Y is empty and by the fact that I and Z cannot be empty simultaneously, we have three possible cases which define three types of realizing sets for bind(G). Namely,

Type I : $I \neq \emptyset$ and $Z = Y = \emptyset$;
Type II : $I = \emptyset$ and $Y \neq \emptyset \neq Z$;
Type III : $I \neq \emptyset \neq Z$ and $Y = \emptyset$.

□

(3.3) *A realizing set of Type I for* bind$(G) > 1$ *is a maximum independent set.*

Proof. Let X be a realizing set of Type I. By $Z = Y = \emptyset$, X is maximal independent and by the minimality of $|\Gamma(X)|/|X|$, X is a maximum independent set. □

(3.4) *Let X be a realizing set of Type II. Then the induced subgraph $\langle Y \rangle_G$ by Y in G is non-trivial and connected.*

Proof. Let $y \in Y$. Then $\Gamma y \not\subseteq C$, otherwise, for the set $X' = X \cup \{y\}$, we have $|\Gamma(X')|/|X'| < \text{bind}(G)$. Thus, $|Y| > 1$ and $\langle Y \rangle_G$ has non-trivial components. Let $D_1, \ldots, D_t$ be the components of $\langle Y \rangle_G$ with $V(D_i) = V_i$ and obviously, $|V_i| \geqslant 2$. For the set $X' = X \cup V_1$ (the set V_i can be chosen arbitrarily) we have

$$(|\Gamma(X)|/|X|) \leqslant (|\Gamma(X')|/|X'|) \leqslant (|\Gamma(X)| + |V_1|)/(|X| + |V_1|).$$

This implies $|\Gamma(X)| - |X| \leqslant 0$, a contradiction. □

The above result implies the following corollary

(3.5) *Let X be a realizing set of Type II for* bind$(G) > 1$, *i.e., $X = Z$ and $C = \Gamma(Z)$, and let $D_1, \ldots, D_t$ be the components of $G - C$ with $V(D_i) = V_i$, $|V_1| \geqslant \ldots \geqslant |V_t| \geqslant 2$. Then $Z = \bigcup_{i=1}^{t-1} X_i$.*

A few additional facts.

(3.6) *For any realizing set X we have*
   (a) *if $Z \neq \emptyset$, then $|Z| \geqslant 2$;*
   (b) *C is a cut-set of G or $G - C$ is trivial.*

4. Some bounds for bind(G)

Let be a given graph G and let $\delta(G) = d(v)$. The set $X = V \setminus \{\Gamma(v)\}$ belongs to F_G. By this we have the following upper bound [21]:

$$\text{bind}(G) < \frac{n-1}{n - \delta(G)}.$$

An improvement of this bound can be found in the following way. Let i be the number of isolated vertices in the graph $G - \Gamma(v)$. Then it is easy to see that

$$\text{bind}(G) < \frac{n-i}{n - \delta(G)}.$$

Because the computation of the binding number is rather complicated in the general case, any partial results in calculating bind(G) for some classes of graphs or in discovering some new bounds involving different parameters of graphs are welcome (see [3], [6], [21]).

Now we present a lower bound, at present the only one known, involving the connectivity number and the hallian index.

A graph G is *Hallian* if $|\Gamma(X)| \geqslant |X|$ for any set $X \subseteq V$ or equivalently, if its vertices can be covered by a set of vertex disjoint independent edges or odd cycles. Other equivalent conditions are summarized in [2] and [5]. A graph G is *k-Hallian* if for any set A of vertices of order at most k the subgraph of G induced by the set $V \setminus A$ is Hallian. The largest k such that G is k-Hallian is called the *Hallian index* of G and is denoted by $h(G)$. The *vertex connectivity* $\kappa(G)$ of a graph G is the minimum number of vertices whose removal results in a disconnected or in a trivial graph. Clearly, $h(G) \leqslant \delta(G) - 1$ and $\kappa(G) \leqslant \delta(G)$.

(4.1) [3] *If a graph G of order n is k-Hallian and l-connected, and $r = \min\{k, l\}$, then*
$$\mathrm{bind}(G) \geqslant \frac{n - \delta(G) + r}{n - \delta(G)}$$

Combining the upper bound given above and (4.1) we obtain the following corollary

(4.2) *If a graph G of order n has $h(G) = \delta(G) - 1$ and $\kappa(G) \geqslant h(G)$, then*
$$\mathrm{bind}(G) = \frac{n-1}{n - \delta(G)}.$$

Applications of properties of Hallian graphs and these results can be found in [2], [3], [4], [5], [15]. In these papers infinite families of graphs are described with an extremal value of the binding number and with the realizing set of Type III. However, the condition of (4.2) for the extremal value of the binding number is not necessary. For the graph $G = C_6 + C_6$, the join of two cycles, the realizing set is $X = V \setminus \Gamma(v)$, where $d(v) = \delta(v)$, and $\mathrm{bind}(G) = (n-1)/(n - \delta(G))$ but $\kappa(G) \geqslant h(G) = \delta(G) - 2$. A characterization of all graphs with this extremal binding number seems to be a difficult problem.

5. Examples, Problems and Concluding Remarks

(5.1) Let T be a plane tree with at least three vertices and with no vertex of degree 2. Let $C = (v_1, \ldots, v_m, v_1)$ be a cycle, where $v_1, \ldots, v_m$ denote all leaves of T in a cyclic order. A *Halin* graph H is a plane graph with $V(H) = V(T)$ and $E(H) = E(T) \cup E(C)$.

For a Halin graph H the following holds:
$$\kappa(H) = 3 \quad \text{and} \quad h(H) = 1 \quad \text{or} \quad h(H) = 2.$$

Thus (for details see [20]) if $h(H) = 1$, then the realizing set X for $\mathrm{bind}(H)$ is maximum independent, i.e., of Type I, otherwise, $X = V \setminus \Gamma(v)$, where v is a vertex of degree 3, i.e., X is of Type III.

Some other graphs with the binding number greater than 1 and with an independent realizing set are known, but this class of graphs has not been characterized. Moreover, at present, no general sufficient conditions for the existence of independent realizing sets for $\mathrm{bind}(G) > 1$ are known. For these graphs we have the following equality: $\mathrm{bind}(G) = \bigl(n - \alpha(G)\bigr)/\alpha(G) = \mu^{-1}(G) - 1$, where $\alpha(G)$ and $\mu(G)$ are the independence number and the independence ratio, respectively.

(**5.2**) Let $G = K_2 + 3K_n$ ($n \geq 3$) denote the join of vertex disjoint one K_2 and 3 copies of K_n. For the graph G we have: $\kappa(G) = 2$, $h(G) = n = \delta - 1$, $\text{bind}(G) = (n+1)/n$, where the realizing set is formed by all vertices of two copies of K_n and is of Type II. It is easy to see that in that manner more general constructions of infinite families of graphs preserving the type of the realizing set can be given.

References

[1] I. Anderson, *Perfect Matchings of a Graph*, J. Comb. Theory Ser. B **10** (1971), 183–186.

[2] M. Borowiecki, *On Some Classes of Hallian Graphs and Digraphs*, Graphs, Hypergrahs and Matroids III, Zielona Góra, 1989, pp. 17–28.

[3] M. Borowiecki, D. Michalak, *Some Properties of Hallian Graphs*, Zastosowania Matematyki **XIX** (1987), 363–370.

[4] _____, *On Hallian Digraphs and Their Binding Numbers*, Combinatorics and Graph Theory, Banach Center Publications 25, PWN Warsaw 1989, pp. 33–37.

[5] _____, *Hallian Graphs and Some of Their Applications*, Recent Studies in Graph Theory, Vishwa International Publications, Gulbarga 1989, pp. 14–28.

[6] O. Favaron, M. C. Heydemann, J. C. Meyer, D. Sotteau, *A Parameter Linked with G-Factors and Binding Number*, Rapports de Recherche 351 L.R.I. Orsay.

[7] A. Faragó, *f-Independence Number of Graphs*, Combinatorics: 7th Hung. Colloq. Eger, July 5–10, 1987, Amsterdam, 1988, pp. 221–226.

[8] D. R. Guichard, *Binding Number of the Cartesian Product of Two Cycles*, Ars Comb. **19** (1985), 175–178.

[9] F. Harary, *Graph Theory*, Addison-Wesley, Mass. 1969.

[10] W. Jiangang, T. Songlin, L. Jiugiang, *The Binding Number of Product Graphs*, Graph Theory, Singapore 1983, Lect. Notes in Math., Springer, Berlin, 1984, pp. 119–128.

[11] V. G. Kane, S. P. Mohanty, *Binding Number, Cycles and Complete Graphs*, Combinatorics and Graph Theory, LMN 885, Springer-Verlag, Berlin, 1981, pp. 290–296.

[12] V. G. Kane, S. P. Mohanty, R. S. Hales, *Product Graphs and Binding Number*, Ars Comb. **11** (1981), 201–224.

[13] V. G. Kane, S. P. Mohanty, E. G. Straus, *Which Rational Numbers are Binding Numbers?*, J. Graph Theory **5** (1981), 379–384.

[14] P. Katerinis, D. R. Woodall, *Binding Numbers of Graphs and the Existence of k-Factors*, Quart. J. Math. Oxford (2) **38** (1987), 221–228.

[15] M. Kwasnik, D. Michalak, *The Join of Graphs and the Binding Minimality*, Časopis Pěst. Mat. **114** (1989), 262–275.

[16] _____, *On the Binding Number of Line Graphs and of Product Graphs*, Discuss. Math. **VIII** (1986), 17–29.

[17] D. Michalak, *The Binding Number of k-Trees*, to appear.

[18] Shi Ronghua, *The Binding Number of a Graph and Its Triangle*, Acta Math. Appl. Sinica **2** (1985), 79–86.
[19] Shi Ronghua, Letter, 1988.
[20] M. Skowrońska, *The Binding Number of Halin Graphs*, Discrete Applied Mathematics 22 (1988/89), 93–97.
[21] D. R. Woodall, *The Binding Number of a Graph and Its Anderson Number*, J. Comb. Theory, Ser. B **15** (1973), 225–255.

Mieczysłav Borowiecki
Higher College of Engineering,
Institute of Mathematics, Physics and Chemistry,
Podgórna 50, 65-246 Zielona Góra, Poland

Fourth Czechoslovakian Symposium on
Combinatorics, Graphs and Complexity
J. Nešetřil and M. Fiedler (Editors)
© 1992 Elsevier Science Publishers B.V. All rights reserved.

Note on Algorithmic Solvability of Trahtenbrot-Zykov Problem

PETER BUGATA

Let G be an undirected graph without loops and multiple edges. As usual, let $V(G)$ be its vertex and $E(G)$ its edge set. The *neighbourhood* of a vertex x in G (denoted by $N(x, G)$) is the subgraph of G induced by the set of all vertices adjacent to x. By the *neighbourhood set* of a graph G we mean the set $N(G) = \{N(x;G); x \in V(G)\}$ with isomorphic graphs considered as identical.

At the Smolenice Symposium (in 1963) Zykov posed the following problem: For which finite graph H does there exist a (finite) graph G with $N(G) = \{H\}$?

The origins of the interest for this problem lie in Trahtenbrot's investigations in automata theory yielding the name *Trahtenbrot-Zykov problem* (in sequel T-Z problem). There are two modifications of this problem: finite, if the graph G is required to be finite, and the infinite one. The solution of both of them is known only for special classes of graphs such as paths, cycles, graphs homeomorphic to a star (for a survey see [4]). In [3] Bulitko proved

Theorem 1. *There exists no algorithm which, given a finite graph H, will determine whether there exists a graph G with $N(G) = \{H\}$.*

His proof is based on the algorithmic unsolvability of the "domino problem" (Kahr, Moore and Hao Wang, 1962). Another proof can be found in [1] together with a proof of the weaker theorem on the algorithmic solvability of the finite modification of the T-Z problem.

Theorem 2. *There exists no algorithm which, given a finite set N of finite graphs, will determine whether there exists a finite graph G with $N(G) = N$.*

The method in [1] consists of two steps. In the first one we show a reduction of the well-known Post problem to the problem of existence of vertex-labelled graphs with prescibed neighbourhoods; the second contains a way of transforming labelled graphs into unlabelled ones.

So far no proof is known for an analogue of Bulitko's theorem for the finite modification of the T-Z problem. However, we succeeded in obtaining a similar result for digraphs.

The *neighbourhood* of a vertex x in a digraph $\vec{G}$ is the subgraph of $\vec{G}$ induced by the set of all terminal vertices of edges whose initial vertex is x. Similarly as in the undirected case, we can define the *neighbourhood set* of a digraph and pose the T–Z problem.

The T–Z problem for digraphs has been solved for some special classes, for example for oriented trees (see [2]). In the directed case, both modifications of the T–Z problem are algorithmically unsolvable even in the class of all symmetric digraphs.

Theorem 3. *There exists no algorithm which, given a finite symmetric digraph $\vec{H}$, will determine whether there exists a digraph (finite digraph) $\vec{G}$ with $N(\vec{G}) = \vec{H}$.*

Proof of Theorem 3

Let uv denote the edge of an undirected graph connecting vertices u and v and let (x, y) be the edge of a digraph with initial vertex x and terminal vertex y. A digraph $\vec{G}$ is said to be *symmetric (asymmetric)*, if for every $(x, y) \in E(\vec{G})$ the ordered pair (y, x) is (is not) an edge of $\vec{G}$.

By a *universal* vertex of the undirected graph G we mean its vertex of degree $|V(G)| - 1$. A graph obtained from a cycle by adding one universal (in the new graph) vertex is called a *wheel*. The symbol $\bigcup_{i=1}^{n} H_i$ denotes the disjoint union of the graphs $H_1, \ldots, H_n$.

Let H be an undirected graph. We define the digraph $\vec{D}(H)$ as the digraph with the vertex set $V(H)$ such that (x, y) is an edge of $\vec{D}(H)$ if and only if $xy \in E(H)$.

Lemma 1. *If there exists a graph (finite graph) G such that $N(G) = \{H_1, \ldots, H_n\}$, then there exists a digraph (finite digraph) $\vec{D}$ such that $N(\vec{D}) = \{\vec{D}(H_1), \ldots, \vec{D}(H_n)\}$.*

Proof. Take $\vec{D} = \vec{D}(G)$. □

Lemma 2. *If there exists a digraph (finite digraph) $\vec{D}$ such that $N(\vec{D}) = \{\vec{H}_1, \ldots, \vec{H}_n\}$, then there exists a digraph (finite digraph) $\vec{G}$ with $N(\vec{G}) = \{\bigcup_{i=1}^{n} \vec{H}_i\}$.*

Proof. Let $\vec{D}$ be a digraph with $N(\vec{D}) = \{\vec{H}_1, \ldots, \vec{H}_n\}$.

Let $\vec{G}_{x,i}$ for $x \in V(\vec{D})$ and $i \in \{1, 2, 3\}$ be disjoint copies of $\vec{D}$. For every $\vec{G}_{x,i}$ we choose vertices $v_{x,i}^{(1)}$ and $v_{x,i}^{(2)}$ such that $N(v_{x,i}^{(1)}, \vec{G}_{x,i})$ is isomorphic to $\vec{H}_1$ and $N(v_{x,i}^{(2)}, \vec{G}_{x,i})$ is isomorphic to $\vec{H}_2$. We construct a digraph $\vec{G}'$: Its vertex set is the union of the vertex sets of all graphs $\vec{G}_{x,i}$ and the edge set we obtain from the union of the edge sets of all $\vec{G}_{x,i}$ in the following way: for every $z \in V(\vec{G}_{x,i})$ such that $N(z, \vec{G}_{x,i})$ is isomorphic to $\vec{H}_1$ ($\vec{H}_2$) we add edges (z, u) for $(u \in V(N(v_{z,j}^{(2)}, \vec{G}_{z,j}))$ ($u \in V(N(v_{z,j}^{(1)}, \vec{G}_{z,j}))$), where $j \equiv (i+1) \bmod 3$. It is not difficult to see that $N(\vec{G}') = \{\vec{H}_1 \bigcup \vec{H}_2, \vec{H}_3, \ldots, \vec{H}_n\}$. By the repetition of this step we obtain the graph $\vec{G}$ with $N(\vec{G}) = \{\bigcup_{i=1}^{n} \vec{H}_i\}$. □

Lemma 3. Let $\vec{G}$ be a digraph (finite digraph) such that $N(\vec{G}) = \{\bigcup_{i=1}^{n} \vec{D}(H_i)\}$, where H_i, $i = 1, \ldots, n$ is a connected undirected graph with a non-empty edge set. Then there exists an undirected graph (finite graph) G with a non-empty edge set such that $N(G) \subseteq \{H_1, \ldots H_n\}$.

Proof. Let $\vec{G}$ satisfy the assumption of the lemma. We construct an undirected graph G':

$V(G') = V(\vec{G})$;
$xy \in E(G') \iff ((x,y) \in E(\vec{G})\ \&\ (y,x) \in E(\vec{G}))$.

Since $N(\vec{G}) = \{\bigcup_{i=1}^{n} \vec{D}(H_i)\}$, the set $E(G')$ is non-empty. We show that for every non-isolated vertex x of G', $N(x, G')$ is isomorphic to the disjoint union of some graphs from the set $\{H_1, \ldots, H_n\}$.

Let K be a component of $N(x, G')$ which is not isomorphic to any of the graphs $H_1, \ldots, H_2$. The connectivity of these graphs implies the existence of vertices y, z of $N(x, \vec{G})$ such that $(y,x), (y,z), (z,y) \in E(\vec{G})$ and $(z,x) \notin E(\vec{G})$. Then $N(y, \vec{G})$ contains the vertices x and z such that $(x,z) \in E(\vec{G})$ and $(z,x) \notin E(\vec{G})$— a contradiction to the structure of $N(\vec{G})$.

Now we define an undirected graph G:

$$V(G) = \bigcup_{x \in V(G')} \{x_K\ ;\ K \text{ is a component of } N(x, G')\};$$

for $x, y \in V(G')$ and components K, L of $N(x, G'), N(y, G')$, respectively

$$(x_K, y_L) \in E(G) \iff x \in L\ \&\ y \in K.$$

We show that $N(G) \subseteq \{H_1, \ldots, H_n\}$. Let x_K be an arbitrary vertex of G. If $y_L \in N(x_K, G)$, then $(x_K, y_L) \in E(G)$ and consequently $y \in K$. Let the map $\eta: V(N(x_K, G)) \to V(K)$ be defined by: $\eta(y_L) = y$.

It is not difficult to see that η is an isomorphism of the graphs $N(x_K, G)$ and K. This implies that the neighbourhood of an arbitrary vertex of G is isomorphic to H_1 for some $i \in \{1, \ldots, n\}$. $\square$

We prove the algorithmic unsolvability of the T–Z problem for digraphs using a method developed in [1]. In the quoted paper we assign a finite set system $\mathcal{M}(S)$ to every instance S of Post's problem (a so called Post system). The system $\mathcal{M}(S)$ contains finite sets of wheels on at least 5 vertices whose vertices are labelled by the elements of a finite alphabet $\Sigma \supseteq \{z\}$ (with cardinality depending on the size of S).

Every system $\mathcal{M}(S)$ has the following properties:

1. Every $M \in \mathcal{M}(S)$ contains a graph with a z-labelled universal vertex.
2. If $M \in \mathcal{M}(S)$, $M' \subseteq M$ and M' contains a graph with a z-labelled universal vertex, then $M' \in \mathcal{M}(S)$, too.

Furthermore, for an arbitrary one-to-one mapping Φ from Σ to the set of all integers greater than one and for every $M \in \mathcal{M}(S)$ we define the set M_Φ as the set of all unlabelled graphs obtained from the graphs of M in the following way:

We replace every non-universal vertex (labelled $q \in \Sigma$) by the complete graph on $\Phi(q)$ vertices and every universal vertex (labelled $q_0 \in \Sigma$) by the complete graph on $\Phi(q_0) - 1$ vertices (see [1]). The vertices belonging to different complete graphs are adjacent if and only if the vertices corresponding to them in the original graph are adjacent. These graphs are called generalized wheels.

In [1] we supposed Φ has the following property:

$$\forall q, q', q'' \in \Sigma: \quad \Phi(q) + \Phi(q') > \Phi(q''). \tag{1}$$

According to Corollary 1 and Corollary 2 of [1] we obtain:

Lemma 4. *There exists no algorithm which, given a Post system S, will determine whether there exists a graph (finite graph) G such that $N(G) = M_\Phi$ for some $M \in \mathcal{M}(S)$.*

Next we suppose that Φ has in addition to (1) the property:

$$\forall q \in \Sigma - \{z\}: \quad \Phi(z) > \Phi(q) + 1. \tag{2}$$

Lemma 5. *Let S be an arbitrary Post system. There exists a graph (finite graph) G with $N(G) \in \{M_\Phi; M \in \mathcal{M}(S)\}$ if and only if there exists a digraph (finite digraph) $\vec{G}$ such that $N(\vec{G}) = \{\bigcup_{H \in M_\Phi} \vec{D}(H)\}$ for some $M \in \mathcal{M}(S)$.*

Proof. a) If there exists a graph G with $N(G) = M_\Phi$ for some $M \in \mathcal{M}(S)$, then according to Lemma 1 and Lemma 2 there exists a digraph $\vec{G}$ with $N(\vec{G}) = \{\bigcup_{H \in M_\Phi} \vec{D}(H)\}$.

b) Conversely, if there exists a digraph $\vec{G}$ with $N(\vec{G}) = \{\bigcup_{H \in M_\Phi} \vec{D}(H)\}$, then according to Lemma 3 there exists a graph G with non-empty edge set such that $N(G) \subseteq M_\Phi$. With respect to the properties of $\mathcal{M}(S)$ it is sufficient to prove that $N(G)$ contains a graph with $\Phi(z) - 1$ universal vertices. We show that the construction in the proof of Lemma 3 guarantees G is of the required structure.

As $N(\vec{G}) = \{\bigcup_{H \in M_\Phi} \vec{D}(H)\}$, a graph $\vec{D}(L)$, where L is a generalized wheel with $\Phi(z) - 1$ universal vertices, is an induced subgraph of $\vec{G}$. Thus L is an induced subgraph of G' from the proof of Lemma 3. Let x be an arbitrary universal vertex of L. $N(x, G')$ contains a generalized wheel on $\Phi(z) - 2$ vertices as an induced subgraph. With respect to the structure of M_Φ for $M \in \mathcal{M}(S)$ and the property (2) of Φ it is obvious that a generalized wheel with $\Phi(z) - 1$ universal vertices is a component of $N(x, G')$. If we denote this component as K, then $N(x_K, G)$ is isomorphic to this wheel. □

Proof of Theorem 3 we obtain using Lemma 4 and Lemma 5.

Problem. Is it true that both modifications of the T–Z problem are algorithmically unsolvable in the class of all asymmetric digraphs?

References

[1] P. Bugata, *On algorithmic solvability of Trahtenbrot-Zykov problem*, KAM Series: Discrete Mathematics and Combinatorics, Charles University, Prague, 1990, preprint.
[2] P. Bugata, A. Nagy, *All oriented trees are realizable*, to appear.
[3] V. K. Bulitko, *On graphs with given vertex-neighbourhoods*, Trudy Mat. Inst. Steklov.
[4] P. Hell, *Graphs with given neighbourhoods I*, Problémes Combinatoires et Théorie des Graphes (Proc. Colloq. Orsay 1976), C. N. R. S., Paris, 1978, pp. 219–223.
[5] A. A. Zykov, *Problem 30*, Theory of Graphs and Its Applications (Proc. Symp. Smolenice 1963), Prague, 1964, pp. 164–165.

Peter Bugata
Department of Geometry and Algebra,
P. J. Šafárik University,
041 54 Košice, Czechoslovakia

Fourth Czechoslovakian Symposium on
Combinatorics, Graphs and Complexity
J. Nešetřil and M. Fiedler (Editors)
© 1992 Elsevier Science Publishers B.V. All rights reserved.

Cartesian Dimensions of a Graph

GUSTAV BUROSCH, PIER VITTORIO CECCHERINI

Four kinds of dimensions $\mu_i(G)$ of a finite graph G are defined and studied, $i = 1, 2, 3, 4$. Dimension $\mu_1(G)$ is the "Sabidussi dimension" and $\mu_2(G)$ coincides with the "isometric dimension" of G defined by R. L. Graham and P. M. Winkler. We list some properties of these dimensions and calculate their value in some particular cases: complete graphs, cycles, trees, (3,4)-connected graphs, and n-partite graphs $K_{m_1,m_2,\ldots,m_n}$. The dimension of a product is always the sum of those of the factors; this leads for instance to the value of the dimension of a hypercube and of an n-dimensional grid $P_{m_1} \square P_{m_2} \square \cdots \square P_{m_n}$. All proofs are omitted for shortness and can be found in [2], where some other results are also given.

1. Introduction

Any graph $G = (V, E)$ under consideration will be *simple*, *finite* and *connected*. There are many possibilities of defining a *dimension* of G. One direction is based on the concept of cartesian product of graphs, but there are other reasonable possibilities, see for instance F. Harary and R. A. Melter [9]. We will only deal with concepts of dimension on the basis of cartesian products.

The "Sabidussi dimension" $\mu_1(G)$ will be defined in §2 as the number of the prime factors which appear in the prime factor decomposition of G, cf. [11], [12].

The other dimensions under consideration use *irreducible embeddings* of G into cartesian products. Different kinds of such embeddings shall give different concepts of the dimensions $\mu_i(G)$, $i \in I = \{2,3,4\}$. The dimension $\mu_i(G)$, $i \in I$, will be the maximum number of factors of a cartesian product $P = \square G_k$ such that there exists an irreducible embedding $\alpha_i \colon G \hookrightarrow P$.

Cases $i = 2, 3, 4$ correspond respectively to the cases when $\alpha_i(G)$ is an *isometric embedding*, an *induced embedding* or a *general embedding*, i.e. when G is isomorphic to an *isometric subgraph* or an *induced subgraph* or simply a *subgraph* of P, respectively.

In the present paper all proofs are omitted for shortness and can be found in [2], where some other results are also given.

2. The Sabidussi dimension of a graph

The *distance* $d_G(x,y) = d(x,y)$ between vertices x, y of G is the number of edges of a shortest xy-path. Clearly (V, d) is a *metric space*.

Let $\{G_i : i \in I = \{1, 2, \ldots, n\}\}$ be a family of graphs with distances d_i, $i \in I$. The (cartesian) product $\Box G_i$ is the graph defined by:

$$V(\Box G_i) = \prod V(G_i), \quad \{x, y\} \in E(\Box G_i) \quad \text{iff} \quad \sum d_i(x_i, y_i) = 1,$$

where $x = (x_1, x_2, \ldots, x_n)$, $y = (y_1, y_2, \ldots, y_n)$. The distance d of $\Box G_i$ turns out to be: $\forall x, y \in V(\Box G_i) : d(x, y) = \sum d_i(x_i, y_i)$.

We shall also consider the product of a family of graphs consisting of only one member: if $I = \{1\}$, then $\Box G_i = G_1$.

The *trivial graph* $U = U(x)$ has only one vertex x and no edges. A graph G is called *prime* if it is not trivial and it is not the product of non-trivial graphs. According to a celebrated result by G. Sabidussi [11] and by V. G. Vizing [12], every connected non-trivial graph of "finite type" has, up to isomorphisms, a *unique prime factor (cartesian) decomposition*.

The *Sabidussi dimension* $\mu_1(G)$ of a finite connected non-trivial graph G is the number of factors of the prime factor decomposition of G. For the trivial graph U we shall assume $\mu_1(U) = 0$.

Obviously one has the following *product theorem*:

$$\mu_1(\Box_i G_i) = \sum_i \mu_1(G_i).$$

3. Embeddings of graphs into cartesian products

In order to introduce the other types of dimensions of a graph, we need some facts concerning embeddings.

A graph $G' = (V', E')$ is a *subgraph* of $G = (V, E)$ if $V' \subseteq V$ and $E' \subseteq E$. Obviously for any vertices x, y of G', $d_G(x, y) \leq d_{G'}(x, y)$ holds. Moreover the subgraph G' is called *induced* (in G) if any edge of G between vertices of G' belongs to G', too. Finally the subgraph G' is called *isometric* (in G) if for any vertices x, y of G', $d_G(x, y) = d_{G'}(x, y)$.

An *embedding* (resp. an *induced embedding*, or an *isometric embedding*) α of a graph G into a graph H is an isomorphism of G onto a subgraph (resp. an induced subgraph or an isometric subgraph) H' of H. We shall write

$$\alpha : G \hookrightarrow H, \quad \alpha : G \hookrightarrow_{\text{induced}} H, \quad \alpha : G \hookrightarrow_{\text{isometric}} H,$$

resp., and we shall say that G is *embeddable* (resp. *induced* or *isometric embeddable*). Note that any isometric embedding is an induced one, too.

The embedding $\text{id}_G : G \hookrightarrow G$ is *trivial* and exists in any case.

Note that any embedding $\alpha : G \hookrightarrow G_1 \Box G_2 \Box \cdots \Box G_m$ induces embeddings of type $\beta : G \hookrightarrow G_1 \Box G_2 \Box \cdots \Box G_m \Box G_{m+1} \Box \ldots \Box G_{m+n}$ where $G_{m+1}, \ldots, G_{m+n}$ are arbitrary graphs; one can assume $\beta(x) = (y_1, \ldots, y_{m+n})$ where $(y_1, \ldots, y_m) = \alpha(x)$ and $y_i \in V(G_i)$ is any fixed vertex, $i = m+1, \ldots, m+n$.

An embedding $\alpha : G \hookrightarrow \Box_{i \in I} G_i$, $I = \{1, 2, \ldots, n\}$ is called *irreducible* if for any $h \in I$ there exist vertices $x, y \in V(G)$ such that $\beta(x) = (x_1, \ldots, x_n)$, $\beta(y) = (y_1, \ldots, y_n)$ with $x_h \neq y_h$.

Let us mention that I. Havel and J. Morávek [8] characterized those graphs which are embeddable into a hypercube.

4. The cartesian dimensions of a graph

Let G be a finite connected non-trivial graph. The cartesian dimension $\mu_4(G)$, $\mu_3(G)$ and $\mu_2(G)$ resp., is the maximum number of factors of an irreducible embedding

$$\alpha: G \hookrightarrow G_1 \square G_2 \square \cdots \square G_m \qquad (1)$$

or

$$\alpha: G \hookrightarrow_{\text{induced}} G_1 \square G_2 \square \cdots \square G_m \qquad (2)$$

or

$$\alpha: G \hookrightarrow_{\text{isometric}} G_1 \square G_2 \square \cdots \square G_m \qquad (3)$$

respectively. For the trivial graph U we shall assume $\mu_2(U) = \mu_3(U) = \mu_4(U) = 0$.
Note that directly from the definitions it follows that, if $G \not\cong U$, then

$$1 \leqslant \mu_1(G) \leqslant \mu_2(G) \leqslant \mu_3(G) \leqslant \mu_4(G). \qquad (4)$$

Below we will see that $\mu_4(G) \leqslant |E(G)|$.

The previous definition of $\mu_2(G)$ agrees with the concept of the *isometric dimension* of G given by R. L. Graham and P. M. Winkler [7]. They assume that α is *irredundant*, i.e., that it is an irreducible embedding with the additional property that for any $h = 1, 2, \ldots, n$ and any $x \in V(G_h)$ there exists $y \in V(G)$ such that $\alpha(y) = (y_1, \ldots, y_n)$ with $y_h = x$. Indeed any irreducible embedding can be made irredundant by discarding each unused vertex.

R. L. Graham and P. M. Winkler stated the following *uniqueness theorem*.

Theorem [6, 7]. *Every connected graph G has a unique canonical irreducible isometric embedding $\alpha: G \hookrightarrow_{\text{isometric}} G_1 \square G_2 \square \cdots \square G_k$ in which every factor G_i has only the trivial embedding $\text{id}_{G_i}: G_i \hookrightarrow G_i$. For any other irredundant isometric embedding $G \hookrightarrow_{\text{isometric}} H_1 \square H_2 \square \cdots \square H_m$ there is a surjection $\varphi: \{1, \ldots, k\} \mapsto \{1, \ldots, m\}$ between the index sets and irredundant isometric embeddings $H_i \hookrightarrow_{\text{isometric}} \square_j : \{G_j : \varphi(j) = i\}$ for which everything commutes; that is the canonical irredundant isometric embedding can be factored through any other.*

From this theorem one obtains the following *product theorem* which was not explicitely mentioned in the papers of Graham and Winkler, [7], [13]:

$$\mu_2(\square_i G_i) = \sum_i \mu_2(G_i).$$

Below we will state the product theorem also for μ_3 and μ_4.

On the other hand, the *uniqueness theorem fails* for embeddings in the general case. Indeed the graph G in Fig. 1 has the three irredundant embeddings

$$G \hookrightarrow (K_4 - k) \square (K_4 - k), \quad G \hookrightarrow C_3 \square C_3, \quad G \hookrightarrow K_{2,3} \square K_2$$

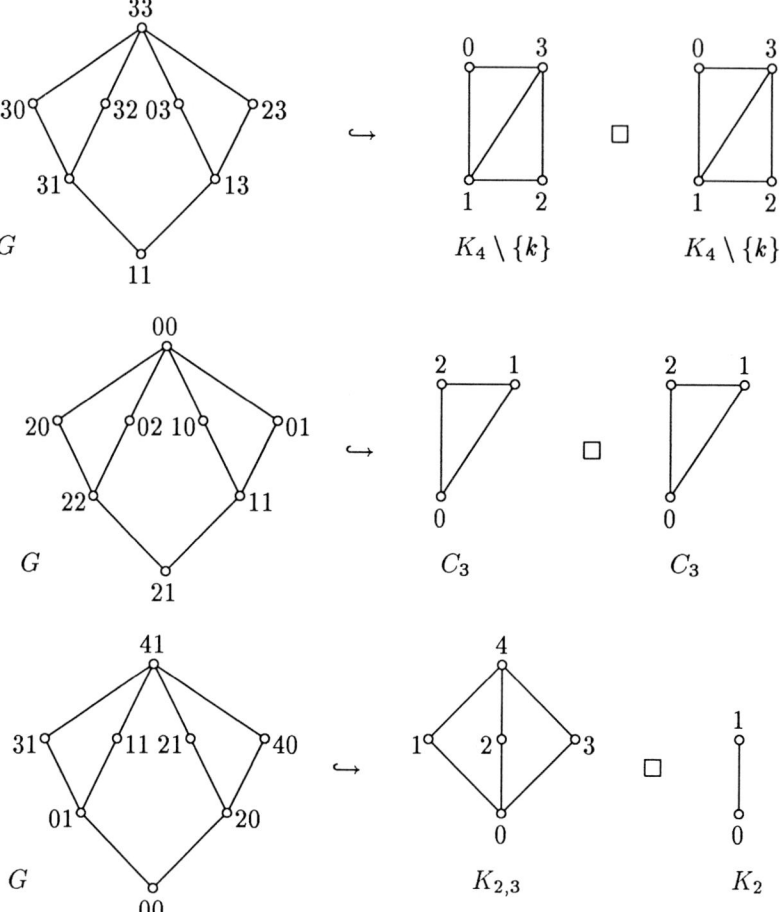

FIGURE 1

in which every factor has only the trivial embedding but no two factors in different embeddings are isomorphic.

In order to define the canonical embedding, R. L. Graham and P. M. Winkler introduced, following an idea of D. Z. Djoković [4], a relation θ on $E(G)$:

$$\{x,y\}\theta\{x',y'\}: \leftrightarrow d_G(x,x') + d_G(y,y') \neq d_G(x,y') + d_G(x',y)$$

which is well defined, reflexive and symmetric, but not generally transitive; denote by $E_1, \ldots, E_k$ the equivalence classes of the transitive closure θ' of θ. Let G_i be the graph obtained from G by identifying any two vertices iff they are connected by the path consisting of edges belonging to $E \setminus E_i$. This defines a canonical map $\alpha_i: V(G) \to V(G_i)$, and gives the required embedding

$$\alpha: G \hookrightarrow G_1 \,\square\, G_2 \,\square\, \cdots \,\square\, G_k: x \mapsto (\alpha_1(x), \ldots, \alpha_k(x)).$$

R. L. Graham and P. M. Winkler considered the value $\mu_2(G)$ in the special cases when G is a tree or a cycle. They obtained that $\mu_2(G) \leqslant |E(G)|$, with equality iff G is a tree, and that $\mu_2(C_{2n+1}) = 1$, $\mu_2(C_{2n}) = n$. Moreover they noted that any triangulated planar graph G has only the trivial isometric embedding id_G.

5. Results for μ_3 and μ_4

One can see [2] that for each inequality in (4) there exists a graph $G^{(i)}$ for which $\mu_i(G^{(i)}) < \mu_{i+1}(G^{(i)})$, $(i = 1, 2, 3)$. Indeed

$$1 = \mu_1(P_2) < \mu_2(P_2) = 2, \quad 1 = \mu_2(C_7) < \mu_3(C_7) = 2 < \mu_4(C_7) = 3.$$

The main concept used in [2] for evaluating the dimension of a graph is the *direction of an edge* associated with an embedding into a product.

Let $\alpha \colon G \hookrightarrow G_1 \square G_2 \square \cdots \square G_n$ be any embedding of a graph $G = (V, E)$ into the product of n graphs G_i, $i = 1, \ldots, n$. The *edge coloring* associated with the embedding α is a map

$$\alpha^* \colon E \to \{1, 2, \ldots, n\},$$

where $\alpha^*(\{x,y\}) = h$ iff $\alpha(x) = (x_1, x_2, \ldots, x_n)$, $\alpha(y) = (y_1, y_2, \ldots, y_n)$, $(x_h, y_h) \in E(G_h)$ and $x_i = y_i$ for $i \neq h$. The color $\alpha^*(e)$ of an edge $e \in E$ will be also called the *direction* of e.

A graph G will be called (3,4)-*connected* if for any two edges e and e' of G there exists a sequence $e = e_1, e_2, \ldots, e_q = e'$ such that e_i, e_{i+1} are either two edges of a C_3 or opposite edges of a C_4 in G.

Note that there are two similar concepts: (a) "strongly triangulated" graphs in the sense of R. Nowakowski and I. Rival [10] when any two vertices are joined by a sequence of triangles with consecutive ones sharing an edge; (b) graphs such that any two edges can be connected by a sequence of triangles with consecutive ones sharing an edge, cf. R. L. Graham [5]. Obviously (b) implies both (a) and the (3,4)-connectedness. Fig. 2 shows an example of a (3,4)-connected graph G verifying (a) but not (b); Fig. 3 (resp. Fig. 4) shows an example of a (3,4)-connected (resp. strongly triangulated) graph which is not strongly triangulated (resp. (3,4)-connected).

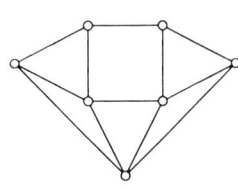
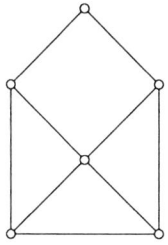
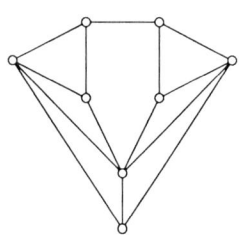

FIGURE 2 FIGURE 3 FIGURE 4

Let us consider the graph G of Fig. 2. As all edges in a triangle must have the same color and two opposite edges in a C_4 must have the same color too, it is clear that all edges of G must have the same color and thus $\mu_4(G) = 1$.

Up to now we have no general methods for calculating $\mu_3(G)$ and $\mu_4(G)$. Our only way is—as in the previous example—to use the knowledge of the dimensions of some subgraphs of G. Especially it is of interest to know subgraphs of dimension 1.

Proposition 1. *For any $(3,4)$-connected graph G, it is $\mu_3(G) = \mu_4(G) = 1$.*

Proposition 2. *For any graph G, it is $\mu_i(G) \leqslant |E(G)|$, $i = 3, 4$ where the equality holds iff G is a tree.*

Proposition 3. *For any cycle of length $\geqslant 4$, it is*

$$\mu_3(C_{2n+1}) = n - 1, \quad \mu_4(C_{2n+1}) = n, \quad \mu_3(C_{2n}) = \mu_4(C_{2n}) = n.$$

Proposition 4. *Let $n \geqslant 2$ and $m_1, m_2, \ldots, m_n \geqslant 1$. The dimension of the complete n-partite graph $G = K_{m_1, m_2, \ldots, m_n}$ is given by:*

$$\mu_i(G) = \begin{cases} 2 & \text{if } G = K_{2,2} \\ n & \text{if } G = M_{1,n} \\ 1 & \text{otherwise} \end{cases} \quad (i = 3, 4).$$

Proposition 5. *For any $n \geqslant 2$ and any graphs $G_1, G_2, \ldots, G_n$,*

$$\mu_i(G_1 \square G_2 \square \cdots \square G_n) = \sum \mu_i(G_t), \quad i = 3, 4.$$

Corollary. *For the n-dimensional grid $P_{m_1} \square P_{m_2} \square \cdots \square P_{m_n}$, $m_1, m_2, \ldots, m_n \geqslant 1$ (in particular for the n-dimensional hypercube $(P_1)^n$),*

$$\mu_i(P_{m_1} \square P_{m_2} \square \cdots \square P_{m_n}) = m_1 + m_2 + \ldots + m_n, \quad i = 3, 4.$$

References

[1] G. Burosch, P. V. Ceccherini, *Cartesian dimensions of graphs*, preprint, 1990.
[2] ———, *On the cartesian dimensions of graphs*, Proc. Int. Conf. on Graphs and Hypergraphs, Varenna, June 1991, submitted.
[3] P. V. Ceccherini, *An intrinsical definition of cartesian product of graphs*, preprint, 1989.
[4] D. Z. Djoković, *Distance preserving subgraphs of hypercubes*, J. Comb. Th. (B) **14**, 263–267.
[5] R. L. Graham, *Isometric embeddings of graphs*, Preprint.
[6] R. L. Graham, P. M. Winkler, *Isometric embeddings of graphs*, Proc. Nat. Acad. Sci. USA **81**, 7259–7260.

[7] R. L. Graham, P. M. Winkler, *On isometric embeddings of graphs*, Trans. Amer. Math. Soc. **288**, 527–736.

[8] I. Havel, J. Morávek, *B-valuations of graphs*, Czechoslovak Mathematical Journal **22**, 338–351.

[9] F. Harary, R. A. Melter, *On the metric dimension of a graph*, Ars Combin. **2**, 191–195.

[10] R. Nowakowski, I. Rival, *Retract rigid cartesian products of graph*, Discrete Math. **70**, 169–184.

[11] G. Sabidussi, *Graph multiplication*, Math. Zeitschr. **72**, 446–457.

[12] V. G. Vizing, *The cartesian product of graphs*, Vychislitel'nye Sistemy **9**, 30–43.

[13] P. M. Winkler, *The metric structure of graphs: Theory and Applications*, Survey in Combinatorics 1987 Pap. 11th Br. Combin. Conf., London Math. Soc. Lect. Notes Ser. 123, London, England, 1987, pp. 197–221.

Gustav Burosch
Fachbereich Mathematik Universität Rostock,
D-2500 Rostock,
Federal Republic of Germany

Pier Vittorio Ceccherini
Dipartimento di Mathematica "G. Castelnuovo",
Università di Roma "La Sapienza",
P. le Aldo Moro 2, 00185 Roma
Italy

Fourth Czechoslovakian Symposium on
Combinatorics, Graphs and Complexity
J. Nešetřil and M. Fiedler (Editors)
Elsevier Science Publishers B.V.

The Steiner Minimal Tree Problem in L_p^2

DIETMAR CIESLIK

The Steiner Minimal Tree Problem has been considered in the plane with Euclidean and rectilinear norm (see [7] and [8]). We intend to discuss this problem for all planes with p-norm.

Denote by L_p^2 the affine plane with norm

$$\|(t_1, t_2)\|_p = (|t_1|_2^p + |t_2|^p)^{\frac{1}{p}} \quad \text{for } 1 \leqslant p < \infty$$

and

$$\|(t_1, t_2)\|_\infty = \max\{|t_1|, |t_2|\}.$$

L_p^2 is a two-dimensional Banach space with unit ball $B(p) = \{x : \|x\|_p \leqslant 1\}$. Let N be a finite set in L_p^2. The Steiner Minimal Tree Problem is to find a connected graph $G = (V, E)$ which interconnects the points of N with minimal length. That means $V \supseteq N$ and

$$l_p(T) := \sum_{vv' \in E} \|v - v'\|_p = \min!$$

where

$$\underline{vv'} = \{tv + (1-t)v' : 0 \leqslant t \leqslant 1\}.$$

Such a graph must be a tree which is called a *Steiner Minimal Tree* (SMT) for N in L_p^2. The points from $V - N$ are called *Steiner points*.

If we do not allow Steiner points, that means $V = N$, then we get a *Minimal Spanning Tree* (MST) for N in L_p^2.

The practical motivation to consider such trees in L_p^2 is given by constructing transport and communication networks. It is a problem of modelling to decide which plane is the best for a given situation (see [11] and [12]).

1. The combinatorial structure of shortest trees

Let N be a finite set in L_p^2. It is easy to find an MST for N:

Observation 1. *A minimal spanning tree in the weighted graph $G = (N, \binom{N}{2}, D)$, where $D(v, v') = \|v - v'\|_p$, is an MST for N in L_p^2.*

It is well-known that a minimal spanning tree in a graph can be found in polynomially bounded time (see [1] and [10]). Hence an MST is simple to determine in L_p^2. But this is not true for SMT.

Denote by $g(v)$ the degree of the vertex v in the tree T.

Observation 2. *Let $T = (V, E)$ be an SMT for N. Then it holds*
(a) $g(v) \geqslant 1$ for all v in V;
(b) *without loss of generality*, $g(v) \geqslant 3$ for all v in $V - N$;
(c) *with respect to (b)*, $\mathrm{card}(V - N) \leqslant \mathrm{card}\, N - 2$.

The *proof* of (a) is obvious. The triangle inequality of the norm implies (b). And (c) is given by

$$2 \cdot (\mathrm{card}\, V - 1) = 2 \cdot \mathrm{card}\, E = \sum_{v \in V} g(v)$$

$$\sum_{v \in N} g(v) + \sum_{v \in V-N} g(v) \geqslant \mathrm{card}\, N + 3 \cdot \mathrm{card}(V - N).$$

□

Observation 3. *Let $T = (V, E)$ be an SMT for N in L_p^2. Then it holds $g(v) \leqslant 4$ for all v in V.*

The proof follows by application of strict convexity of $B(p)$ for $1 < p < \infty$ (see [2]) and for $p = 1, \infty$ by a statement in [8]. □

Especially vertices with degree four can be met in L_1^2 and L_∞^2. But in L_2^2 all vertices of an SMT have degree less than four.

2. Reductions

Lemma 1. *Let $N = \{v_1, \ldots, v_4\}$ be a set with four elements in L_p^2. If we search a point q with $\sum_{v \in N} \|v - q\|_p = \min!$, it is sufficient to look for a solution by*
(a) *If $N = \{v_1, \ldots, v_4\}$ is collinear in this order, then $q = v_2$ or $q = v_3$.*
(b) *Let $\mathrm{conv}\, N = \mathrm{conv}\{v_1, v_2, v_3\}$ be a triangle, then $q = v_4$.*
(c) *Let $\mathrm{conv}\, N$ be a quadrangle with v_1, v_2, v_3, v_4 in this order on its boundary, then $q \in \overline{v_1 v_3} \cap \overline{v_2 v_4}$.*

For a *proof* see [2]. □

It easy to see that in an SMT the cases (a) and (b) are impossible for a Steiner point of degree four, if N is the set of its neighbours. Hence we remove such vertices in an arbitrary given tree by

Procedure 1. *Let $T = (V, E)$ be a tree for N, v in $V - N$ with $g(v) = 4$ and $v_1, \ldots, v_4$ be the neighbours of v.*

Define $G = (V - \{v\}, E - \{vv_i : i = 1, \ldots, 4\})$. G is a forest with four components $G_1, \ldots, G_4$, where G_i contains the vertex v_i.
Define $G_{ij} = G_i \cup G_j \cup (\emptyset, \{v_i v_j\})$ for $i, j = 1, \ldots, 4, i \neq j$.

If we repeat this procedure we get a family of trees without Steiner points of degree four. A suitable composition of some trees in this family constitutes an SMT, if we minimize the lengths.

Lemma 2. *Let $T = (V, E)$ be a tree for N. Then it holds $g(v) = 1$ for all v in N and $g(v) = 3$ for all v in $V - N$ iff $\operatorname{card}(V - N) = \operatorname{card} N - 2$.*

This lemma is simple to prove.

Procedure 2. *Let $T = (V, E)$ be a tree for N, v in N with $g(v) > 1$.*
Define $G = (V - \{v\}, E - \{vv' : v' \text{ is a neighbour of } v\})$. G is a forest with $g(v)$ components $G_1, \ldots, G_{g(v)}$, where $G_i = (V_i, E_i)$.
Define $G_{(i)} = (V_i \cup \{v_i\}, E_i \cup \{v_i v' : v' \text{ is a neighbour of } v \text{ in } G \text{ and } v' \text{ is in } V_i\})$ for $i = 1, \ldots, g(v)$. (v_i not in V).

If we repeat this procedure we get a family of trees in which for every tree $T = (V, E)$ for N it holds $g(v) = 1$ for all v in N and $g(v) = 3$ for all v in $V - N$ and hence by Lemma 2 $\operatorname{card}(V - N) = \operatorname{card} N - 2$. Such trees are called *full trees* (see [7]).

3. A solution method

As a consequence of the last statements, it is sufficient to look for solution methods for full trees. Let $T = (V, E)$ be a full tree for N with $V = \{v_1, \ldots, v_{2n-2}\}$ and $N = \{v_1, \ldots, v_n\}$ ($\operatorname{card} N = n \geqslant 3$). Let $(a_{ij})_{i,j=1,\ldots,2n-2}$ be a matrix with $a_{ij} = 1$ iff v_i is adjacent to v_j in T (the adjacency matrix). Then it is necessary to minimize the function

$$F_p(T) = F_p(v_{n+1}, \ldots, v_{2n-2}) :=$$
$$= \sum_{i=1}^{n} \sum_{j=n+1}^{2n-2} a_{ij} \|v_i - v_j\|_p + \sum_{i=n+1}^{2n-3} \sum_{j=i+1}^{2n-2} a_{ij} \|v_i - v_j\|_p.$$

It is possible to do this by well-known methods in nonlinear programming (see [11]).
Now we can find an SMT for a finite set N in L_p^2 by the following algorithm:

Procedure 3. *Let N be a finite set in L_p^2.*
1. *Generate all trees $T = (V, E)$ for N with $\operatorname{card} N \leqslant \operatorname{card} V \leqslant 2 \cdot \operatorname{card} N - 2$ and $1 \leqslant g(v) \leqslant 4$ for all v in N and $3 \leqslant g(v) \leqslant 4$ for all v in $V - N$.*
2. *Reduce every tree by Procedures 1 and 2 to a family of full trees.*
3. *Minimize the function $F_p(.)$ for every full tree.*
4. *Construct an SMT from suitable full trees of minimal length.*

The first step of Procedure 3 uses exponential time. Moreover it was shown that in the cases $p = 1, 2$ and ∞, the Steiner Minimal Tree Problem is NP-hard (see [5] and [6]). Since an MST is easy to find (see Observation 1), such a tree can be used as an approximative solution for an SMT. Then we are interested in the defect

$$m(p) := \inf_{N \text{ finite}} \frac{l_p(T)}{l_p(T')}$$

where T is an SMT and T' an MST for N in L_p^2.

It can be shown that

$$\frac{1}{\sqrt[4]{6}} \leqslant m(p) \leqslant \sqrt[4]{\frac{2}{3}} \quad \text{for all } 1 \leqslant p \leqslant \infty.$$

For the proof and for some better bounds for $m(p)$ depending on p see [3]. Moreover it is known that $m(1) = m(\infty) = \frac{2}{3}$ (see [9]) and $\frac{4}{5} \leqslant m(2) \leqslant \frac{\sqrt{3}}{2}$ (see [4] and [7]). □

References

[1] D. Cheriton, R. E. Tarjan, *Finding minimum spanning trees*, SIAM J. Computing **5** (1976), 724–742.

[2] D. Cieslik, *Das Steinerproblem für Bäume minimaler Länge in der Banach-Minkowski-Ebene*, Preprint Mathematik no. 15 (1986), Greifswald.

[3] _____, *The Steiner-ratio in Banach-Minkowski-planes*, Contemporary Methods in Graph Theory (R. Bodendieck, eds.), Mannheim, 1990, to appear.

[4] D. Z. Du, F. K. Hwang, *A new bound for the Steiner ratio*, Trans. Am. Math. Soc. **278** (1983), 137–148.

[5] M. R. Garey, R. E. Graham, D. S. Johnson, *The complexity of computing Steiner minimal trees*, SIAM J. Appl. Math. **32** (1977), 826–834.

[6] M. R. Garey, D. S. Johnson, *The rectilinear Steiner tree problem is NP-complete*, SIAM J. Appl. Math. **32** (1977), 826–834.

[7] E. N. Gilbert, H. O. Pollak, *Steiner minimal trees*, SIAM J. Appl. Math. **16** (1968), 1–29.

[8] M. Hanan, *On Steiner's problem with rectilinear distance*, SIAM J. Appl. Math. **14** (1966), 255–265.

[9] F. K. Hwang, *On Steiner minimal trees with rectilinear distance*, SIAM J. Appl. Math. **30** (1976), 104–114.

[10] J. B. Kruskal, *On the shortest spanning subtree of a graph and the traveling salesman problem*, Proc. Am. Math. Soc. **7** (1956), 48–50.

[11] R. F. Love, J. G. Morris, G. O. Weselowsky, *Facilities Location-Models and Methods*, North-Holland, 1989.

[12] J. McGregor Smith, *Generalized Steiner network problems in engineering design*, Design optimization (1985), 119–161.

Dietmar Cieslik
Fachbereich Mathematik,
Ernst-Moritz-Arndt-Universität,
Jahnstr. 15a, 2200 Greifswald, Germany

Fourth Czechoslovakian Symposium on
Combinatorics, Graphs and Complexity
J. Nešetřil and M. Fiedler (Editors)
© 1992 Elsevier Science Publishers B.V. All rights reserved.

On k-Connected Subgraphs of the Hypercube

TOMÁŠ DVOŘÁK

The *hypercube of dimension* n, denoted by Q_n, is the graph of 2^n vertices labelled by binary vectors of length n, an edge joining two vertices whenever the corresponding vectors differ in exactly one coordinate. We call a graph *cubical* if it is a subgraph of Q_n for some n. Cubical graphs have been intensively studied in the last 20 years and the topic has recently found new applications in computer science (see [4]).

Each cubical graph is necessarily bipartite, but this condition is far from sufficient. For the sake of characterization of cubical graphs it is reasonable to introduce the following concept:

A *C-valuation* of the graph G is its edge coloring such that

(i) in each cycle no color occurs an odd number of times
(ii) in each open path there is a color that occurs an odd number of times.

A graph is cubical iff there exists its C-valuation ([3]).

In the following we shall restrict ourselves to cubical graphs. By a C_n-*valuation* of G we understand such a C-valuation of G that exactly n colors are used. The smallest n for which G is a subgraph of Q_n is called the *cubical dimension* of G and denoted by $\mathrm{cd}(G)$. If G is connected, $\mathrm{cd}(G)$ is the smallest n such that there exists a C_n-valuation of G (the color of an edge in the C_n-valuation corresponds to the coordinate in which its endvertices differ). To state a simple example, let P_k be the path of $k+1$ vertices, then $\mathrm{cd}(P_k) = \lceil \log_2(k+1) \rceil$. This follows from the fact that the hypercube is hamiltonian, as it is the n-th Cartesian power of the complete graph K_2.

Following [5], we shall define the *maximal cubical dimension* of G, denoted $\mathrm{maxdim}(G)$, to be the largest n such that G is a subgraph of Q_n and for each $i \in \{1,\ldots,n\}$ there are two adjacent vertices in G which differ in the i-th coordinate. Clearly, $\mathrm{maxdim}(G)$ is the largest n for which a C_n-valuation of G exists.

While to the problem to determine the cubical dimension of various classes of graphs much attention has been paid, not too much is known about the maximal dimension. One of the problems that arise here is to determine an upper bound for $\mathrm{maxdim}(G)$. The first known result in this respect is

$$\mathrm{maxdim}(G) \leqslant |V(G)| - 1$$

for any cubical graph G ([2]).
From the results of [1] follows (although explicitely it is stated there only for cd) that for 2-vertex-connected graphs this bound can be improved:

Theorem 1 ([1]). *Let G be a 2-vertex-connected graph. If G is cubical then*

$$\text{maxdim}(G) \leqslant \frac{1}{2}|V(G)|.$$

The assumption of vertex connectivity is substantial. It cannot be replaced by edge connectivity, as can be demonstrated on graphs G_k obtained by gluing together k cycles of length 4 in one vertex. Clearly $|V(G_k)| = 3k+1$ and $\text{maxdim}(G_k) = 2k$.

In the following we shall investigate the maximal cubical dimension of k-edge-connected graphs and use this to obtain upper bounds for the cubical dimension of graphs with minimal degree at least 2. First we prove three auxiliary propositions:

Lemma 1. *Let H be a cubical graph and c one of the colors used in a given C-valuation of H. If all edges of color c are removed, the resulting graph is disconnected.*

Proof. Let $\{x, y\} \in E(H)$ be of blue color. If the graph obtained by removing all blue edges were connected there would be a path P between x and y in H with no blue edges. P and $\{x, y\}$ form a cycle in which the blue color appears exactly once. This contradicts condition (i) in the definition of C-valuation. $\square$

Lemma 2. *Every cubical graph with minimum degree $\delta(G)$ has at least $2^{\delta(G)}$ vertices.*

Proof. By induction on $|V(G)|$. We may suppose without loss of generality that G is connected. Omitting the trivial case, assume G is C-valued, $|V(G)| \geqslant 2$. Removing all edges of a chosen color disconnects G by Lemma 1. Each of the resulting components has minimal degree at least $\delta(G) - 1$ and thus has at least $2^{\delta(G)-1}$ vertices by induction. This implies $|V(G)| \geqslant 2^{\delta(G)}$. $\square$

Lemma 3. *Let T be a tree. Let A, B be the classes of its bipartition. Then*

$$\sum_{u \in A} \bigl(\deg(u) - 1\bigr) = |B| - 1.$$

Theorem 2. *Let G be a cubical k-edge-connected graph ($k \geqslant 2$). Then*

$$\text{maxdim}(G) \leqslant \frac{2^{k-1}}{2^k - 1}(|V(G)| - 1).$$

Proof. Suppose G is C_n-valued and consider its block-cutvertex graph $bc(G)$ with vertex set $V(bc(G)) = B \cup A$, where $B = \{B_1, \ldots, B_l\}$ is the set of blocks (maximal 2-vertex-connected subgraphs) and A the set of cutvertices of G. Then

$$\sum_{i=1}^{l} |V(B_i)| = |V(G)| + \sum_{u \in A} \Bigl(\deg_{bc(G)}(u) - 1\Bigr) = |V(G)| + l - 1 \qquad (*)$$

according to Lemma 3. As the blocks of a k-edge-connected graph are again k-edge-connected, $\delta(B_i) \geq k$. This and Lemma 2 gives

$$l2^k \leq \sum_{i=1}^{l} |V(B_i)|$$

hence

$$l \leq \frac{1}{2^k-1}(|V(G)|-1). \qquad (**)$$

Using Theorem 1 for each block, $(*)$ and $(**)$ we obtain

$$n \leq \sum_{1}^{l} \frac{1}{2}|V(B_i)| \leq \frac{2^{k-1}}{2^k-1}(|V(G)|-1).$$

$\square$

Now turn to the cubical dimension of graphs where each vertex has degree at least 2. If such a graph G is 2-edge-connected, then Theorem 2 implies $cd(G) \leq \frac{2}{3}(|V(G)|-1)$. If, on the other hand, such a graph is not 2-edge-connected, we will show that even a slightly better upper bound is valid. Let $b(G)$ denote the number of bridges and $e(G)$ the number of endvertices (vertices of degree 1) of G. Before proving the main theorem consider the slightly technical

Lemma 4. Let T be a tree of diameter $d(T)$, assume $e(T) \geq 2$. Then

$$cd(T) \leq \lceil \log_2(d(T)+1) \rceil + e(T) - 2.$$

Proof. By induction on $e(T)$. For $e(T) = 2$, T is a path and $cd(T) = \lceil \log_2(d(T)+1) \rceil$ as noticed before. Assume now $e(T) \geq 3$ and choose an edge $\{x_1, x_2\}$ such that $\deg(x_1) \geq 3$. The removal of $\{x_1, x_2\}$ splits the graph into two connected components T_1 and T_2, each having at most $e(T) - 1$ endvertices. Color both T_1 and T_2 by induction using the same set of colors. To complete the C-valuation of T, assign a new color to $\{x_1, x_2\}$. The number of colors used is

$$\max_{i \in \{1,2\}} (\lceil \log_2(d(T_i)+1) \rceil + e(T_i) - 2) + 1 \leq \lceil \log_2(d(T)+1) \rceil + e(T) - 2.$$

$\square$

Now we can proceed with the

Theorem 3. Let G be a cubical graph with minimal degree $\delta(G) \geq 2$ and $b(G)$ bridges. If $b(G) \geq 1$ then

$$cd(G) \leq \frac{2}{3}(|V(G)| - b(G) - 1) + \lceil \log_2(b(G)+1) \rceil - 2.$$

Proof. We shall construct a C-valuation of G as follows: First color each of the maximal 2-edge-connected subgraphs $G_1, \ldots, G_l$ of G with the same set of colors;

the number of colors used is at most $\max_{1 \leq i \leq l} \frac{2}{3}(|V(G_i)| - 1)$ by Theorem 2. Let $c_1, \ldots, c_m$ be all the cutvertices of G. Consider the graph G^* obtained from G by contraction of $G_1, \ldots, G_l$. Color G^* with a new set of $\operatorname{cd}(G^*)$ colors; $\operatorname{cd}(G^*) \leq \lceil \log_2(d(G^*) + 1) \rceil + e(G^*) - 2$ by Lemma 4. It remains to transmit the colors from the edges of G^* onto the corresponding bridges of G. The number of colors used in this C-valuation of G is at most

$$\max_{1 \leq i \leq l} \frac{2}{3}(|V(G_i)| - 1) + \lceil \log_2(d(G^*) + 1) \rceil + e(G^*) - 2.$$

Using the facts that $|V(G_i)| \geq 4$ (otherwise G_i contains a triangle, which contradicts the assumption that G is cubical) and that $l \geq e(G^*)$ (since each endvertex of G^* must be a G_i) leads to

$$\max_{1 \leq i \leq l} |V(G_i)| \leq |V(G)| - 4(l-1) - m = |V(G)| - |V(G^*)| - 3l + 4 \leq$$
$$\leq |V(G)| - |V(G^*)| - 3e(G^*) + 4.$$

Hence using $|V(G^*)| = b(G) + 1$

$$\operatorname{cd}(G) \leq \max_{1 \leq i \leq l} \frac{2}{3}(|V(G_i)| - 1) + \lceil \log_2(d(G^*) + 1) \rceil + e(G^*) - 2 \leq$$
$$\leq \frac{2}{3}(|V(G)| - |b(G)| - 1) - e(G^*) + \lceil \log_2(d(G^*) + 1) \rceil$$

which with $d(G^*) \leq b(G)$ and $e(G) \geq 2$ leads to the desired upper bound. □

The bound of Theorem 3 is reached by each graph H_k (shown on Figure 1 for $k = 3$).

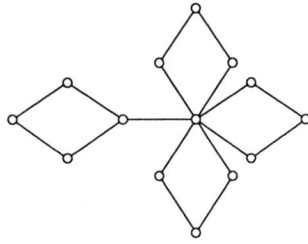

FIGURE 1

Since $\lceil \log_2(b(G) + 1) \rceil - \frac{2}{3}b(G) - 2 \leq -\frac{4}{3}$, Theorems 2 and 3 imply

Corollary. Let G be a cubical graph with minimal degree $\delta(G) \geq 2$ and (G) bridges. Then

(i) $\operatorname{cd}(G) \leq \frac{2}{3}(|V(G)| - 1)$ ([4])
(ii) if $b(G) \geq 1$ then $\operatorname{cd}(G) \leq \frac{2}{3}(|V(G)| - 3)$.

Problem. Is there for k-connected graphs ($k \geqslant 3$) a better upper bound than that in Theorem 1? Clearly, from Theorem 1 it follows that $\operatorname{maxdim}(G) \leqslant \frac{1}{2}(|V(G)|-k+ +2)$ for any cubical k-vertex-connected graph G. On the other hand, all examples we know fulfil $\operatorname{maxdim}(G) \leqslant \frac{1}{3}(|V(G)|-1)$ for $k=3$ and $\operatorname{maxdim}(G) \leqslant \frac{|V(G)|}{2^{k-2}} + k-5$ for $k \geqslant 4$ (The construction of graphs with maxdim reaching the above bounds is the following: For $k = 3$ (and any $r \geqslant 2$) construct such a graph from two copies of Q_3 connected by a "chain" of $P_2 \times P_r$ (see Fig. 2 for case $r = 2$). The case $k \geqslant 4$ is similar; take two copies of Q_k and connect them by a "chain" of $Q_{k-2} \times P_r$).

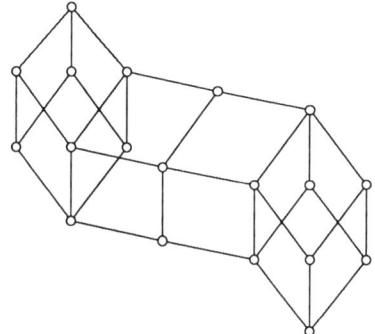

FIGURE 2

References

[1] F. Afrati, C. H. Papadimitriou, G. Papageorgiou, *The complexity of cubical graphs*, Inform. and Control **66** (1985), 53-60.
[2] M. R. Garey, R. L. Graham, *On cubical graphs*, J. Combin. Theory Ser. B **18** (1973), 263–267.
[3] I. Havel, J. Morávek, *B-valuations of graphs*, Čas. pěst. mat. **22** (1972), 338–351.
[4] M. Livingston, Q. F. Stout, *Embeddings in Hypercubes*, Math. Comput. Modelling **11** (1988), 222–227.
[5] A. S. Wagner, *Embedding Trees in the Hypercube*, Techn. Report 204/87, University of Toronto, 1987.

Tomáš Dvořák
Mathematical Institute,
Czechoslovak Academy of Sciences,
Žitná 25,
115 67 Praha
Czechoslovakia

On Some of My Favourite Problems in Various Branches of Combinatorics

P. ERDŐS

I published many papers with related titles. Unfortunately I have not many really new problems, but I will try to spend more time with those questions where something new has happened relatively recently.

1. Consider the graph of the n-dimensional cube Q_n. It has 2^n vertices and $n2^{n-1}$ edges. Denote by $f(n; C_4)$ the smallest integer for which every subgraph of $f(n; C_4)$ edges contains a C_4. I conjectured more than 20 years ago that for every $\varepsilon > 0$ and $n > n_0(\varepsilon)$

$$f(n; C_4) < \left(\frac{1}{2} + \varepsilon\right) n2^{n-1}. \tag{1}$$

As far as I know (1) is still open and I offer 100 dollars for a proof or disproof. Perhaps $f(n; C_4)$ can be determined exactly and perhaps one could try to guess and later prove the exact value for $f(n; C_4)$. Perhaps for sufficiently large c

$$f(n; C_4) < n2^{n-2} + c2^n. \tag{2}$$

It is easy to see that (2) fails for small c. To guess the exact formula for $f(n; C_4)$ it might be worthwhile to determine $f(n; C_4)$ for small values of n.

Let H be any graph which can be embedded into the r-dimensional cube for some r. Denote by $f(n; H)$ the smallest integer for which every subgraph of Q_n containing $f(n; H)$ edges contains H. I conjectured long ago that

$$\frac{f(n; C_6)}{n2^{n-1}} \to 0. \tag{3}$$

Fan Chung disproved (3). In fact she showed that Q_n is the union of four subgraphs neither of which contains a C_6. Thus $f(n; C_6) > n2^{n-3}$ and Graham thought that perhaps

$$f(n; C_6) = n2^{n-3} + 1.$$

Fan Chung could only prove that for ε small

$$f(n; C_6) < \left(\frac{1}{2} - \varepsilon\right) n2^{n-1}.$$

One could try to investigate the following more general problem: What property of H determines the size of $f(n; H)$?

Perhaps it is more reasonable to ask: What property of H detemines

$$\lim_{n \to \infty} \frac{f(n; H)}{n 2^{n-1}}?$$

Perhaps there is an Erdős-Stone-Simonovits type theorem here. Let me remind the reader of our old theorem: Denote by $F_T(G; n)$ the smallest integer for which every graph of n vertices and $F_T(G; n)$ edges contains G as a subgraph. This is of course the classical Turán type problem and the letter T is there to remind the reader of this. We proved that if G is k-chromatic then

$$F_T(G; n) = \frac{1}{2}\left(1 - \frac{1}{k-1}\right) n^2 + o(n^2)$$

and in fact $o(n^2)$ can be replaced in (4) by $o(n^{2-\varepsilon})$. In other words the order of magnitude of $F_T(G; n)$ is determined by the chromatic number for $k > 2$. It is not clear if an analogous result holds for Q_n and if the answer is affirmative what is the property which replaces the chromatic number. By the way as far as I know nobody characterized the graphs which can be embedded in some Q_n.

For two-chromatic graphs many very interesting Turán-type problems remain. Simonovits and I proved

$$F_T(Q_3, n) < c n^{8/5}$$

and we conjectured

$$F_T(Q_3, n) > c' n^{8/5}$$

and in fact we are sure that

$$\lim \frac{F_T(Q_3, n)}{n^{8/5}} = c, \qquad 0 < c < \infty$$

exists, but we could not even prove

$$\frac{F_T(Q_3, n)}{n^{3/2}} = \infty.$$

We stated with some trepidation the following much more general conjecture. Let G be a bipartite graph and let $r \geqslant 3$ be the largest integer for which G has an induced subgraph, each vertex of which has degree $\geqslant r$. Then

$$F_T(G; n) > n^{2-\frac{1}{r-1}+\varepsilon}.$$

If G has no such induced subgraph then

$$F_T(G; n) < c n^{2-\frac{1}{r-1}}.$$

Thus in particular if $r = 3$ our conjecture implies that if G has the vertices x; $y_1, \ldots, y_k$; $z_1, \ldots, z_{\binom{k}{2}}$; x is joined to all the y's and every z is joined to two y's, distinct z's are joined to distinct pairs of y's, then

$$F_T(G; n) < c_k n^{3/2}.$$

Recently Füredi obtained a very simple and ingenious proof of this conjecture, but our general conjecture remains open even for $r = 3$. Füredi's paper is not yet published.

Chung and Trotter considered the following problem. Define the graph $G_{n,r}$ as follows (r is large and $n \to \infty$). It consists of a cycle C_n plus two vertices are joined if their distance is $\leqslant r$. $G_{n,r}$ has n vertices and rn edges. Now let $f(n,r)$ be the smallest integer for which every subgraph of $G_{n,r}$ of $f(n,r)$ edges contains a triangle. They proved

$$(2 - \sqrt{2})rn < f(n,r) < \tfrac{1}{11}(5 + \sqrt{3})rn$$

and of course conjectured that there is a constant c for which

$$f(n,r) = (c + o(1))rn.$$

V. T. Sós suggested that perhaps every subgraph of $G_{n,r}$ having $(c+\varepsilon)rn$ edges will contain for large r and $n \to \infty$ all three-chromatic graphs and that an Erdős-Stone type theorem might perhaps hold. As far as I know this interesting question has not been investigated.

B. Bollobás, Extremal graph theory, London Math. Soc. Monograph 11, Academic Press, London, 1978.

P. Erdős and A. F. Stone, On the structure of linear graphs, Bull. Amer. Math. Soc. 52 (1946), 1087–1091.

P. Erdős and M. Simonovits, Some extremal problems in graph theory, Proc. Coll. Balatonfüred (1969) North Holland 1970, 377–390.

P. Erdős and M. Simonovits, A limit theorem in graph theory, Studia Sci. Math. Hung. 1 (1969), 91–97.

M. Simonovits, Extremal graph theory, Graph Theory 2 (L. Beineke, R. Wilson, eds.), Academic Press, 1983, 161–200.

F. Chung and W. Trotter, Triangle Free Graphs of Restricted Bandwith, Progress In Graph Theory (Waterloo 1982), Academic Press, Toronto (1984), p. 175–190.

Z. Füredi, On a Turán type problem of Erdős, Combinatorica 11, 1 (1991), 75–79.

2. One of my favourite conjectures is our conjecture with Faber and Lovász which goes back to 1972. Let G_i, $1 \leqslant i \leqslant n$, be n edge disjoint complete graphs of size n. Is it true that $\bigcup_{i=1}^{n} G_i$ has chromatic number n? I offer 500 dollars for a proof or disproof of this conjecture. Hindman proved it for $n \leqslant 10$ and recently it was proved by Chang and Lawler that $\bigcup_{i=1}^{n} G_i$ has chromatic number $\leqslant \frac{3}{2}n - 2$. Very recently Jeff Kahn proved that the chromatic number of $\bigcup_{i=1}^{n} G_i$ is $n(1 + o(1))$. (Kahn's paper is not yet published.)

One could try to estimate the maximum chromatic number of the edge disjoint union of $\bigcup_{i=1}^{m} G_i$, $m > n$, where the G_i are complete graphs or one could try to drop the edge disjointness etc; but it is not certain if one gets any really nice conjectures.

A few years ago Nu and Hsu conjectured that if n vertex disjoint triangles T_i, $1 \leqslant i \leqslant n$, are given and their $3n$ vertices are joined by a hamiltonian cycle which does not contain any edges of the triangles, then this $G(3n; 6n)$ has an independent set of n vertices. My only contribution was that I conjectured that the graph is three-chromatic. It is surprising and annoying that these conjectures are perhaps really difficult. I tried but so far did not succeed in finding a related more general conjecture which is simple and natural.

W. I. Chang and E. Lawler, Edge coloring of hypergraph, and a conjecture of Erdős, Faber, Lovász. Combinatorica 8 (1980), 293–295.

J. Kahn, Coloring nearly-disjoint hypergraphs with $n + o(n)$ colors, J. Comb. Th. A (to appear).

3. Nearly 20 years ago Lovász and I conjectured that if $|A_i| = n$, $1 \leqslant i \leqslant m(n)$, $A_i \cap A_j \neq \emptyset$ is a system of n-tuples and $m(n)$ is the smallest integer for which the family cannot be represented by fewer than n elements (i.e. if $|S| < n$ there always is an i for which $A_i \cap S \neq \emptyset$) then

$$\frac{m(n)}{n} \to \infty. \qquad (1)$$

Very recently Jeff Kahn disproved (1). Clearly the $p^2 + p + 1$ lines of size $p + 1$ of a finite geometry can not be represented by fewer than $p + 1$ points, but the number of sets is very large. We in fact proved that there are $n^{\frac{3}{2}+\varepsilon}$ lines ($n = p + 1$) which cannot be represented by fewer than n points and we conjectured that there are $cn \log n$ lines which cannot be represented by fewer than n points. Very recently Jeff Kahn proved this conjecture. I think he also proved that $cn \log n$ is best possible if one considers a subset of lines. In general, it is not even known that $\frac{m(n)}{n} > 3$. (1) can perhaps be strengthened in the following way. To every $\varepsilon > 0$ there is a c_ε ($c_\varepsilon \to \infty$ as $\varepsilon \to 0$) for which there are $c_\varepsilon n$ sets $A_i, |A_i| = n, A_i \cap A_j = \emptyset$ which cannot be represented by fewer than $(1 - \varepsilon)n$ elements.

P. Erdős and L. Lovász, Problems and results on 3 chromatic hypergraphs and some related questions, Infinite and finite sets Coll. Keszthely 1973, 609–627.

S. J. Dow, D. A. Drake, Z. Füredi and J. Larson, A lower bound for the cardinality of a maximal family of mutually intersecting sets of equal size, Proc. 10-th Southeastern Conference, Boca Raton 1985. Congressus Num. 48 (1985), 47–98.

4. I conjectured many years ago that there is an absolute constant C so that in every finite geometry there is a blocking set (i.e. a set which meets every line) which meets every line in at most C points. Unfortunately I never had any idea how to prove or disprove this conjecture. It might be of some interest to prove that C cannot be too small. Also perhaps the following more general result could hold: Let $|S| = n$, $A_i \subset S$, $|A_i| > c\sqrt{n}$ is a family of subsets satisfying $|A_i \cap A_j| < 2$. Then there is a subset $B \subset S$ for which for every i

$$1 \leqslant |A_i \cap B| \leqslant C$$

where C can (in fact must) depend on c and $C \to \infty$ as $c \to 0$, i.e. in a partial geometry there is a blocking set which meets every line in few points.

Jean Larson and I posed the following problem: Is it true that for every $n > n_0(c)$ there is a family of subsets $A_i \subset S$, $|S| = n$, $A_i > n^{\frac{1}{2}-c}$ for which every pair of elements x, y of S is in exactly one A_i? Singhi and Shrikhande proved that it is very likely that for infinitely many n such a family does not exist.

P. Erdős and Jean Larson, On pairwise balanced block designs with the sizes of blocks as uniform as possible, Annals of Discrete Math. 15 (1982),129–134.

S. S. Shrikhande and N. M. Singhi, On a problem of Erdős and Larson, Combinatorica 5 (1985), 351–358.

5. I asked a few years ago the following question: Let $a_1 < a_2 < \ldots < a_k$ be any Sidon sequence of integers (i.e. the sums $a_i + a_j$ are all distinct). Is it then true that the sequence can be embedded into a perfect difference set? In other words: Is there a prime p and a sequence $a_1 < a_2 < \ldots < a_k < a_{k+1} < \ldots < a_{p+1}$ for which every non-zero residue $\mathrm{mod}(p^2 + p + 1)$ can be uniquely expressed in the form $a_i - a_j$, $1 \leqslant i < j \leqslant p+1$ and our Sidon sequence occurs among the a's? A much weaker conjecture would be: Is there a Sidon sequence

$$a_1 < a_2 < \ldots < a_k < a_{k+1} < \ldots \leqslant a_n, \quad \frac{a_n}{n^2} = 1 + o(1)?$$

Unfortunately I could not even answer this much weaker conjecture.

A very old conjecture of Turán and myself states that if $f(n)$ denotes the number of solutions of $n = a_i + a_j$ and $f(n) > 0$ for all $n > n_0$ then $\limsup f(n) = \infty$. I offered long ago (and still offer) 500 dollars for a proof or disproof. Perhaps if

$$\sum_{\substack{f(n)>0 \\ n<x}} 1 > cx \qquad (1)$$

holds for every $x > x_0$ then $\overline{\lim} f(n) = \infty$ already follows. It is well known that there is a Sidon sequence for which (1) holds for infinitely many x. I wondered if $f(n) > 0$ for all $n > n_0$ implies

$$\frac{1}{x} \sum_{n<x} f(n)^2 \to \infty \qquad (2)$$

but Ruzsa found a counterexample. His paper will soon appear in Monatshefte der Mathematik. It is not clear if $f(n) > 0$ for all $n > n_0$ implies

$$\frac{1}{x} \sum f(n)^r \to \infty$$

for sufficiently large r. After Ruzsa's result one would expect that the answer is negative.

I proved that if $a_n < cn^2$ for all $n > n_0$ then for infinitely many n, $f(n) > 1$ (i.e. for a Sidon sequence $\limsup a_n/n^2 = \infty$.) Is it true that $a_n < cn^2$ implies $\limsup f(n) = \infty$? It is not hard to prove that the number of solutions of $a_i - a_j = n$ is unbounded.

Let $1 \leqslant a_1 < \ldots < a_k \leqslant n$ be a Sidon sequence. Denote by $f(n)$ the maximum of the smallest integer which cannot be written in the form $a_i - a_j$ (i.e. there always is an $m \leqslant f(n)$ for which $m = a_i - a_j$ has no solution, where the maximum is taken over all Sidon sequences of the above form). Is it true that $f(n) < n^{\frac{1}{2}+\varepsilon}$, perhaps in fact $f(n) < cn^{\frac{1}{2}}$? If true this would be best possible.

H. Halberstam and K. F. Roth, Sequences, Springer Verlag.

6. During the meeting held in memory of G. Dirac, Nešetřil and I formulated some problems. Does there exist a graph G every edge of which occurs in at most three triangles and for which $G \to (K_3)^2$? In other words if we color the edges of G by two colors there always is a monochromatic triangle. The answer is almost certainly negative and perhaps the proof of this will not be difficult. We also asked: If every edge of G occurs in at most four triangles and G contains no K_6 can $G \to (K_3)^2$ hold? Clearly many related questions can be asked.

Perhaps our second problem is more interesting. Two edges e_1 and e_2 in G are called strongly independent if they have no common vertex and there is no path of length three containing both e_1 and e_2. We conjectured that if G has maximum degree n and has more than $\frac{1}{4}5n^2$ edges then G has two strongly independent edges. It is easy to see that if true, this is best possible. Chung-Trotter and independently Gyárfás-Tuza proved this conjecture. Their joint paper appeared recently.

We further stated the following Vizing type conjecture: Let G have maximum degree n. Then the edges can be coloured by at most $\frac{1}{4}5n^2$ colors so that edges which have distance $\leqslant 1$ get different colors. If true this conjecture is easily seen to be best possible, but perhaps the conjecture is a bit too optimistic.

F. R. K. Chung, A. Gyárfás, Z. Tuza and W. T. Trotter, The maximum number of edges in $2K_2$-free graphs of bounded degree. Discrete Math. 81 (1990) 2, 129–135.

L. D. Andersen, The strong chromatic index of a cubic graph is at most 10, Topological, Algebraical and Combinatorial structures (ed. J. Nešetřil), Annal Discr. Math. (to appear).

7. At the international conference held at Warsaw in 1983 Pisier told me a very striking conjecture. A sequence $a_1 < a_2 < \ldots$ is called independent if all

the finite sums $\sum \varepsilon_i a_i$, $\varepsilon_i = 0$ or 1 are distinct (e.g. the powers of 2 are clearly independent). Pisier now conjectures that $b_1 < b_2 < \ldots$ is the union of a finite number of independent sequences if there is an $\varepsilon > 0$ so that for every n, every set of n, b's contains a subsequence of εn independent numbers. The condition is clearly necessary. The problem is whether it is also sufficient. Pisier was led to this beautiful conjecture by its application to Sidon sets. Nešetřil, Rödl and I published a long paper on Pisier type problems which I hope will appear soon. Practically all our results are negative. Herewith I state only two problems which we found very interesting: Is it true that $b_1 < b_2 < \ldots$ is the union of a finite number of Sidon sequences if every set of n, b's contains a Sidon sequence of cn terms? It follows from a deep result of Nešetřil and Rödl that the answer is negative for $c < \frac{1}{4}$. It is trivially affirmative for $c > \frac{3}{4}$ and the question is open for $\frac{1}{4} \leqslant c \leqslant \frac{3}{4}$. It is annoying that we could not decide it for $c = \frac{3}{4}$.

We have a similar question about decomposing a graph as the union of triangle free graphs.

Perhaps I can now state my first serious conjecture which goes back nearly 60 years and which is still alive and unsolved.

Let $1 \leqslant a_1 < a_2 < \ldots < a_n$ be an independent sequence. Is it then true that there is an absolute constant c for which $a_n > 2^{n-c}$? Conway and Guy found independent sequences with $a_n < 2^{n-2}$ but nobody found such a sequence with $a_n < 2^{n-3}$. I offer 500 dollars for a proof or disproof of $a_n > 2^{n-c}$.

The following problem is stated in my paper with Nešetřil and Rödl: Does there exist an infinite sequence of integers $x_1 < x_2 < \ldots$ which does not contain $n+1$ terms together with all their $2^{n+1} - 1$ subsums (i.e. there is no subsequence $x_{i_1}, \ldots, x_{i_{n+1}}$ for which all the sums

$$\sum_{r=1}^{n+1} \varepsilon_r x_{i_r}, \quad \varepsilon_r = 0 \text{ or } 1$$

are all x's), but if we color the x's by 2 colors (or more generally by t colors) then there is an infinite subsequence $x_{u_1} < x_{u_2} < \ldots$ so that all the n-fold sums

$$\sum_{r=1}^{n} \varepsilon_r x_{u_{l_r}}, \quad \varepsilon_r = 0 \text{ or } 1, \quad l_1 < l_2 < \ldots < l_n,$$

have the same color.

We stated that for $r = 2$ this follows from Ramsey's Theorem. At the moment we cannot prove this, Nešetřil and Rödl prove only that all the consecutive sums $x_{u_i} + x_{u_{i+1}}$ have the same color.

Even this weaker statement for consecutive sums is open for $r > 2$.

P. Erdős, Problems and results in additive number theory, Coll. Théorie des Nombres, Bruxelles (1955), 127–137.

R. Guy, Unsolved problems in number theory, Springer Verlag.

P. Erdős, F. Nešetřil, V. Rödl, On Pisier Type Problems and Results, In: Mathematics of Ramsey Theory, Springer Verlag (1990), 214–231.

8. Hajnal and I have the following conjecture: Let H be a graph and $G(n)$ ($n \to \infty$) a graph of n vertices which does not contain H as an induced subgraph. Is it then true that G contains a trivial subgraph of size $\geq n^\varepsilon$? (I.e. G contains either an independent set or a complete graph of size n^ε). We could only prove this with $e^{c\sqrt{\log n}}$ instead of n^ε. Of course $\varepsilon = \varepsilon(H)$ and no doubt for suitable H, ε will be small. We could prove our conjecture for a large class of graphs but could not prove it for a C_5.

P. Erdős and A. Hajnal, Ramsey type theorems, Discrete Applied Math. 25 (1989), 37–52.

9. Let $n_1 < n_2 < \ldots$ be an infinite sequence of integers and $G^{(r)}(n_i)$ a sequence of r-uniform hypergraphs of n_i vertices. We say that the edge density of this sequence is α if α is the largest real number for which there is a sequence $m_i \to \infty$, $m_i \leq n_i$ so that for infinitely many i $G_{n_i}^{(r)}$ has a subgraph $G^r(m_i; (\alpha+o(1))\binom{m_i}{r})$ (i.e. it has a subgraph of m_i vertices and $(\alpha + o(1))\binom{m_i}{r}$ hyperedges).

The Erdős-Stone-Simonovits theorem implies that for $r = 2$ the only possible values of the density of $G^2(n_i)$ are 0, 1 or $\frac{1}{2}(1 - \frac{1}{k})$. For $r \geq 3$ I proved that if the density is positive it is $\geq r!/r^r$ i.e. it cannot have a density α, $0 < \alpha < r!/r^r$. I then conjectured that for every r there is an absolute constant c_r for which if the density of $G^{(r)}(n_i)$ is $> r!/r^r + \varepsilon$ than it is fact larger than $r!/r^r + c_r$. I called this hypothetical c_r the jumping constant.

This modest looking conjecture seems to present great difficulties and is still open, but after the recent results of Frankl and Rödl it is doubtful that it is true. I conjectured the following stronger conjecture: The possible values of the density form a well ordered denumerable set. This conjecture was disproved by Frankl and Rödl and it is in fact possible that the density can take any value between $r!/r^r$ and 1.

P. Frankl and V. Rödl, Hypergraphs do not jump. Combinatorica 4 (1984), 139–159.

P. Erdős, On extremal problems of graphs and generalized graphs, Israel Journal of Math. 2 (1965), 183–190.

10. The following problem seems to me to be very interesting. Let $f(n)$ be the largest integer for which every $G(n)$ contains an induced regular subgraph of $f(n)$ vertices. I think this problem is due to Fajtlowicz, Staten and myself. The interesting and startling problem is whether the largest such graph is trivial, i.e. either the complete graph or an independent set. A stronger conjecture is that $f(n)/\log n \to \infty$. Bollobás proved that $f(n) < n^{\frac{1}{2}}$.

11. Let $G(n)$ be a graph which does not contain a complete bipartite graph $K(c \log n, c \log n)$ for some $c > c_0$. The probability method gives that such graphs

exist. In a recent paper Hajnal and I proved that such a graph contains exponentially many non-isomorphic induced subgraphs. We conjectured that the same holds if our $G(n)$ does not contain a trivial subgraph of size $c\log n$ (i.e. it does not contain a complete graph or an independent set of size $c\log n$.)

Alon and Bollobás have the following very nice problem: Let $G(n)$ be a graph which contains no trivial subgraph of size $c\log n$. Is it then true that it contains cn^2 induced subgraphs no two of which have the same number of edges and vertices? I do not see that this holds even if we assume that G does not contain a $K(c\log n, c\log n)$. By the way, perhaps in the Alon-Bollobás problem cn^2 can be replaced by $n^{\frac{5}{2}}$; the random graph shows that it does not hold with $n^{\frac{5}{2}+\varepsilon}$. V. T. Sós and I proved the Alon-Bollobás conjecture with $n^{\frac{3}{2}}$ instead of n^2.

P. Erdős and H. Hajnal, On the number of distinct induced subgraphs of a graph, Graph theory and combinatorics 1988, Annals of discrete Math. 43, 145–154.

N. Alon and B. Bollobás, Graphs with a small number of induced subgraphs, ibid. 23–30.

12. Now I state three problems which we formulated with colleagues in Memphis.

1. Let $f(n)$ be the largest integer for which there is a graph $G(n)$ of n vertices every vertex of which has degree $\geqslant f(n)$ and $G(n)$ has no C_4. Is it true that $f(n+1) \geqslant f(n)$ for $n > n_0$? If this fails too is it true that there is an absolute constant c for which for every $m > n$, $f(m) > f(n) - c$? We had no success at all with these questions. Clearly many generalizations are possible, which we leave to the reader.

2. Is it true that every $G(n; 2n-1)$ has an induced subgraph of $m < n(1-\varepsilon)$ vertices every vertex of which has degree $\geqslant 3$? Faudree found a proof with $m < n - c\sqrt{n}$. Bollobás, Häggkvist and others found independently a proof with the same bound. Perhaps after all this is the correct bound.

3. I stated im my paper at the Kalamazoo meeting (1980) that we proved that for $m \geqslant 2n+1$, every $G\bigl(m; \binom{2n+1}{2} - \binom{n}{2} - 1\bigr)$ is the union of a bipartite graph and a graph each vertex of which has degree $< n$. Faudree had a simple proof for $m = 2n+1$, we could extend it for $m = 2n+2$ and $m = 2n+3$ but we could never do the general case. We proved many related conjectures, but had no success with this one.

P. Erdős, R. F. Faudree, C. C. Rousseau, R. H. Schelp, Subgraphs of Minimum Degree k, Discrete Math. 85 (1990), 53–58.

P. Erdős, R. F. Faudree, A. Gyárfás, R. H. Schelp, Cycles in graphs without proper subgraphs of minimum degree 3, Ars Combin. 25 (1988), B, 195–201.

13. Here are two old problems of Hajnal, Szemerédi and myself which deal with infinite graphs: Is there a function $f(n)$ for which every graph of chromatic number $> \aleph_0$ contains an n-chromatic graph of $\leqslant f(n)$ vertices? We proved that

$f(n)$ must tend to infinity faster than a k-times iterated exponential. Perhaps this is best possible, but perhaps there is no such $f(n)$. This problem was somewhat neglected and I offer 250 dollars for settling it.

Our second problem states: Is it true that for every $g(n) \to \infty$ as slowly as we please there always is a graph of infinite chromatic number every n chromatic subgraph of which can be made bipartite by omitting fewer than $g(n)$ edges. I offer 100 dollars for deciding this question. Rödl showed that such an r-graph $G^{(r)}$ exists for every $r \geqslant 3$.

P. Erdős, A. Hajnal and E. Szemerédi, On almost bipartite large chromatic graphs, Annals of Discrete Math. 12 (1982), 117–123.

V. Rödl, Nearly bipartite graphs with large chromatic number, Combinatorica 2 (1982), 377–389.

14. The following old and beautiful problem of Walter Taylor is still open: Let G be any graph of chromatic number $\aleph_1$. Is it true that for every cardinal number m there is a graph H of chromatic number m all finite subgraphs of which are contained in G_2?

Let A be a class of finite graphs. What properties must A have that for every infinite cardinal m there should be a graph G of chromatic number m every finite subgraph of which should be in A? It is not clear that A can be characterized in a simple way. Hajnal and I showed that A must contain every bipartite graph and Hajnal, Shelah and I showed that A must contain all large cycles. All these results have been superseded by results of Hajnal and Komjáth. If A_1 and A_2 are two such classes, is $A_1 \cap A_2$ also such a class?

Let G_1 and G_2 be two graphs of uncountable chromatic number. Is there a four-chromatic graph G which is contained in both G_1 and G_2? It is not impossible that there is in fact an $\aleph_0$-chromatic graph contained in both G_1 and G_2.

P. Erdős, A. Hajnal and S. Shelah, On some general properties of chromatic numbers, Topics in Topology, Proc. Coll. Math. Soc. Bolyai 8 (1979), 243–255.

A. Hajnal and P. Komjáth, What must and what need not be contained in a graph of uncountable chromatic number, Combinatorica 4 (1984), 47–52.

To end this paper I now state two old problems.

15. This problem is due to Chao Ko, R. Rado and myself. Let

$$|S| = 4n, \quad A_i \subset S, \quad |A_i| = 2n, \quad |A_i \cap A_j| \geqslant 2, \quad 1 \leqslant i \leqslant f(n).$$

Is it true that

$$\max f(n) = \frac{\binom{4n}{2n} - \binom{2n}{n}^2}{2}? \tag{1}$$

It is well known and easy to see that (1) if true is best possible. Our triple paper contains many problems, all but (1) have been settled. I offered 250 pounds for a proof or disproof of (1). Cooper and P. Frankl have several more general conjectures, but as far as I know (1) is still open.

P. Erdős, Chao Ko and R. Rado, Intersection theorems for systems of finite sets, Quarterly Journal of Math. 12 (1961), 313–320.

16. Finally our old problem with R. Rado on Δ-systems. A family of sets B_i, $1 \leq i \leq r$ is called a Δ-system if the intersection of any two of them is the same (i.e. it equals the intersection of all of them.) Let now

$$|A_i| = n, \quad 1 \leq i \leq f(n;k)$$

be the smallest integer for which every family of $f(n;k)$ sets of size n contains a Δ-system of k sets. We proved

$$2^n < f(n;3) < 2^n n!$$

Abbott and Hanson proved $f(n;3) \geq 10^{\frac{n}{2}}$ and Joel Spencer showed that for every $\varepsilon > 0$ and $n > n_0(\varepsilon)$, $f(n;3) < (1+\varepsilon)^n n!$. I offer 1000 dollars for a proof or disproof of

$$f(n;3) < c^n. \tag{2}$$

No doubt for every k there is a c_k for which

$$f(n;k) < c_k^n.$$

A family of sets B_i, $1 \leq i \leq r$ is called a weak Δ-system if the size of $|B_i \cap B_j|$ is the same for every $1 \leq i < j \leq r$. Milner, Rado and I investigated weak Δ-systems. Denote by $g(n;k)$ the smallest integer for which any family of $g(n;3)$ sets of size n contains a subfamily of k sets which form a weak Δ-system. Presumably $g(n;k)$ is much smaller than $f(n;k)$, but we could not prove

$$g(n;3) < c^n. \tag{3}$$

We proved that $g(n;3) > c_1^n$, but have no conjecture about $\lim(g(n;3))^{\frac{1}{n}}$. One final remark. In our paper we consider problems of infinite Δ-systems and it is perhaps surprising that we could solve all the infinite problems for the Δ-systems and weak Δ-systems but so far the finite problems (2) and (3) remain intractable.

P. Erdős and R. Rado, Intersection theorems for system of sets, I and II Journal London Math. Soc. 35 (1960), 85–90, 44 (1969), 467–479.

P. Erdős, E. Milner and R. Rado, Intersection theorems for systems of sets III. J. Austral. Math. Soc. 18 (1979), 22–40.

P. Erdős
Mathematical Institute,
Hungarian Academy of Sciences,
H-1364 Budapest, P.O.B. 127,
Hungary

Fourth Czechoslovakian Symposium on
Combinatorics, Graphs and Complexity
J. Nešetřil and M. Fiedler (Editors)
© 1992 Elsevier Science Publishers B.V. All rights reserved.

Realizability of Some Starlike Trees

D. FRONČEK

A graph H is called e-realizable if there exists a graph G such that the neighbourhood of any edge of G (i.e. the subgraph of G induced by all vertices adjacent to at least one end vertex of this edge) is isomorphic to H. We show that certain starlike trees (graphs homeomorphic to stars) with short branches are not e-realizable.

By the *neighbourhood of a vertex* x (or *v-neighbourhood*) in a finite undirected graph G without loops and multiple edges (denoted $N_G(x)$) we mean the subgraph of G induced by the set of all vertices adjacent to x (the neighbours of x).

Analogously by the *neighbourhood of an edge* $f = xy$ (or *e-neighbourhood*) in G (denoted $N_G(xy)$) we mean the subgraph of G induced by all vertices adjacent to at least one of the vertices x, y but different from them.

A given graph H is called *v-realizable* (*e-realizable*) if there exists a graph G in which the neighbourhood of any vertex (edge) is isomorphic to H. Then the graph G is called *v-realization* (*e-realization*) of H.

By the *starlike tree* $S_{k_1,k_2,\ldots,k_m}$ we mean a graph homeomorphic to the star $S_m = K_{1,m}$, the branches of which are of lengths $k_1, k_2, \ldots, k_m$, $m \geq 3$, $k_i \geq 1$. If all the branches are of the same length k then we say the graph is a *regular starlike tree* and denote it $S_{m(k)}$. The vertex of degree m is called the *central vertex*.

Brown and Connelly [1] showed which starlike trees are v-realizable. In this paper we show that certain starlike trees with at least 5 branches of small lengths are not e-realizable.

We can easily see that if G is an e-realization of some connected graph H with at least two vertices and G contains a triangle then at least one edge of any triangle belongs to at least two triangles. Let $\langle x_1, x_2, x_3 \rangle = C_3$ (the symbol $\langle x_1, x_2, \ldots, x_n \rangle$ denotes the graph induced by the set of vertices $\{x_1, x_2, \ldots, x_n\}$). Then $N_G(x_1 x_2)$ contains at least one vertex adjacent to x_3 which is adjacent to at least one of x_1, x_2 and then either $x_1 x_3$ or $x_2 x_3$ (or both) belong to more than one triangle.

Now suppose that G is an e-realization of a starlike tree with $m \geq 5$ branches, at least one of them of length greater than 1. (Zelinka [2] observed that each star S_m is e-realizable by $K_{2,m+1}$). Let $y_1 y_2$ be an edge belonging to $k \geq 3$ triangles $\langle y_1, y_2, x_1 \rangle, \ldots, \langle y_1, y_2, x_k \rangle$.

If $k < m$ then there exist at least $m - k + 2$ vertices adjacent either to y_1 or to y_2 (but not to both). Let u_1 be a neighbour of y_1, then there exist neighbours $v_1, v_2, \ldots, v_{m-k}$ of y_2 adjacent to u_1 because y_2 is the central vertex of $N_G(y_1 u_1)$. Analogously, each neighbour v_i of y_2 is adjacent to $m - k$ vertices $u_{i_1}, \ldots, u_{i_{m-k}}$. If $m - k \geqslant 2$ then the graph induced by all edges $u_i v_j$ is regular of degree at least 2, which is a contradiction. If $m - k = 1$ then $N_G(y_2 v_i)$ (of each existing v_i) contains just one vertex u_i and analogously each $N_G(y_1 u_j)$ contains just one vertex v_j. Thus there exist the edges $u_i v_i$ for each existing u_i, v_i. $N_G(y_1 y_2)$ is connected and then each vertex x_j is adjacent either to some u_i or v_i because there exists no edge $x_r x_s$—in the opposite case $N_G(y_1 u_i)$ contains $C_3 = \langle x_r, x_s, y_2 \rangle$. In both cases $N_G(u_i v_i)$ contains $C_3 = \langle x_r, y_1, y_2 \rangle$, which is a contradiction.

Now suppose that $y_1 y_2$ belongs to exactly two triangles $\langle y_1, y_2, x_1 \rangle$ and $\langle y_1, y_2, x_2 \rangle$ and vertices $u_1, u_2, \ldots, u_r$ are adjacent to y_1 and $v_1, v_2, \ldots, v_s$ are adjacent to y_2 (possibly either $r = 0$ or $s = 0$). Let there exist an edge $u_{i_1} v_{j_1}$. Then y_2 is the central vertex of $N_G(y_1 u_{i_1})$ and y_1 is the central vertex of $N_G(y_2 v_{j_1})$ and hence u_{i_1} has to be adjacent to $m - 2$ vertices $v_{j_1}, \ldots, v_{j_{m-2}}$ and v_{j_1} has to be adjacent to $u_{i_1}, \ldots, u_{i_{m-2}}$, which is a contradiction and there exists no edge $u_i v_j$. But then the central vertex of $N_G(y_1 y_2)$ belongs to more than two triangles. If it is u_i then it belongs to triangles $\langle y_1, u_i, u_{i_1} \rangle, \ldots, \langle y_1, u_i, u_{m-1} \rangle$, if it is v_i the case is similar, if it is x_i then it belongs to $\langle y_1, y_2, x_i \rangle$ and either to $\langle y_1, x_i, v_{i_1} \rangle, \langle y_1, x_i, v_{i_2} \rangle$ or $\langle y_1, x_i, u_{i_1} \rangle, \langle y_1, x_i, u_{i_2} \rangle$, which is the case proved above. So we can see that in $N_G(y_1 y_2)$ no edge $u_i v_j$ exists and $m = k$.

It is clear that there exists no edge $x_i x_j$—in the opposite case $N_G(y_1 x_p)$ ($p \neq i, j$) contains $C_3 = \langle y_2, x_i, x_j \rangle$. Suppose that x_1 is the central vertex of $N_G(y_1 y_2)$. Then it is adjacent to at least three vertices of the set $\{u_1, u_2, \ldots, u_r\}$ or $\{v_1, v_2, \ldots, v_s\}$, say u_1, u_2, u_3. Then y_1 is adjacent to $y_2, x_1, \ldots, x_m, u_1, u_2, u_3$ and it is of degree $m + 1$ in $N_G(y_2 x_1)$, which is a contradiction.

Thus one of vertices u_i, v_j has to be the central vertex of $N_G(y_1 y_2)$. Let it be u_1. If it is adjacent to two vertices x_i, x_j then $N_G(y_1 x_p)$ ($p \neq i, j$) contains $C_4 = \langle y_2, x_i, u_1, x_j \rangle$, which is again a contradiction. Hence u_1 is adjacent to at least $m - 1$ vertices of the set $\{u_1, u_2, \ldots, u_r\}$, say $u_2, \ldots, u_m$. But in this case $N_G(x_1 y_1)$ contains u_1 of degree $m - 1 \geqslant 4$ and also y_2 of degree $m - 1$ and $N_G(x_1 y_1)$ is not a starlike tree.

So we have proved that there exists no edge belonging to two triangles and then G is a graph without triangles.

Lemma 1. *Let G be an e-realization of a starlike tree. Let x_0 be the central vertex of $N_G(y_1 y_2)$ adjacent to y_1. Then y_2 is the central vertex of $N_G(x_0 y_1)$.*

Proof. If x_0 is the central vertex of $N_G(y_1 y_2)$ adjacent to y_1 then there are vertices x_1, x_2, x_3 which are adjacent to both x_0 and y_2 and thus y_2 is the central vertex of $N_G(x_0 y_1)$. □

Lemma 2. *Let $S_{k_1, k_2, \ldots, k_m}$ be a starlike tree with at least five branches, the lengths of which are $k_1 = 3$, $2 \leqslant k_i \leqslant 3$ for $i = 2, 3, \ldots, m$. Then $S_{k_1, k_2, \ldots, k_m}$ is not e-realizable.*

Proof. Let G be an e-realization of $S_{k_1,k_2,\ldots,k_m}$ and let x_0 be the central vertex of $N_G(y_1y_2)$ adjacent to y_1. Let the branches of $N_G(y_1y_2)$ be $\langle x_0, x_{11}, x_{12}, x_{13}\rangle$, $\langle x_0, x_{21}, \ldots, x_{2k_2}\rangle$, ..., $\langle x_0, x_{m1}, \ldots, x_{mk_m}\rangle$. In this case the vertices x_{i2} are adjacent to y_1 and x_{i1} and x_{i3} to y_2 for each $i = 1, 2, \ldots, m$.

So $N_G(y_1x_{12})$ contains $P_4 = \langle x_{13}, y_2, x_{11}, x_0\rangle$. But none of the vertices x_{13}, y_2, x_{11}, x_0 can be the central vertex of $N_G(y_1x_{12})$. If it is y_2 then according to Lemma 1 x_{12} is the central vertex of $N_G(y_1x_0)$. If it is x_0 then x_{12} is the central vertex of $N_G(y_1x_0)$. If it is x_{11} or x_{13} then y_1 is the central vertex of $N_G(x_{12}x_{11})$ or $N_G(x_{12}x_{13})$ and either x_{11} or x_{13} is adjacent to all vertices x_{i2} and thus its degree in $N_G(y_1y_2)$ is at least m, which is a contradiction.

Hence either x_{13} or x_0 is at a distance of at least 4 from the central vertex of $N_G(y_1x_{12})$, which is again a contradiction and the proof is over. $\square$

Lemma 3. *Let $S_{m(k)}$ be a regular starlike tree with $m \geq 5$ branches of length 2. Then $S_{m(k)}$ is not e-realizable.*

Proof. Let G be an e-realization of $S_{m(k)}$ and y_1y_2 be its edge. Let x_0 be the central vertex of $N_G(y_1y_2)$ adjacent to y_1 and $\langle x_0, x_{i1}, x_{i2}\rangle$ $(1 \leq i \leq m)$ be the branches of $N_G(y_1y_2)$. Then y_1 is adjacent to all vertices x_{i2} and y_2 to all vertices x_{i1}. Because $N_G(y_1x_{12})$ contains $P_3 = \langle x_0, x_{11}, y_2\rangle$ and similarly like in Lemma 2 none of them can be the central vertex of $N_G(y_1x_{12})$ then $N_G(y_1x_{12})$ contains a branch of length greater than 2, which is a contradiction. $\square$

From Lemma 2 and Lemma 3 our main result immediately follows.

Theorem. *Starlike trees $S_{k_1,k_2,\ldots,k_m}$ with more than 4 branches of length 2 or 3 are not e-realizable.*

References

[1] M. Brown, R. Connelly, *On graphs with a constant link I*, Proof Techniques in Graph Theory (F. Harary, ed.), Academic Press, 1969; *On graphs with a constant link II*, Discrete Math. **11** (1975), 199–232.

[2] B. Zelinka, *Edge neighbourhood graphs*, Czech. Math. Journ. **36** (1986), 44–47.

D. Fronček
Silesian University at Opava,
Department of Mathematics,
Bezručovo nám. 13, 746 01 Opava,
Czechoslovakia

Fourth Czechoslovakian Symposium on
Combinatorics, Graphs and Complexity
J. Nešetřil and M. Fiedler (Editors)
© 1992 Elsevier Science Publishers B.V. All rights reserved.

The Construction of All Configurations $(12_4, 16_3)$

Harald Gropp

It is proved that the number of non-isomorphic configurations $(12_4, 16_3)$ is 574. This result is obtained by line deletions of the 80 Steiner systems $S(2,3,15)$ and a special discussion of the 5 non-colourable cubic graphs on 12 vertices.

1. Introduction and notation

> Zum Schlusse möchte ich noch bemerken, dass man, wie eben die bisherigen Arbeiten über die Konfiguration $(12_4, 16_3)$ zeigen, versuchsweise zu neuen Konfigurationen gelangen kann, dass es jedoch wünschenswert wäre, irgendein Ordnungsprinzip einzuführen, das gestatten würde, in irgendeinem Maasse Übersicht über die ganze Materie zu gewinnen.

In the last paragraph of his paper [2] B. Bydžovský discusses the question of getting a survey about all configurations $(12_4, 16_3)$.

Definition 1.1. *A configuration (v_r, b_k) is a finite incidence structure with the following properties:*
1. *There are v points and b lines.*
2. *There are k points on each line and r lines through each point.*
3. *Two different lines intersect each other at most once and two different points are connected by a line at most once.*

In this paper only configurations with parameters $v = 12$, $b = 16$, $k = 3$, $r = 4$ are discussed. The reader is referred to further papers about configurations (e.g. [6]). Quite naturally the following regular graph is closely related to each configuration and plays an important role in this paper.

Definition 1.2. *The configuration graph of a configuration (v_r, b_k) has the v points as vertices, and two vertices are connected by an edge if they are not collinear in the configuration.*

The following lemma is obvious.

Lemma 1.3. *The configuration graph of a configuration (v_r, b_k) is a d-regular graph on v vertices where d is the so called deficiency $\bigl(d = v - r(k-1) - 1\bigr)$.*

The configurations $(12_4, 16_3)$ seem to be the only non-symmetric (i.e. $b > v$) configurations with a long and interesting history, apart from Steiner systems. The first configuration $(12_4, 16_3)$ was constructed by O. Hesse [8] in 1848. Until 1948 four more configurations were known constructed by J. de Vries in 1889, B. Bydžovský in 1939, J. Metelka in 1944 and M. Zacharias in 1948. For references see [14]. After this period of single constructions all configurations with certain properties were constructed in several papers of the Czechoslovaks B. Bydžovský, J. Metelka, V. Metelka, and J. Novák, altogether more than 200. Unfortunately, all these papers have not been very known. A reason may be that they are published in German or Czech language in German or Czechoslovak journals (see e.g. [10], [11], [12], [13]).

The limited length of this paper does not allow to include too many details here. Especially, all aspects of the geometrical realizability which are discussed in many of the older papers are not covered here at all. It is my aim to concentrate on the construction of all non-isomorphic configurations $(12_4, 16_3)$ in a purely combinatorial sense. A more detailed paper on the history of configurations $(12_4, 16_3)$ and their properties, however, is in preparation [7].

In section 2 the construction method is explained. The 5 non-colourable cubic graphs with 12 vertices and "their" configurations are discussed in section 3. Section 4 contains the final result together with a few statistical data about the 574 configurations $(12_4, 16_3)$. The reader is again referred to [7] for more information.

2. Methods and background

2.1 Known enumeration results for configurations with $k = 3$

Until now the following configurations $(12_4, 16_3)$ have been enumerated:

There are unique configurations 7_3 and 8_3. There are 3 configurations 9_3, 10 configurations 10_3, 31 configurations 11_3, 229 configurations 12_3, 2036 configurations 13_3, and 21399 configurations 14_3. Moreover, 5 more numbers for non-symmetric configurations (including Steiner systems) are known: There is a unique configuration $(9_4, 12_3)$, 5 configurations $(12_5, 20_3)$, 2 configurations $(13_6, 26_3)$, 787 configurations $(14_6, 28_3)$, and 80 configurations $(15_7, 35_3)$. References can be found in [6].

2.2 Line deletion of a Steiner system

The 80 configurations $(15_7, 35_3)$ or Steiner systems $S(2, 3, 15)$ were constructed by Cole, Cummings and White in 1919 and can be found in [9]. By deleting 3 collinear points and all lines through these points a configuration $(12_4, 16_3)$ is obtained. Vice versa, a given configuration $(12_4, 16_3)$ can be extended to a Steiner system $S(2, 3, 15)$ by 3-colouring the 18 edges of its configuration graph. This method is the inverse method of the so-called line deletion of a Steiner system (see [6]) and has been used in 1899 to determine all Steiner systems $S(2, 3, 13)$.

2.3 Configurations (10_3) and Steiner systems $S(2, 3, 13)$

This analogous relation between configurations (10_3) and Steiner systems $S(2, 3, 13)$ is described in [3]. A few results are mentioned here.

The Construction of All Configurations $(12_4, 16_3)$

There are 21 cubic graphs with 10 vertices. 19 of them are connected. Exactly 2 of them are not 3-colourable, a bridge graph (i.e. a graph where the deletion of a vertex yields a non-connected graph) and the Petersen graph.

3. The non-colourable graphs

3.1 The cubic graphs with 12 vertices

There are exactly 94 cubic graphs with 12 vertices, 85 of which are connected and can be found in Baraev, Faradzhev [1]. The remaining 9 graphs are unions of smaller graphs. [1] also contains tables of properties of these graphs. Exactly 5 of them are not 3-colourable, the 4 graphs nos. 1, 2, 3, and 4 which are bridge graphs an graph no. 63 which can be obtained from the Petersen graph by replacing one of its vertices by a triangle.

These 5 graphs have to be checked whether the 48 edges of their complement can be partitioned into 16 edge-disjoint triangles. Since they are not 3-colourable the configurations $(12_4, 16_3)$ which have one of these 5 graphs as configuration graph are not obtained by the method used below.

3.2 The bridge graphs

Lemma 3.1. *There is no configuration $(12_4, 16_3)$ such that its configuration graph is one of the 4 bridge graphs.*

Proof. Partition the vertex set into 3 subsets. Let $A = \{1, 2, 3, 4\}$, $B = \{5, 6\}$, and $C = \{7, 8, 9, 10, 11, 12\}$. Each of the 4 bridge graphs has 5 edges in A, 1 edge in B, 8 edges in C, and 2 edges between A and B as well as B and C. Each line of the configuration covers 3 edges of the complementary graph. Hence there are at most 6 lines containing 1 point of A, B, and C each and at most 8 lines containing more than one point of a subset. This is a contradiction, since 16 lines are needed. □

3.3 Graph no. 63

Graph no. 63 in the list of [1] is described by the following 18 edges on the point set

$$\{1, 2, 3, 4, 5, 6, 7, 8, 9, 10, 11, 12\}:$$
$$1, 2; 1, 3; 1, 4; 2, 3; 2, 5; 3, 6;\ 4, 7; 4, 8; 5, 9;\ 5, 10; 6, 11; 6, 12;$$
$$7, 9; 7, 11; 8, 10; 8, 12; 9, 12; 10, 11.$$

Its automorphism group has order 12 and can be generated be 2 elements s and t:

$$s = (1, 2, 3)(4, 5, 6)(7, 9, 12, 8, 10, 11),\ t = (1, 2)(4, 5), (7, 9), (8, 10)(11, 12).$$

In order to construct all configurations whose configuration graph is graph no. 63 the point set is partitioned into 4 subsets of size 3.

Let $A = \{1,2,3\}$, $B = \{4,7,8\}$, $C = \{5,9,10\}$, and $D = \{6,11,12\}$. There are 3 configuration lines with 2 points in a set: $\ell_1 = \{7,8,x\}$, $\ell_2 = \{9,10,y\}$, $\ell_3 = \{11,12,z\}$.

The other 13 lines contain points of 3 of the 4 sets. Denote by α, β, γ, and δ the number of lines which do not contain a point of A, B, C, D resp. The numbers of remaining edges which have to be covered by lines of the configuration are: 8 between A and B, A and C, A and D each and 7 between B and C, B and D, C and D. The lines ℓ_1, ℓ_2, ℓ_3 each cover an even number of edges between two subsets. This implies: $\beta \equiv \gamma \equiv \delta \bmod(2)$ and $\alpha \not\equiv \beta \bmod(2)$. Considering the possibilities for x, y, z there remain 2 types: Either $\alpha = 1$, $\beta = \gamma = \delta = 4$ (Type 1) or $\alpha = 4$, $\beta = \gamma = \delta = 3$ (Type 2).

3.3.1 Type 1

Lemma 3.2. *There is exactly 1 configuration $(12_4, 16_3)$ belonging to type 1.*

Cfg. 570:

$\{7,8,5\}, \{9,10,6\}, \{11,12,4\}, \{4,5,6\}, \{2,4,9\}, \{3,4,10\}, \{1,7,10\}, \{3,8,9\},$
$\{2,7,6\}, \{3,7,12\}, \{1,8,6\}, \{2,8,11\}, \{3,5,11\}, \{1,5,12\}, \{1,9,11\}, \{2,10,12\}.$

Proof. By using the automorphism group of the graph it is possible to choose $x = 5$, $y = 6$, $z = 4$. It follows that $BD = \{4,5,6\}$. The remaining 12 non-edges in BC, BD, and CD can be combined with the points 1, 2 and 3 in a unique way yielding configuration no. 570. □

3.3.2 Type 2

Lemma 3.3. *There are exactly 4 non-isomorphic configurations $(12_4, 16_3)$ belonging to type 2.*

Cfg. 571:
$\{4,5,11\}, \{4,9,6\}, \{8,5,6\}, \{7,10,12\}, \{1,7,8\}, \{3,9,10\}, \{2,11,12\}, \{2,4,10\},$
$\{3,7,5\}, \{2,8,9\}, \{3,4,12\}, \{2,7,6\}, \{3,8,11\}, \{1,5,12\}, \{1,9,11\}, \{1,10,6\}.$
Cfg. 572:
$\{4,5,11\}, \{4,9,6\}, \{8,5,6\}, \{7,10,12\}, \{1,7,8\}, \{2,9,10\}, \{3,11,12\}, \{3,4,10\},$
$\{3,7,5\}, \{3,8,9\}, \{2,4,12\}, \{2,7,6\}, \{2,8,11\}, \{1,5,12\}, \{1,9,11\}, \{1,10,6\}.$
Cfg. 573:
$\{4,5,11\}, \{7,5,6\}, \{4,10,12\}, \{8,9,6\}, \{1,7,8\}, \{2,9,10\}, \{3,11,12\}, \{3,4,9\},$
$\{3,7,10\}, \{3,8,5\}, \{2,4,6\}, \{2,7,12\}, \{2,8,11\}, \{1,5,12\}, \{1,9,11\}, \{1,10,6\}.$
Cfg. 574:
$\{4,5,11\}, \{7,5,6\}, \{4,10,12\}, \{8,9,6\}, \{2,7,8\}, \{2,9,10\}, \{2,11,12\}, \{3,4,9\},$
$\{3,7,10\}, \{1,8,5\}, \{2,4,6\}, \{1,7,12\}, \{3,8,11\}, \{3,5,12\}, \{1,9,11\}, \{1,10,6\}.$

Proof. At first, the 4 lines which do not contain a point of set A are constructed. It turns out that the points 4, 5, and 6 must occur twice and the other 6 points occur once on these 4 lines. The following subtypes are obtained:

Type 2a. The line $\{4, 5, 6\}$ occurs. This leads to a contradiction.

Type 2b. The points 4, 5, and 6 are pairwise connected on 3 different lines. The fourth line is parallel to these 3 lines. Again by using the automorphism group of the graph to find isomorphisms between solutions exactly 2 non-isomorphic configurations can be constructed: no. 571 and no. 572.

Type 2c. Only 2 of the 3 lines connecting 4, 5, and 6 do not contain a point of A. With respect to the automorphism group the 4 lines outside of A are unique. They can be extended to the whole configuration in 2 non-isomorphic ways and yield configurations no. 573 and no. 574. □

4. The results

4.1 The main theorem

As a result of sections 2 and 3 there is the following theorem.

Theorem 4.1. *There are exactly 574 non-isomorphic configurations $(12_4, 16_3)$, 5 of them have graph no. 63 as configuration graph.*

Proof. The 5 graphs nos. 1, 2, 3, 4, and 63 give rise to exactly 5 configurations. The configurations which have a 3-colourable configuration graph are obtained by line deletion of the 80 Steiner systems $S(2, 3, 15)$. Together they produce exactly 569 non-isomorphic configurations (compare section 3). □

The biggest problem is to decide whether two configurations produced by deleting a line of a Steiner system $S(2, 3, 15)$ are isomorphic. A normal form of a configuration $(12_4, 16_3)$ is defined in a similar way as described in [4]. Moreover, the following point types defined by the Czechoslovak authors mentioned above are used and are very helpful (for details see, e.g., [13]).

Given a fixed point X in a configuration $(12_4, 16_3)$. There are 5 possible incidence structures induced on the 3 points X_1, X_2, X_3 which are not connected with X. If possible, the intersection structure of the lines which contain none of these 4 points is used to distinguish subtypes of the 5 structures above. Altogether, 10 types of points are considered.

For each produced configuration the 12 point types are computed. In a second step iterated point types are used to compute the normal form of the configuration. If too many points have the same iterated point type the configuration is treated without computer in order to save computing time. Hence a small number of configurations is checked for isomorphism without computer.

4.2 Some statistical remarks

Some statistical data about the 574 configurations $(12_4, 16_3)$ are described below since it is not possible to give here a list of all the configurations or of their individual properties. Again the reader is referred to [7].

The 574 configurations have 76 different configuration graphs. 18 cubic graphs do not occur as a configuration graph. The maximum number of configurations having the same graph is 24 (graphs no. 51 and 57).

The possible sizes of blocking sets are 4, 5, and 6 (compare [5]). There are exactly 16 configurations with a blocking set of size 4, 282 with a blocking set of size 5, and 517 with a blocking set of size 6. Hence there are 57 blocking set free configurations $(12_4, 16_3)$.

By the way, the existence of a blocking set of size 4 means that there are 4 pairwise non-connected points in the configuration. This fact is mentioned by B. Bydžovský (in the same paper mentioned in the beginning [2]) in the following sentences which shall conclude my paper since they contain a good characterization of how the problem of determining all configurations $(12_4, 16_3)$ could be finally solved. In fact, a detailed analysis of point types as well as connections between design theory, graph theory and configuration theory have solved this problem.

> Die früher bekannten Konfigurationen haben mindestens einen Vierer von gegenseitig getrennten Punkten; die in diesem Aufsatze behandelten Konfigurationen besitzen keine solche Punktgruppe. Diese und ähnliche Beobachtungen könnten zur Lösung der oben angedeuteten allgemeinen Aufgabe einigermassen beitragen.

References

[1] A. M. Baraev, I. A. Faradzhev, *Postroenie i issledovanie na EVM odnorodnyh i odnorodnyh dvudolnyh grafov*, Algoritmicheskie issledovaniya v kombinatorike, Nauka, 1978, pp. 25–60.

[2] B. Bydžovský, *Über zwei neue ebene Konfigurationen $(12_4, 16_3)$*, Czech. Math. Journal **4** (1954), 193–218.

[3] J. W. DiPaola, H. Gropp, *Hyperbolic graphs from hyperbolic planes*, Congr. Numerantium **68** (1989), 23–44.

[4] H. Gropp, *Configurations and Steiner systems $S(2,4,25)$ II—Trojan configurations n_3*, to appear.

[5] _____, *Blocking sets in configurations n_3*, Blocking Sets Conf. Giessen, 1989.

[6] _____, *Non-symmetric configurations with deficiencies 1 and 2*, Combinatorica'90, Gaeta, May, 1990.

[7] _____, *Configurations $(12_4, 16_3)$, their history and properties*, in preparation.

[8] O. Hesse, *Über Curven dritter Ordnung und die Kegelschnitte, welche diese Curven in drei verschiedenen Puncten berühren*, J. reine angew. Math. **36** (1848), 143–176.

[9] R. A. Mathon, K. T. Phelps, A. Rosa, *Small Steiner triple systems and their properties*, Ars Combinatorica **15** (1983), 3–110.

[10] J. Metelka, *O rovinných konfiguracích $(12_4, 16_3)$*, Časopis pro pěstování matematiky **80** (1955), 133–145.

[11] V. Metelka *Rovinné konfigurace $(12_4, 16_3)$ s D-body*, Časopis pro pěstování matematiky **82** (1957), 385–439.

[12] _____, *Über gewisse ebene Konfigurationen* $(12_4, 16_3)$, *die auf den irreduziblen Kurven dritter Ordnung endlich Gruppoide bilden und über die Konfigurationen* C_{12}, Časopis pro pěstování matematiky **95** (1970), 25–53.

[13] _____, *O jistých rovinných konfiguracích* $(12_4, 16_3)$ *obsahujících B, C, a E-body a konfiguracích singulárních*, Časopis pro pěstování matematiky **105** (1980), 219–255.

[14] M. Zacharias, *Konstruktionen der ebenen Konfigurationen* $(12_4, 16_3)$, Math. Nachrichten **8** (1952), 1–6.

Harald Gropp,
Mühlingstr. 19,
D-6900 Heidelberg, Germany

Fourth Czechoslovakian Symposium on
Combinatorics, Graphs and Complexity
J. Nešetřil and M. Fiedler (Editors)
© 1992 Elsevier Science Publishers B.V. All rights reserved.

(p, q)-realizability of Integer Sequences with Respect to Möbius Strip

MIRKO HORŇÁK

Let σ be a compact connected surface and (C_0, C_1) an imbedding of a connected pseudograph in σ such that any member of the set $\mathbf{C}_2$ of all connected components of $\sigma - \bigcup_{c_1 \in \mathbf{C}_1} c_1$ is homeomorphic to the open disk. The ordered triple $D = (C_0, C_1, C_2)$, where $C_2 = \bigcup_{c \in \mathbf{C}_2} \{\bar{c}\}$, is a *cell-decomposition* (shortly a *decomposition*) of σ and elements of $C_i(D) = C_i$ are *i-dimensional cells* of D (vertices, edges, faces for $i = 0, 1, 2$). Cells c, c' of different dimension are said to be *incident* if one of the following conditions is fulfilled: $c \in c'$, $c' \in c$, $c \subseteq c'$, $c' \subseteq c$. Let $I(D)$ be the set of all pairs of cells of D being incident and $B_i(D)$ the set of all *i-dimensional boundary cells* (formed by boundary points of σ only), $i = 0, 1$. The *degree of* $c_i \in C_i(D), i \in \{0, 2\}$, is defined by

$\deg_D(c_0) = |\{c_1 \in C_1(D) : \{c_0, c_1\} \in I(D)\}|$
$\quad + |\{c_1 \in C_1(D) : \{c_0, c_1\} \in I(D), |\{c'_0 \in C_0(D) : \{c'_0, c_1\} \in I(D)\}| = 1\}|,$
$\deg_D(c_2) = |\{c_1 \in C_1(D) : \{c_1, c_2\} \in I(D)\}|$
$\quad + |\{c_1 \in C_1(D) - B_1(D) : \{c_1, c_2\} \in I(D), |\{c'_2 \in C_2(D) : \{c_1, c'_2\} \in I(D)\}| = 1\}|.$

For integers m, n set $[m, n] = \bigcup_{i=m}^{n} \{i\}$, $[m, \infty) = \bigcup_{i=m}^{\infty} \{i\}$ and for $n \in [1, \infty)$ let $(m)_n$ be the unique $m' \in [3, n+2]$ such that $m' \equiv m \pmod{n}$. For finite sequences $A_i = (a_{i1}, \ldots, a_{il_i}), i = 1, 2$, let $A_1 A_2$ denote the sequence $(a_{11}, \ldots, a_{1l_1}, a_{21}, \ldots, a_{2l_2})$; according to the associativity of this product of sequences we have a natural generalization $\prod_{i=1}^{k} A_i$ for finite sequences A_i with $i \in [1, k]$ as well as A_1^k for a non-negative integer k (A_1^0 is the empty sequence ( )). For $n \in [0, \infty)$ and $m \in [n, \infty)$ put

$$Z_{m,n} = \left\{ \prod_{i=1}^{j}(a_i) \in \bigcup_{k=m}^{\infty} [2, \infty)^k : |\{l \in [1, j] : a_l \geqslant 3\}| \geqslant n \right\}.$$

Let D be a decomposition of a surface with non-empty connected boundary. Its boundary structure can be combinatorially described by the set $\bar{V}(D)$ of all sequences $\prod_{i=1}^{n}(\deg_D(v_i))$ such that $v_i \in B_0(D)$ and, provided $v_{n+1} = v_1$, vertices v_i and v_{i+1} are joined by a boundary edge of D for each $i \in [1, n]$; evidently, the set $\bar{V}(D) \subseteq Z_{1,0}$ can be reconstructed from any of its members. If $p, q \in [1, \infty)$, a sequence $A \in Z_{1,0}$ is said to be (p,q)-*realizable* $((p,q)$-*realizable with respect to Möbius strip*, shortly $-(p,q)$-*realizable*) if there exists a decomposition D of the disk (of the Möbius strip) such that $A \in \bar{V}(D)$, for every $c_2 \in C_2(D)$

$$\deg_D(c_2) \equiv 0 \pmod{p}$$

and for every $c_0 \in C_0(D) - B_0(D)$

$$\deg_D(c_0) \equiv 0 \pmod{q};$$

D is said to be a (p,q)-*realization* $(-(p,q)$-*realization*) of A. Some partial results for the problem of (p,q)-realizability were found by Medyanik [7] ($p = 3$, $q = 3$, 4, 5), Fleischner and Roy [1] ($p = 3$, $q = 2$) and Jaeger ($p = 2$, $q \in [3, \infty)$). In Horňák [3, 4, 5] the problem has been solved completely for the set $R_0 = \{(p,q) \in [1, \infty)^2 : (p-2)(q-2) < 4\}$. An independent solution has been presented in Gallai [2]. Here the problem of $-(p,q)$-realizability will be solved for all $(p,q) \in R_0$.

For $(p,q) \in [1, \infty)^2$ let binary relations $T_1(p), T(p,q)$ on $Z_{0,0}$ and $T_2(p,q)$ on $Z_{2,0}$ be defined as follows:

$$T_1(p) = \{(A, B) \colon (\exists A_1, A_2 \in Z_{0,0})(\exists n \in [0, \infty))(A = A_1 A_2 \wedge B = A_1(2)^{np} A_2)\},$$

$$T_2(p,q) = \{(A, B) \colon (\exists A_1, A_2 \in Z_{0,0})(\exists a_1, a_2 \in [2, \infty))(\exists m \in [0, \infty))$$

$$(\exists n \in [\lceil \tfrac{m+2}{p} \rceil, \infty))(\exists \prod_{i=1}^{m}(k_i) \in [1, \infty)^m)$$

$$(A = A_1(a_1)[\prod_{i=1}^{m}(k_i q)](a_2)A_2 \wedge B = A_1(a_1+1)(2)^{np-m-2}(a_2+1)A_2)\},$$

$$L(p,q) = \{\prod_{i=1}^{l}(A_i) \in \bigcup_{j=1}^{\infty} Z_{0,0}^j \colon (\forall i \in [1, l-1])(A_i, A_{i+1}) \in T_1(p) \cup T_2(p,q)\},$$

$$T(p,q) = \{(A, B) \colon (\exists \prod_{i=1}^{l}(A_i) \in L(p,q))(A_1 = A \wedge A_l = B)\}.$$

Theorem 1. *If A_1, A_2, A_2', $A_3 \in Z_{0,0}$, $(p,q) \in [1, \infty)^2$, $(A_2, A_2') \in T(p,q)$ and the sequence $A_1 A_2 A_3$ is (p,q)-realizable with respect to Möbius strip, so is the sequence $A_1 A_2' A_3$.*

The *proof* proceeds as that of Theorem 3.1 in [3] and uses the fact that pasting together a Möbius strip and a disk in such a way that their intersection is equal to the intersection of their boundaries and homeomorphic to a line segment results in a Möbius strip.

The solution of the problem of (p,q)-realizability for $(p,q) \in R_0$ can be presented as follows: In [3] for $(p,q) \in R_1 = \{(r,s) \in R_0 \colon \min(\{r\} \cup \{s\}) = 3\}$ and in [4] for $(p,q) \in R_0 - R_1$ there are defined sets $S_{p,q} \subseteq Z_{3,3}$ of

$$\frac{4pq}{2p + 2q - pq}$$

(for $\min(\{p\} \cup \{q\}) \geq 2$) or

$$\frac{1}{2}(3 + (-1)^{\max(\{p\} \cup \{q\})})$$

(for $\min(\{p\} \cup \{q\}) = 1$) sequences $(p,q;i)$, $i = 1, \ldots, s(p,q)$, containing $((0)_q)^{(0)_p}$ and it is proved that for any $A \in Z_{0,0}$ there exists exactly one $B \in S_{p,q}$ denoted as $G(p,q,A)$ such that $(A,B) \in T(p,q)$. Thus $Z_{0,0}$ can be decomposed into subsets called (p,q)-sets; let $[p,q,A]$ be the (p,q)-set containing $A \in Z_{0,0}$, i.e.

$$[p,q,A] = \{B \in Z_{0,0} : G(p,q,B) = G(p,q,A)\}.$$

If on the set $G_{p,q}$ of all (p,q)-sets a product of $[p,q,A]$ and $[p,q,B]$ is defined by $[p,q,A][p,q,B] = [p,q,AB]$, then, as shown in [5], $G_{p,q}$ with corresponding binary operation is isomorphic to the group with the abstract definition $S^p = T^2 = (ST)^q = E$. In this group $[p,q,((0)_q)], [p,q,((1)_q)]$ and $\varepsilon_{p,q} = [p,q,((0)_q)^{(0)_p}]$ are in the roles of S, T, E, respectively, and

$$[p,q,(a)] = [p,q,((0)_q,(1)_q)^{a(q-1)}((0)_q)]$$

for $a \in [2, \infty)$.

Theorem 2. If $(p,q) \in R_0$ and $A = \prod_{i=1}^{k}(a_i) \in Z_{1,1}$, then the following conditions are equivalent:
1. A is (p,q)-realizable.
2. $[p,q,A] = \varepsilon_{p,q}$.
3. $\prod_{i=1}^{k}[(ST)^{-a_i}S] = E$.

For the *proof* see Theorem 5.1 of [3], Theorem 4 of [4] and the proof of Theorem 2.2 of [5].

For $A \in S_{p,q}$ in [3,5] there is defined a sequence $I(p,q,A) \in Z_{3,3}$ fulfilling $[p,q,I(p,q,A)] = [p,q,A]^{-1}$ (see the proof of Theorem 2.1 of [5]).

Theorem 3. If $A \in Z_{1,0}$, $A' \in Z_{2,2}$, $(p,q) \in R_0$, the (p,q)-sets $[p,q,A]$ and $[p,q,A']$ are conjugate and the sequence A is (p,q)-realizable with respect to Möbius strip, so is the sequence A'.

Proof. There exists $\beta \in G_{p,q}$ such that $\beta[p,q,A]\beta^{-1} = [p,q,A']$. If $B \in S_{p,q} \cap \beta$, then

$$\beta[p,q,A]\beta^{-1} = [p,q,BAI(p,q,B)],$$
$$G(p,q,A') = G(p,q,BAI(p,q,B)),$$
$$\{(BAI(p,q,B), G(p,q,A')), (A', G(p,q,A'))\} \subseteq T(p,q),$$

hence Theorem 4.2 of [3] and Theorem 3 of [4] yield $(BAI(p,q,B), A') \in T(p,q)$. It is easy to see that Lemma 4.9b) and consequently Theorem 4.3 of [3] are true for $(p,q) \in R_0 - R_1$ as well. Now the proof follows from a non-orientable analogue of the generalized version (for $(p,q) \in R_0$) of Theorem 4.3 (based on our Theorem 1). □

Due to Theorem 3 our analysis can be limited to a small number of (p,q)-sets representing conjugacy classes. Let $c(p,q)$ be the number of conjugacy classes of (p,q)-sets; as in [5] they are denoted $\langle p,q,i\rangle$, $i = 1, \ldots, c(p,q)$, where $\langle p,q,1\rangle = \{\varepsilon_{p,q}\}$, and representatives of $\langle p,q,j\rangle$, $j = 2, \ldots, c(p,q)$, in terms of S, T are:

$(3,3)/S$, S^2, T; $(3,4)/ST$, T, $(ST)^2$, S; $(3,5)/ST$, $(ST)^2$, T, S;
$(4,3)/S$, T, S^2, ST; $(5,3)/S^2$, S, T, ST;
$(p,2)$, $p \in [2,\infty)/S^k$, $k = 1, \ldots, \lceil\frac{1}{2}(p-1)\rceil$, T, for $p \equiv 0 \pmod 2$ also ST;
$(2,q)$, $q \in [3,\infty)/(ST)^k$, $k = 1, \ldots, \lceil\frac{1}{2}(q-1)\rceil$, S, for $q \equiv 0 \pmod 2$ also $T4$;
$(p,1)$, $p \in [1,\infty)/S$—only for $p \equiv 0 \pmod 2$ (for $p \equiv 1 \pmod 2$) $c(p,1) = 1$);
$(1,q)$, $q \in [2,\infty)/T$—only for $q \equiv 0 \pmod 2$.

Let D be a decomposition of the Möbius strip σ and let P be a path (possibly closed) of $(C_0(D), C_1(D))$ joining $u, v \in B_0(D)$ in such a way that $\{u\} \cup \{v\}$ is the set of all vertices of P belonging to $B_0(D)$. For $k \in [0,1]$ the path P is said to be a k-path of D if the union of edges of P has k' crossings with the central curve of σ, $k' \equiv k \pmod 2$.

Proposition 4. *If D is a decomposition of the Möbius strip, then there exists a 1-path of D.*

Proof. Suppose the set of all 1-paths of D is empty. Then using a continuous deformation of $(C_0(D), C_1(D))$ we can obtain a decomposition $\tilde{D}$ of the Möbius strip σ isomorphic to D such that all its 0-paths have no crossing or touch points with respect to the central curve l of σ (and have, of course, no 1-paths). Since the connected component of $\sigma - \bigcup_{c_1 \in C_1(\tilde{D})} c_1$ containing l clearly cannot be homeomorphic to the open disk, $\tilde{D}$ is not a decomposition of σ—a contradiction. □

Corollary 5. *If $(p,q) \in [1,\infty)^2$ and a sequence $\prod_{i=1}^{k}(a_i) \in Z_{0,0}$ is (p,q)-realizable with respect to the Möbius strip, then $\sum_{i=1}^{k} a_i \geq 2k + 2$.*

Proof. Let D be a decomposition of the Möbius strip such that $\bar{V}(D)$ contains the sequence $\prod_{i=1}^{k}(\deg_D(v_i)) = \prod_{i=1}^{k}(a_i)$ and let P be a 1-path of D joining v_m to v_n for some $m, n \in [1,k]$. Since $\deg_D(v_i) \geq 2$ for each $i \in [1,k]$ and

$$\sum_{i \in \{m\} \cup \{n\}} \deg_D(v_i) \geq \sum_{i \in \{m\} \cup \{n\}} 2 + 2$$

(the first as well as the last edge of P are not boundary edges of D), the desired inequality follows. □

1-paths play an important role in our solution—cutting the Möbius strip decomposed by a $-(p,q)$-realization D of $B \in Z_{1,1}$ along a 1-path of D results in a (p,q)-realization D' of $B' \in Z_{0,0}$. If a vertex $v \in C_0(D) - B_0(D)$ splits into vertices $v'_1, v'_2 \in B_0(D')$, then $\deg_{D'}(v'_1) + \deg_{D'}(v'_2) = \deg_D(v) + 2 \equiv 2 \pmod{q}$. That is why it will be useful to know how the mapping

$$A = \prod_{i=1}^{k}(a_i) \to C(q, A) = \prod_{i=1}^{k}((2-a_i)_q)$$

acts on (p,q)-sets.

Proposition 6. *If $(p,q) \in R_0$, $A, B \in Z_{0,0}$ and $[p,q,A] = [p,q,B]$, then*

$$[p,q,C(q,A)] = [p,q,C(q,B)].$$

Proof. Since $((\ ),(2)^p) \in T(p,q)$, we have $[p,q,(\ )] = [p,q,(2)^p] = \varepsilon_{p,q}$. Moreover, $[p,q,((0)_q)^{(0)_p}] = \varepsilon_{p,q}$, $[p,q,(a)] = [p,q,(b)]$ whenever $a,b \in [2,\infty)$, $a \equiv b$ $\pmod{q}$, $C(q, \prod_{i=1}^{l}(b_i)) = \prod_{i=1}^{l}C(q,(b_i))$ for every $\prod_{i=1}^{l}(b_i) \in Z_{0,0}$, $C(q,((0)_q)) = (q+2)$ and $C(q,(2)) = ((0)_q)$. Recall that $[p,q,A] = [p,q,B]$ if and only if $G(p,q,A) = G(p,q,B)$; then according to the definition of $T(p,q)$ two possibilities are sufficient to be considered:
 1. $A = B = (\ )$,
 2. $A = (a_1)(q)^m(a_2)$ with $a_1, a_2 \in [2, q+1]$ and $m \in [0, p-1]$, $B = (a_1 + 1)(2)^{2p-m-2}(a_2+1)$.

In the second case (case 1 is trivial) for $(p,q) \in R_1$ there are pq^2 pairs (A,B) to be checked while for $(p,q) \in R_0 - R_1$ the required statement follows immediately from Lemma 4 of [4]. □

Theorem 7. *If $(p,q) \in R_0$, $m \in [1, c(p,q)]$, $A = \prod_{i=1}^{k}(a_i) \in Z_{1,1}$, $\sum_{i=1}^{k} a_i \geq 2k+2$ and $[p,q,A] \in \langle p,q,m \rangle$, then the sequence A is (p,q)-realizable with respect to Möbius strip if and only if the triple (p,q,m) fulfils one of the following conditions:*
 1. $(p,q) \in \{(3,3),(3,4),(4,3)\}$, $m \in \{1,4,5\}$,
 2. $(p,q) \in \{(3,5),(5,3)\}$, $m \in \{1,2,3,5\}$,
 3. $2 \in \{p\} \cup \{q\}$, $p+q \equiv 0 \pmod 2$, $m \in \bigcup_{i=1}^{\lceil \frac{1}{4}(p+q)\rceil}\{2i-1\}$,
 4. $2 \in \{p\} \cup \{q\}$, $5 \geq p+q \equiv 1 \pmod 2$, $m \in [1, \frac{1}{2}(p+q-1)]$,
 5. $1 \in \{p\} \cup \{q\}$, $m = 1$.

Proof. (a) Assume (p,q,m) does not fulfil any of the mentioned conditions, but the sequence A is $-(p,q)$-realizable.
 (aa) $(p,q) \in R_1$: First let $(p,q,m) = (4,3,2)$. Since $[4,3,(3,2)^2(2)] \in \langle 4,3,2 \rangle$, by Theorem 3 there exists a $-(4,3)$-realization D of $(3,2)^2(2)$. According to Proposition 4 consider a 1-path P of D joining subsequently vertices $v_0, \ldots, v_{l+1}$ with $v_0, v_{l+1} \in B_0(D)$. Cut the Möbius strip decomposed by D along P and denote the

resulting disk-decomposition by D'; if v_i splits into v'_{i1}, v'_{i2} for $i \in [0, l+1]$, then $\deg_{D'}(v'_{i1}) = \deg_{D'}(v'_{i2}) = 2$ for $i = 0, l+1$ and

$$\deg_{D'}(v'_{i2}) = \deg_D(v_i) + 2 - \deg_{D'}(v'_{i1}) \equiv (2 - \deg_{D'}(v'_{i1}))3 \pmod{3}$$

for $i \in [1, l]$. Clearly, our notation can be chosen such that for $B_j = \prod_{i=1}^{l}(\deg_{D'}(v'_{ij}))$, $j = 1, 2$, the sequence (2) $\prod_{j=1}^{2}[B_j(2)^3]$ belongs to $\bar{V}(D')$; put $G_1 = G(4, 3, B_1)$. As degrees of cells of

$$C_2(D) \cup C_0(D) - \{v_i : i \in [0, l+1]\}$$

do not change passing from D to D', D' is a $(4, 3)$-realization of B. Due to the congruences above we have $[4, 3, B_2] = [4, 3, C(3, B_1)]$, by Proposition 6, $[4, 3, C(3, B_1)] = [4, 3, C(3, G(4, 3, B_1))]$ and then, since $[4, 3, B_1] = [4, 3, G_1]$, for $\alpha = [4, 3, (2)]$ Theorem 2 yields $\alpha[4, 3, G_1]\alpha^3[4, 3, C(3, G_1)]\alpha^3 = \varepsilon_{4,3}$; this is a contradiction, since it can be checked that $\alpha[4, 3, S]\alpha^3[4, 3, C(3, S)]\alpha^3 \neq \varepsilon_{4,3}$ for every $S \in S_{4,3}$.

Analogously we proceed using the following representatives of conjugacy classes of (p, q)-sets:

$$[3, 3, (3, 3, 2)] \in \langle 3, 3, 2 \rangle, \quad [3, 3, (3, 3)] \in \langle 3, 3, 3 \rangle,$$
$$[4, 3, (3, 3, 2)] \in \langle 4, 3, 3 \rangle, \quad [5, 3, (3)^2(2)^4] \in \langle 5, 3, 4 \rangle.$$

The situation is slightly more complicated for

$$[3, 4, (3, 2, 4, 2)] \in \langle 3, 4, 2 \rangle, \quad [3, 4, (3, 4)] \in \langle 3, 4, 3 \rangle, \quad [3, 5, (3, 2, 4, 2)] \in \langle 3, 5, 4 \rangle :$$

If $\deg_D(v_0) = 4$ and $d'_i = \deg_{D'}(v'_{0i})$, $i = 1, 2$, then $d'_1 \in [2, 3]$ and $d'_2 = 5 - d'_1$ so that to obtain a contradiction, $2s(p, q)$ products must be checked to be different from $\varepsilon_{p,q}$.

(ab) $(p, q) \in R_0 - R_1$: We use Lemma 4 of [4] and the following relations:

$$[p, 2, (3)^2(2)^{2i-1}] \in \langle p, 2, 2i \rangle, \quad i = 1, \ldots, \lceil \tfrac{1}{4}(p+2) \rceil$$

(we have to analyze only even p's here),

$$[p, 2, (3, 4)(2)^{p-1}] \in \langle p, 2, \lceil \tfrac{1}{2}(p+3) \rceil \rangle, \quad [p, 2, (3, 4)(2)^p] \in \langle p, 2, \tfrac{1}{2}(p+6) \rangle$$

(only even p's) for $p \in [2, \infty)$ (note that in the last two cases $4s(p, 2)$ products need to be checked, since cyclically from the left to the right the terms 3 and 4 are not separated by the same number of 2's),

$$[2, q, (q+4-2i, 3)] \in \langle 2, q, 2i \rangle, \quad i = 1, \ldots, \lceil \tfrac{1}{4}(q+2) \rceil$$

(only even q's),

$$[2, q, (2q - 1, 2, 3)] \in \langle 2, q, \lceil \tfrac{1}{2}(q + 3) \rceil \rangle, \quad [q, (q, 2, 3)] \in \left\langle 2, q, \tfrac{1}{2}(q + 6) \right\rangle$$

(only even q's) for $q \in [3, \infty)$,

$$[p, 1, (3, 3, 2)] \in \langle p, 1, 2 \rangle$$

for even $p \in [2, \infty)$ and

$$[1, q, (4, 3)] \in \langle 1, q, 2 \rangle$$

for even $q \in [2, \infty)$.

(b) If (p, q, m) is one of the described triples, it is easy to see there exist $a \in [4, \infty)$ and $n \in [0, \infty)$ such that $[p, q, (a)(2)^n] \in \langle p, q, m \rangle$, hence according to Theorem 3 it is sufficient to analyze the case $A \in Z_{1,1} - Z_{2,2}$.

(ba) $(p, q) \in R_1$: As an example consider $[3, 5, (7, 2)] \in \langle 3, 5, 5 \rangle$. We have

$$\begin{aligned}\varepsilon_{3,5} &= [3, 5, (2)(3, 5; 13)(2)^2(5)C(5, (3, 5; 13))] \\ &= [3, 5, (2)(5)^3(4)^2(5)^3(2)^2(5)(2)^3(3, 3, 2, 2, 2)] \\ &= [3, 5, (2)(4)^2(2)^2(5)(3)^2]\end{aligned}$$

and by Theorem 2 there exists a $(3, 5)$-realization D of the sequence $(2)(4)^2(2)^2$ $(5)(3)^2$ decomposing a disk σ with boundary vertices $v_{i,j}$, $i = 1, 2$, $j = 1, \ldots, 8$, fulfilling $\prod_{j=1}^{8}(\deg_D(v_{i,j})) \in \bar{V}(D)$, $i = 1, 2$, and $v_{2,j} = v_{1,j+5}$, $j = 1, 2$, and with boundary edges $e_{i,j}$ joining $v_{i,j}$ to $v_{i,j+1}$, $i = 1, 2$, $j = 1, 2, 3$. Let $\mathcal{F}$ be a continuous mapping defined on σ such that $\mathcal{F}(e_{1,j}) = \mathcal{F}(e_{2,j})$, $j = 1, 2, 3$, and $\bigcup_{j=1}^{3} \mathcal{F}(e_{1,j})$ is the set of all self-intersection points of $\mathcal{F}(\sigma)$; then $\bar{D} = \prod_{i=0}^{2} \left(\bigcup_{c_i \in C_i(D)} \{\mathcal{F}(c_i)\} \right)$ is a decomposition of the Möbius strip $\mathcal{F}(\sigma)$ fulfilling

$$\begin{aligned}\deg_D(\mathcal{F}(v_{1,1})) &= (2 + 2 + 5) - 2 = 7, \\ \deg \bar{D}(\mathcal{F}(v_{1,j})) &= (4 + 3) - 2 = 5, \quad j = 2, 3\end{aligned}$$

(note that $\mathcal{F}(v_{1,j}) = \mathcal{F}(v_{2,j})$, $j = 1, 2, 3, 4$), $\deg \bar{D}(\mathcal{F}(c)) = \deg_D(c)$ for each $c \in C_2(D) \cup C_0(D) - \{v_{1,j} : j \in [1, 8]\}$ and $B_0(\bar{D}) = \{\mathcal{F}(v_{1,j}) : j \in \{1, 5\}\}$ so that it is a $-(3, 5)$-realization of the sequence $(7, 2)$.

If for $a \in [4, q + 3]$ and $n \in [0, p - 1]$ we have $[p, q, (a)(2)^n] \in \langle p, q, m \rangle$, then there exists $r \in [1, s(p, q)]$ such that

$$[p, q, (2)(p, q; r)(2)^{n+1}(a - 2)C(q, (p, q; r))] = \varepsilon_{p,q};$$

we present this information as $(p,q)/(a,n,m,r)$:

$(3,3)/(4,0,4,1)$, $(5,2,1,3)$, $(6,1,4,2)$;
$(3,4)/(4,0,5,1)$, $(4,1,5,14)$, $(4,2,4,4)$, $(6,0,5,20)$, $(6,1,5,2)$, $(6,2,1,3)$;
$(3,5)/(4,0,2,1)$, $(4,1,5,50)$, $(4,2,3,59)$, $(5,0,5,4)$, $(5,1,2,45)$,
    $(5,2,3,40)$, $(6,1,2,2)$, $(6,2,2,55)$, $(7,0,5,32)$, $(7,1,5,13)$,
    $(7,2,1,3)$, $(8,0,2,14)$, $(8,2,2,31)$;
$(4,3)/(4,1,5,2)$, $(4,3,5,24)$, $(5,1,4,3)$, $(5,3,1,1)$, $(6,1,5,9)$, $(6,3,5,11)$;
$(5,3)/(4,1,5,19)$, $(4,2,3,3)$, $(4,3,3,52)$, $(4,4,5,56)$, $(5,0,3,41)$,
    $(5,1,2,24)$, $(5,2,2,57)$, $(5,3,3,20)$, $(5,4,1,2)$, $(6,0,3,44)$,
    $(6,1,3,17)$, $(6,2,5,47)$, $(6,4,5,50)$.

For $a \in [q+4, \infty)$ and $n \in [p, \infty)$ proceed analogously using the facts that $[p,q] = [p,q,(a-q)]$ and $[p,q,(2)^n] = [p,q,(2)^{n-p}]$.

Finally, the $-(p,q)$-realizability of the sequence $(a)(2)^n$ is equivalent to that of $(2)^l(a)(2)^{n-l}$ for $l \in [1,n]$.

(bb) $(p,q) \in R_0 - R_1$: If $p \in [2,\infty)$, $n \in [0, p-1]$ and at least one of p, n is odd, then the sequence $(4)(2)^n$ is $-(p,2)$-realizable—for $A_{ln} = (2)(4)^l(2)^{n+2}(4)^l$, $l \in [0, \infty)$, we have $[p, 2, A_{ln}] = [p, 2, (4)^{2l+n+3}]$, the congruence $2l + n + 3 \equiv 0 \pmod{p}$ is solvable for l and its solution can be found in $[1, p]$ (we need A_{ln} from the set $Z_{1,1}$); in this way all necessary conjugacy classes are covered.

If $q \in [3, \infty)$, $a \in [4, q+3]$ and q is odd or a is even, then the sequence $(a, 2)$ is $-(2, q)$-realizable—by Lemma 4b) of [4] for $b \in [2, \infty)$

$$G(2,q,(2,b,q,2,2,a-2,(2-b)_q, q+2))$$
$$= ((2-b+q-2+2-(a-2)+(2-b)_q - (q+2))_q)(q)^3 = ((2(2-b)-a)_q)(q)^3$$

and the congruence $2(2-b) \equiv a \pmod{q}$ is solvable for b (with a solution in $[2, q+1]$); again this covers all possible cases.

If $p \in [1, \infty)$, $n \in [0, p-1]$ and at least one of p, n is odd, then the sequence $(4)(2)^n$ is $-(p, 1)$-realizable—according to Lemma 4c)d) of [4] we have

$$[p, 1, (2,3)(2)^{n+2}(3)] = \varepsilon_{p,1}.$$

If $q \in [2, \infty)$, $a \in [4, q+3]$ and q is odd or a is even, then (a) is $-(1, q)$-realizable, since by Lemma 4c)e) of [4] $[1, q, (2, q, a-2, q+2)] = \varepsilon_{1,q}$. □

References

[1] H. Fleischner and P. Roy, *Distribution of points of odd degree of certain triangulations in the plane*, Monatsh. für Math. **78** (1974), 385–390.

[2] T. Gallai, *Signierte Zellenzerlegungen II*, Acta Math. Hung. **49** (1987), 185–201.

[3] M. Horňák, *(p, q)-completability of disk-decompositions I*, Math. Nachr. **109** (1982), 281–301.
[4] _____, *(p, q)-completability of disk-decompositions II*, Math. Nachr. **109** (1982), 303–310.
[5] _____, *A relationship between the finite group $S^p = T^2 = (ST)^q = E$ and the (p, q)-completability of disk-decompositions*, Math. Nachr. **110** (1983), 25–36.
[6] F. Jaeger, *On parity patterns of even q-angulations*, J. Comb. Theory (Ser. B), **21** (1976), 206–211.
[7] A. I. Medyanik, *O nekotorykh kombinatornykh svojstvakh primitivnykh vypuklykh mnogogrannikov*, Ukrain. Geom. Sbor. **12** (1972), 94–105.

Mirko Horňák
Department of Geometry and Algebra,
P. J. Šafařík University,
Jesenná 5, 041 54 Košice, ČSFR

Vertex Location Problems

OTO HUDEC

Let $G = (X, W, E)$ be a complete bipartite graph. Let D be the $m \times n$ matrix of distances between each pair of vertices $x_i \in X$, $w_j \in W$, let z be an objective defined on the set M of p-element subsets of W. The vertex location problem is to find a subset $Y \in M$ which minimizes (maximizes) the objective $z(Y)$. For special criteria we obtain the well known p-center and p-median problem. We consider several other optimality criteria and give complexity results for such problems.

1. Introduction

Graph location problems occur when new objects are to be located in order to optimize one or several objectives which are distance-dependent with respect to given vertices of a graph. For given graph location problems, the objects are often idealized as points and may be located anywhere on the graph (including vertices and points on edges). Such problems are motivated by a number of potential applications. Typical problems of this kind arise in the location of ambulance, police or fire stations, facilities, plants, etc.

Up to now the literature has been concentrated on two general types of optimality criteria for graph location problems:

(1) Mini-max Criterion. Let p denote the number of points to be located. The points should be located in such a way that the maximum distance to a vertex from its nearest point is minimized. If the points are restricted to be located at vertices, the problem is called *vertex p-center problem*; if the points may be located on edges or at vertices, the problem is called *absolute p-center problem*. Both problems are known to be NP-hard [4].

(2) Mini-sum Criterion. For a given integer p, p points should be located so as to minimize the sum of distances to each vertex from its nearest point. If the points can only be located at vertices, the problem is called *vertex p-median problem*; if the points may be located on edges and at vertices, the problem is called *absolute p-median problem*. These problems are NP-hard too [5].

The above problems are NP-hard even in the case when the given graph is planar with maximum degree 3. However, results leading to efficient algorithms are presented in the literature when the graph is a tree or $p < 3$ ([2], [4], [5], [6]). We will restrict ourselves to a variety of vertex location problems with objectives consisting of min and max functions.

Definitions and notations

Usually graph location problems are defined on an undirected connected graph $G = (V, E)$ with positive edge lengths. For any two vertices $v_i, v_j \in V$, the distance $d(v_i, v_j)$ is the length of any shortest path joining v_i and v_j. It takes $o(|V|^3)$ steps to compute the entries of the distance matrix D in a general graph and usually D is already given (see [1] for methods of computing). It is convenient to suppose that D is the matrix associated with a complete bipartite graph partitioned into two independent sets X and W, where X represents the set of n vertices (demands) and W the set of possible locations for p points, $p < n$. The edges have lengths corresponding to the entries of D.

Now we can define the vertex location problem as follows:

Let $G = (X, W, E)$ be a complete bipartite graph, let its vertices be partitioned into two independent sets $X = \{x_1, \ldots, x_n\}$, $X = \{w_1, \ldots, w_m\}$. Let $d_{ij} \geqslant 0$ denote the length of the edge joining x_i and w_j, $i = 1, \ldots, n$, $j = 1, \ldots, m$. The $n \times m$ matrix $D = (d_{ij})$ is called the *distance matrix*. For a given integer $p < m$, let W be the set of possible locations for p points. Let $Y = \{y_1, \ldots, y_p\}$ denote any p-element subset of W and M denote the set of p-element subsets of W, obviously $|M| = \binom{m}{p}$. Any $Y \in M$ defines an $n \times p$ submatrix $D(Y)$ of D, where $D(Y)$ consists of p columns of D corresponding to the vertices of Y.

For a fixed Y, let $f \colon \mathbf{R}_p \to \mathbf{R}$ be a given function defined on p values of any row of the matrix $D(Y)$ and let $f(i)$ be the value assigned to the i-th row. Let $g \colon \mathbf{R}_n \to \mathbf{R}$ be a function defined on the set of values $\{f(i) \colon i = 1, \ldots, n\}$. The given functions f, g define an objective z on M by $z(Y) = g(f(1), \ldots, f(n))$. Hence, in the vertex location problem we are to minimize or maximize $z(Y)$ over M.

The above definition extends in a certain way the vertex p-center and vertex p-median problem if $X = W = V$, $m = n$ and $z(Y)$ is the corresponding objective. Further we will consider a modification of this problem in which each point may be located in one of two predetermined vertices. In such a case, for a given integer p, $W = \{u_1, v_1, \ldots, u_p, v_p\}$, $m = 2p$ and for $j = 1, \ldots, p$ the point Y_j may be located either at u_j or v_j. Let the set of possible locations of p points be denoted by M_a, obviously $|M_a| = 2^p$. If the problem is to minimize the minimax objective, we get the *alternative p-center problem*. That problem has been shown to be NP-hard [3].

We limit our attention to the maximization (minimization) of the objectives, where f and g are min or max functions. There are 4 pairs of the dual problems considered either on M or M_a. The complexity of the problems as well as the running time (the same for the dual problems) of polynomial algorithms are presented:

P_1: (minimization, g: min, f: min), to minimize the minimum distance between demands and points.

P_1': (maximization, g: max, f: max), to maximize the maximum distance between demands and points.

Complexity: M: $o(mn)$, M_a: $o(mn)$

P_2: (minimization, g: max, f: max), to minimize the maximum distance between demands and points.

P'_2: (maximization, g: min, f: min), to maximize the minimum distance between demands and points.

Complexity: M: $o(mn) + o(m \log m)$, M_a: $o(mn)$

P_3: (minimization, g: min, f: max), to minimize the minimum distance from a demand to its farthest point.

P'_3: (maximization, g: max, f: min), to maximize the maximum distance from a demand to its nearest point.

Complexity: M: $o(mn \log m)$, M_a: $o(mn)$

P_4: (minimization, g: max, f: min), to minimize the maximum distance from a demand to its nearest point.

P'_4: (maximization, g: min, f: max), to maximize the minimum distance from a demand to its farthest point.

Complexity: M: NP-hard, M_a: NP-hard

References

[1] N. Christofides, *Graph Theory: An Algorithmic Approach*, Academic Press, New York, 1975.

[2] P. Hansen, M. Labbé, D. Peeters, J. F. Thisse, *Single Facility Location on Networks*, Annals of Discr. Math. **31** (1987), 113–146.

[3] O. Hudec, *On Alternative p-Center Problems*, to appear in Zeitschr. Oper. Res..

[4] O. Kariv, S. L. Hakimi, *An Algorithmic Approach to Network Location Problems I: The p-Centers*, SIAM J. Appl. Math. **3** (1979), 513–538.

[5] _____, *An Algorithmic Approach to Network Location Problems II: The p-Medians*, SIAM J. Appl. Math. **3** (1979), 539–560.

[6] N. Megiddo, A. Tamir, *New Results on the Complexity of p-Center Problems*, SIAM J. Comput. **4** (1983), 751–758.

[7] E. Minieka, *The Centers and Medians of a Graph*, Oper. Res. **4** (1977), 641–650.

Oto Hudec
Department of Mathematics, Technical University,
Košice, Czechoslovakia

On Generation of a Class of Flowgraphs

A. J. C. HURKENS, C. A. J. HURKENS, R. W. WHITTY

We present some structure theorems for the class of binary flowgraphs. These graphs show up in the study of the structural complexity of flowcharts. A binary flowgraph is a digraph with distinct vertices s and t such that (1) t is a sink, (2) all vertices other than t have outdegree 2 and (3) for every vertex v there is a path from s to v, and a path from v to t. An irreducible flowgraph (IBF) is a binary flowgraph with no proper subgraph that is a binary flowgraph. We define a simple operation called *generation* that produces an IBF on k vertices from one on $k-1$ vertices. Our main result is that all IBF's can be obtained from an IBF on two vertices by a sequence of generation operations. In some cases the last generation step is uniquely defined and we give some additional results on this matter.

Introduction and definitions

In the following we will consider directed graphs (digraphs). We allow multiple edges and loops. A digraph G is a pair (V, E) or a triple (V, E, z), with vertex set V, edge set E and, if specified, a special vertex z. An edge leaving vertex x and entering vertex y is denoted by xy. For a vertex x the outdegree of x, denoted by $\deg^+(x)$ (indegree of x, $\deg^-(x)$), is the number of edges leaving (entering) x. If S is a set of vertices we use $\deg^+(S)$ ($\deg^-(S)$) for the number of edges leaving (entering) S. A digraph $G = (V, E, z)$ with $|V| \geq 2$ is called a binary graph, if and only if $\deg^+(z) = 0$ and $\deg^+(x) = 2$, for all $x \neq z$. A digraph is called a flowgraph if there exist distinct vertices s and t, such that for every vertex v there is a directed path from s to v, and a directed path from v to t. A binary flowgraph having no binary flowgraph as a proper subgraph is called an *irreducible* binary flowgraph (an IBF). The smallest such graphs are displayed in figure 1.

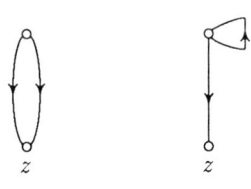

FIGURE 1

A digraph is called *linked to z*, if for every vertex $x \neq z$, there exist two internally vertex disjoint paths from x to z. This notion is used in the following lemma characterizing the class of IBF's. The reader easily verifies

---

Research supported by the Netherlands Organization for the Advancement of Pure Research (N.W.O.), through the Stichting Mathematisch Centrum.
Research supported in part by the U.K. government under the Alvey programme, Project SE069

Lemma 1. *Let G be a binary flowgraph (V, E, z), with $|V| \geq 3$. Then G is irreducible if and only if (i) G is linked to z and (ii) $G - z$ is strongly connected.*

If $G = (V, E, z)$ is a binary graph, and $xy \in E$, then by *subtracting* xy we mean deleting xy, and contracting the remaining edge xy' (which is well defined since $\deg^+(x) = 2$). Note that if $|V| \geq 3$, this procedure yields again a binary graph, which is denoted by $G(xy)$. In this paper we will restrict ourselves to subtracting only edges xy, for which we know that x has indegree 1. For this special case it is not so difficult to define an operation which reconstructs G from $G(xy)$. We will call this *generation*. So, if G' is a digraph with edge uv, then subdividing this edge, by insertion of a new vertex x, into edges ux and xv, and adding a new edge xy, where y is some existing vertex, is called *generating* xy. In this context x is called a *generated vertex*, and xy is called a *generated edge*.

If we have a binary (flow)graph, then clearly generating an edge yields another binary (flow)graph. Starting from an IBF and taking care not to have multiple edges or loops in the resulting graph, generation of an edge yields another IBF. In fact, the main theorem of this paper is that any IBF can be obtained by means of successive generation of edges, starting from one of the graphs of figure 1. We will first state and prove an analogous result on subtracting edges in a binary graph, linked to z, while preserving this property.

The main theorems

Theorem 1. *Let $G = (V, E, z)$ be a binary digraph with $|V| \geq 3$, linked to z, and let x be some vertex of indegree ≤ 1. Then there exists an edge xy such that $G(xy)$ is again linked to z.*

Proof of Theorem 1. Let xy and xy' be the edges leaving x. If $\deg^-(x) = 0$, then subtraction of edge xy (or xy') amounts to deletion of x, which clearly leaves a graph again linked to z. If $\deg^-(x) = 1$, let $p_x x$ be the edge entering x. As G is linked to z, there exist two internally vertex disjoint paths from p_x to z. Only one of these paths passes x and xy', say. We claim that $G(xy)$ is linked to z. To prove this claim suppose to the contrary that $G(xy)$ is not linked to z, let $u \neq z$ be a vertex in $G(xy)$ such that no two internally vertex disjoint paths from u to z exist. Then, by Menger's theorem, there exists either a *vertex* c in $G(xy)$, with $c \notin \{u, z\}$, or an *edge* $c = uz$, such that every u-z-path in $G(xy)$ has to pass c. As G is linked to z there must be a path in G from u to z, not passing c. Clearly, this path must use the edge xy, and therefore pass p_x. But then in $G(xy)$ we can take the beginning of this path to get from u to p_x, after which we have two disjoint alternatives to get from p_x to z. In this way it is possible to get from u to z, while avoiding to pass c. So we see that no such c exists, which contradicts the assumption that $G(xy)$ is not linked to z. □

In the following we will see that if a binary graph $G = (V, E, z)$ is not only linked to z, but also has the property that $G - z$ is strongly connected, then we can find an edge such that subtraction of this edge does not violate either property. While Theorem 1 states that every vertex of indegree 1 is a good candidate as far as the property "linked to z" is concerned, the following examples show that it is not at all obvious to preserve the property of strong connectedness of $G - z$. From now on we will use the expression "generated" edge (vertex) exclusively for those edges (vertices) in an IBF, for which its subtraction (the subtraction of one of its outgoing edges) again yields an IBF. Figure 2 displays an infinite family of graphs $(G_k)_{k \geqslant 2}$, where index k is the number of vertices of indegree 1 (viz. $u_1, \ldots, u_k$), and each graph has only one generated edge, viz. $u_k p_k$.

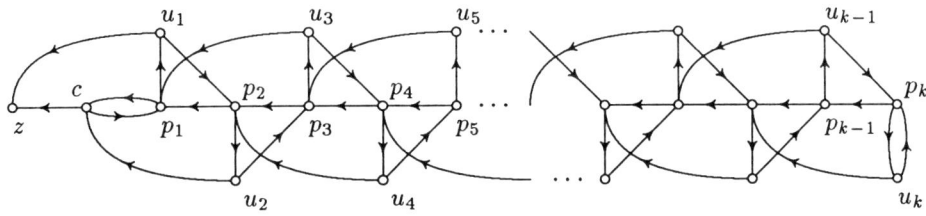

FIGURE 2

It also shows that we have to restrict ourselves to finite graphs. In figure 3 we give an example showing that the number of generated edges can be small, even if $\deg^-(z)$ is arbitrarily large. In the example there are only two generated edges, viz. uv and $u'v'$. The double-headed arrows $\longrightarrow\!\!\!\!\!\rightarrow$ denote edges with their endpoint in z.

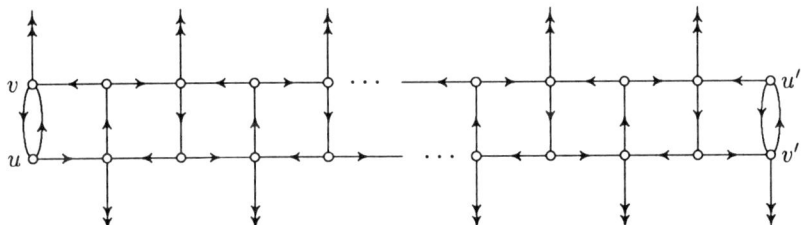

FIGURE 3

However, in the following we will see that if $G = (V, E, z)$ is an IBF with $|V| \geqslant 3$, then there exists at least one generated vertex. Furthermore, if $\deg^-(z) \geqslant 3$, then there exist at least two generated vertices. We deal with the existence of such generated vertices first, and state

Theorem 2. *Let $G = (V, E, z)$ be an IBF with $|V| \geqslant 3$. Then there exists a vertex u of indegree 1 and an edge $uv \in E$, such that $G(uv)$ is an IBF.*

Proof of Theorem 2. Let w be some vertex with $wz \in E$. Let X^1 denote the set of all vertices of G with indegree 1. Consider the digraph $H = (V_H, E_H)$, with $V_H := X^1 - w$, and where $pq \in E_H$ if and only if every path from w to q (in graph G) has to pass p. Now $V_H \neq \emptyset$, since $|X^1| \geqslant 2$. Furthermore H has no directed cycles, for if we would have such a cycle $(u_1 \to u_2 \to \ldots \to u_k \to u_1)$, then none of its vertices u_i would be reachable from w. This would contradict the strong connectedness of $G - z$. So H must have a vertex of outdegree 0, u say. Now u is also a vertex of G. Let $p_u u, uv, uv' \in E$. There exist, in G, two internally vertex disjoint paths from p_u to z, P^1 and P^2 say. Furthermore there exists a path from p_u to w, since $G - z$ is strongly connected. From this it follows that we can assume, w.l.o.g., that one of the paths P^1, P^2 passes w. Moreover, one of the edges leaving u is not used by these paths. Without loss of generality we may assume that this edge is uv. We claim that $G(uv)$ is irreducible.

The proof of Theorem 1 shows that $G(uv)$ is linked to z, so we need only show that $G(uv) - z$ is strongly connected. Since $G - z$ is strongly connected, it suffices to prove that each vertex in G is reachable from p_u without using uv. Define R to be the set of vertices of G that can be reached from p_u without using the edge uv (in graph G). Clearly, $u, w, z \in R$. For any vertex $u' \in X^1 - \{u, w\}$ we know that $uu' \notin E_H$, so by definition, (in graph G) u' is reachable from w without using the edge uv. Hence $u' \in R$. This means that $V - R$ contains no vertex of indegree $\leqslant 1$. Furthermore we have $\deg^-(V - R) \leqslant 1$, as there is at most one candidate entry edge, viz. uv. Note that for $x \in V - R$ we have $\deg^-(x) - \deg^+(x) \geqslant 0$, so it follows that $\deg^+(V - R) \leqslant 1$. As G is linked to z we conclude that $V - R = \emptyset$, so $R = V$, which proves the claim. $\square$

Observing the role of w in the proof above we immediately derive the following

Corollary. *If an IBF $G = (V, E, z)$, with $|V| \geqslant 3$, has a unique generated vertex u, then $uz \notin E$.*

Applications

Theorems 1 and 2 would seem primarily to be motivated by their potential applications to related vertex connectivity problems (e.g. [9]). In fact, however, Theorem 2 has been used by computer scientists for at least ten years, as a technique for analyzing program flowgraphs. Thus a binary graph is associated with the "flow of control" through a program, by equating vertices of outdegree 2 with program decisions or tests, which have True and False outcomes; the root of the binary graph is equated with the termination of the control flow. A precisely analogous association can be made with any deterministic finite automaton on a binary alphabet [2]. In this context, Theorem 2 was first stated in [6] (see also [1]), in the form of an algorithm for constructing all n-test IBF's from those on $n - 1$ tests ($n \geqslant 2$). However, no proof of the correctness of the algorithm was given. A weaker version of Theorem 2, applying to irreducible graphs of regular indegree 1, was discovered independently by J. Fučík (see refernce to [4] in [5]), although we have been unable to see a proof of this. A proof of Theorem 2 has been given in [8], but has been found to contain errors, and in any case, did little to reveal how the theorem works. On the strength of this proof, applications were made to program structuredness [2, 3] and program complexity measurement [7], and the theorem has also been extended to apply to graphs which have the property of irreducibility but are non-binary (i.e. are allowed to have vertices of arbitrary outdegree).

It is clearly of interest in these areas to have a simple revealing proof of Theorem 2 as we have presented it here. Moreover, by careful examination of the proof and by some ad hoc arguments one can derive the following additional results. They show that our examples given in Figures 2 and 3 are in some sense extreme cases.

Additional theorems

For the case that an IBF has only one generated vertex, we can derive some extra information. We find that in this case $\deg^-(z) = 2$, and the number of vertices of indegree 1 is less than $\frac{1}{2}|V|$. This is expressed in the following theorems, the proofs of which we omit.

Theorem 3. Let $G = (V, E, z)$ be an IBF, and let $\deg^-(z) \geq 3$. Then G has at least 2 generated vertices.

Theorem 4. Let $G = (V, E, z)$ be an IBF with $|V| \geq 3$, and with only one generated vertex. Let k be the number of vertices of G with indegree 1. Then $|V| \geq 2k+1$ and, moreover, if G has only one generated edge then $|V| \geq 2k+2$.

References

[1] D. F. Cowell, D. F. Gillies, A. A. Kaposi, *Synthesis and structural analysis of abstract programs*, Computer Journal **23 (3)** (1980), 243–247.

[2] N. E. Fenton, R. W. Whitty, A. A. Kaposi, *A generalised mathematical theory of structured programming*, Theoretical Computer Science **36** (1985), 145–171.

[3] N. E. Fenton, R. W. Whitty, *Characterisations of structured programs*, submitted.

[4] J. Fučík, *Thesis*, Charles University, Prague, 1979. (Czech)

[5] J. Fučík, J. Král, *The hierarchy of program control structures*, Computer Journal **29 (1)** (1986), 24–32.

[6] D. F. Gillies, D. F. Cowell, A. A. Kaposi, *A theory for flowgraph schemas*, Proc. CISS John Hopkins University, Baltimore, USA, 1978, pp. 441–446.

[7] R. E. Prather, *On hierarchical software metrics*, Software Engineering Journal **2 (2)** (1987), 42–45.

[8] R. W. Whitty, *Generation of a class of program flowgraphs*, preprint 1983.

[9] ———, *Vertex-disjoint paths and edge-disjoint branchings in directed graphs*, Journal of Graph Theory **11 (3)** (1987), 349–358.

A. J. C. Hurkens
Department of Mathematics,
Catholic University of Nijmegen,
Nijmegen, The Netherlands

C. A. J. Hurkens
Department of Mathematics and Computing Science,
Eindhoven University of Technology,
Eindhoven, The Netherlands

R. W. Whitty
Department of Mathematical Sciences,
Goldsmiths' College,
New Cross, London SE14 6NW,
United Kingdom

Fourth Czechoslovakian Symposium on
Combinatorics, Graphs and Complexity
J. Nešetřil and M. Fiedler (Editors)
© 1992 Elsevier Science Publishers B.V. All rights reserved.

The Weight of a Graph

Jaroslav Ivančo

The *weight* $w(E)$ of an edge E in a multigraph G is the sum of the degrees of its end vertices; and if G is a pseudograph and E is a loop, then $w(E)$ is defined as twice the degree of its unique end vertex. The *weight* $w(G)$ of a pseudograph G is defined as $\min\{w(E); E \in G\}$.

The pseudograph G triangulates the closed 2-manifold S_g of genus g if G dissects S_g into regions such that each region has an interior homeomorphic to an open disc, and every region meets three different regions along three edges.

A. Kotzig's Theorem [2] states that the weight of every graph which triangulates the sphere S_0 is at most 13; and 13 is the best possible bound.

B. Grünbaum and G. Shephard [1] proved that every graph triangulating the torus S_1 contains an edge of weight at most 15; and 15 is the best possible bound.

J. Zaks [3] proved that the weight of every graph triangulating the closed 2-manifold S_g ($g \geq 1$) is at most $n(g)$, where $n(g)$ is the least odd integer which is greater than $6 + \sqrt{48g+1}$.

If a graph G can be embedded into S_g then the weight of G can be greater than $n(g)$, of course in this case G does not triangulate S_g. If G contains vertices which have degree 1 or 2 then there exists no bound for $w(G)$ depending on g, as can be seen on the complete bipartite graphs $K_{1,r}$ or $K_{2,r}$. If the degree of every vertex of G is greater than 2 (i.e. $\delta(G) > 2$) then such bound for $w(G)$ exists. For such graphs G we establish a strict bound for $w(G)$ depending on the genus of G (i.e. the smallest genus of closed 2-manifolds in which G can be embedded).

First we present the assertion which was proved in [3].

Proposition. (J. Zaks) *Let n denote an odd integer, $n \geq 11$. Let G be a pseudograph triangulating S_g and suppose $w(G) \geq n$. Let v_k denote the number of k-valent vertices of G. Let $e_{i,j}$ denote the number of edges in G having end vertices which are i-valent and j-valent. Then*

$$2e_{5,n-5} + 3e_{4,n-4} + 5e_{3,n-3} \geq 120(1-g) + 10\sum_{i=7}^{n-5}(i-6)v_i + (9n-95)v_{n-4} +$$

$$+ \frac{1}{2}(15n - 165)v_{n-3} + 10 \sum_{i \geq n-2} \left[\frac{i-11}{2}\right] v_i$$

$$(1)$$

Theorem 1. Let G be a graph and suppose $\delta(G) \geq 3$. Let g be the genus of G. Then
$$w(G) \leq \begin{cases} 2g + 13, & \text{if } 0 \leq g \leq 3 \\ 4g + 7, & \text{if } 3 \leq g. \end{cases}$$

Proof. The coefficients of v_i on the right hand side of (1) are all positive and for $i > n/2$ they are at least $5n - 65$; therefore (1) may be reestablished in the following form

$$2e_{5,n-5} + 3e_{4,n-4} + 5e_{3,n-3} \geq 120(1-g) + (5n-65) \sum_{i>n/2} v_i. \tag{2}$$

Now let G be a graph and let g be its genus. Consider an embedding of G in S_g. We dissect every region (which is not a triangle) into triangles by adding new edges the end vertices of which have degrees greater than $\frac{1}{2}(w(G)-1)$ in the graph G. In this way we obtain a pseudograph G^* which triangulates S_g. Evidently, G is a subgraph of G^*, $w(G) \leq w(G^*)$ and if E is an edge of G^* such that $w(E) = w(G^*)$, then E is an edge of G, too.

Let m denote an odd integer such that $w(G^*) \geq m$ (suppose $m \geq 11$) and $v = \sum_{i>m/2} v_i$, where v_i is the number of i-valent vertices of G^* (evidently, $v \geq 3$). If some j-valent vertex V of G^* would have more than $[j/2]$ neighbouring vertices having degrees $< m/2$, then some two consecutive neighbours W_1 and W_2 would have degrees $< m/2$, so $w(W_1 W_2) < m$, contradicting the assumption on G^*. Therefore at most $[j/2]$ neighbours of every j-valent vertex have degrees $< m/2$. All other neighbours have degrees $> m/2$. Since v is the number of vertices which have degrees $> m/2$, thus every j-valent vertex ($j > m/2$) in the pseudograph G^* has degree at most $[j/2] + v - 1$ in the graph G.

If $g = 0$ and $n = 13$, then by (2) we get $2e_{5,8} + 3e_{4,9} + 5e_{3,10} \geq 120$ for the pseudograph G^*. Hence $w(G) \leq w(G^*) \leq 13$.

If $g = 1$ and $n = 15$, then by (2) we get $2e_{5,10} + 3e_{4,11} + 5e_{3,12} \geq 10v > 0$. Hence $w(G) \leq w(G^*) \leq 15$.

If $g \geq 2$ and $n = 15 + 2[12(g-1)/v]$, then by (2) we get

$$2e_{5,n-5} + 3e_{4,n-4} + 5e_{3,n-3} \geq 120(1-g) + 10(1 + [12(g-1)/v])v > 0.$$

Hence $w(G^*) \leq 15 + 2[12(g-1)/v]$.

Let E be an edge of G^* the end vertices of which have degrees i and j where $i + j = w(G^*)$, and suppose that every edge with the weight $w(G^*)$ has the degrees of its end vertices $\geq i$. Then $3 \leq i \leq v$ and $i \leq j = w(G^*) - i > m/2$. E is also an edge of the graph G and its end vertices have degrees at most i and $[j/2] + v - 1$ (or j, if $j \leq [j/2] + v - 1$) in the graph G. Therefore

$$w(G) \leq w(E) \leq i + \min\{j; [j/2] + v - 1\} = \min\{w(G^*); [(w(G^*) - i)/2] + v + i - 1\}. \tag{3}$$

Hence

$$w(G) \leqslant \min\{15 + 2[12(g-1)/v]\,;\, i + v + [(13-i)/2] + [12(g-1)/v]\}.$$

It can easily be seen that this inequality gives the desired bound on $w(G)$ for all i, v ($3 \leqslant i \leqslant v$) except for $g = 3$, $i = v = 7$ and $5 \leqslant i \leqslant v = 8$. These exceptional cases must be treated separately.

Case 1. $g = 3$, $i = v = 7$.

In this case $w(G^*) \leqslant 21$. If $w(G^*) \leqslant 20$, then by (3) we get $w(G) \leqslant 19$. If $w(G^*) = 21$, then by (2) we get $2e_{5,16} + 3e_{4,17} + 5e_{3,18} \geqslant 40$, contradicting the assumption on E (i.e. $7 = i \leqslant 5$).

Case 2. $g = 3$, $i = 5$, $v = 8$.

In this case $w(G^*) \leqslant 21$. If $w(G^*) \leqslant 20$, then by (3) we get $w(G) \leqslant 19$. If $w(G^*) = 21$, then $v_{16} \geqslant 1$ and by (1) ($n = 21$) we get $2e_{5,16} \geqslant 120(1-3) + 10(16 - 6)v_{16} + (5 \cdot 21 - 65)(v - v_{16}) \geqslant 140$. Hence

$$e_{5,16} \geqslant 70. \tag{4}$$

Since at most 8 neighbours of every 16-valent vertex have degrees $< m/2$, thus $e_{5,16} \leqslant 8v = 64$, contradicting (4).

Case 3. $g = 3$, $i = 6$, $v = 8$.

In this case $w(G^*) \leqslant 20$ ($w(G^*) = 21$ contradicts the assumption on E, as in case 1). If $w(G^*) = 20$, then $v_{14} \geqslant 1$ and by (1) ($n = 19$) we get $2e_{5,14} + 3e_{4,15} + 5e_{3,16} \geqslant 120(1-3) + 10(14-6)v_{14} + (5 \cdot 19 - 65)(v - v_{14}) = 50v_{14} > 0$. Hence $w(G^*) \leqslant 19$, in contradiction to the assumption.

Case 4. $g = 3$, $7 \leqslant i \leqslant v = 8$.

By (1) ($n = 19$) we get $2e_{5,14} + 3e_{4,15} + 5e_{3,16} \geqslant 120(1-3) + 10v_7 + 20v_8 + 30v = 10v_7 + 20v_8 > 0$. Hence $w(G) \leqslant w(G^*) \leqslant 19$, which completes the proof. □

Remark 1. In [2], [1] and [3] there are examples of graphs such that $w(G) = 2g + 13$, for $g = 0$, 1 and 2. For $g \geqslant 3$ the graph which we obtain from the complete bipartite graph $K_{3,4g+2}$ by adding three edges joining its $(4g+2)$-valent vertices has genus g and weight $4g + 7$. Therefore the proved bounds are the best possible.

If a graph G contains no triangles then the bound for $w(G)$ is lower, as can be seen in the following assertion.

Theorem 2. Let G be a graph without triangles and suppose $\delta(G) \geqslant 3$. Let g be the genus of G. Then

$$w(G) \leqslant \begin{cases} 8, & \text{if } g = 0 \\ 4g + 5, & \text{if } g \geqslant 1. \end{cases}$$

Proof. It is well known that Euler's formula implies

$$v_3 \geqslant 8(1-g) + \sum_{j \geqslant 5}(j-4)v_j$$

for a graph without triangles. Every k-valent vertex is incident with k edges and thus $kv_k = e_{k,k} + \sum_{i \geqslant 3} e_{i,k}$. Therefore

$$\frac{1}{3}\left(e_{3,3} + \sum_{i \geqslant 3} e_{i,3}\right) \geqslant 8(1-g) + \sum_{j \geqslant 5}\left(1 - \frac{4}{j}\right)\left(e_{j,j} + \sum_{k \geqslant 5} e_{k,j}\right).$$

Hence

$$\frac{2}{3}e_{3,3} + \frac{1}{3}e_{3,4} + \frac{2}{15}e_{3,5} \geqslant 8(1-g) + \sum_{j \geqslant 7}\left(\frac{2}{3} - \frac{4}{j}\right)e_{3,j} + \sum_{i \geqslant j \geqslant 4}\left(2 - 4\frac{i+j}{ij}\right)e_{i,j}. \quad (5)$$

Let h denote the number of edges of G and $e^* = \frac{2}{3}e_{3,3} + \frac{1}{3}e_{3,4} + \frac{2}{15}e_{3,5}$.

If $g = 0$, then the coefficients of $e_{i,j}$ on the right hand side of (5) are all nonnegative; therefore $e^* > 8$, hence $w(G) \leqslant 8$.

If $g \geqslant 1$ and $w(G) \geqslant 4g + 6$, then the coefficients of $e_{i,j}$ on the right hand side of (5) are at least $(8g-6)/(12g+9)$ for all i, j, where $i + j \geqslant 4g + 6$. Therefore $e^* \geqslant (8g-6)h/(12g+9) - 8(g-1) \geqslant 2$, because a graph without triangles must have at least $3(w(G) - 3)$ edges. Hence $w(G) \leqslant 8$, which is a contradiction to the assumption. Therefore $w(G) \leqslant 4g + 5$. □

Remark 2. The complete bipartite graph $K_{3,4g+2}$ has genus g and weight $4g+5$. Therefore the proved bounds are the best possible.

Remark 3. By (5) it can be proved that the inequality

$$w(G) \leqslant [4g/(\delta(G) - 2)] + \delta(G) + 2$$

holds for every graph without triangles, where $\delta(G) \geqslant 3$, $g \geqslant 1$.

References

[1] B. Grünbaum and G. C. Shephard, *Analogues for tilings of Kotzig's Theorem on minimal weights of edges*, in: Theory and Practice of Combinatorics, Annals of Discrete Mathematics **12** (1982), 129–140.

[2] A. Kotzig, *Príspevok k teórii eulerovských polyédrov*, Mat.-fyz. časopis Sloven. akad. vied **5** (1955), 101–113. (Slovak; Russian summary)

[3] J. Zaks, *Extending Kotzig's Theorem*, Isr. J. Math. **45** (1983), 281–296.

Jaroslav Ivančo
Katedra geometrie a algebry,
PF UPJŠ Košice
Poland

Fourth Czechoslovakian Symposium on
Combinatorics, Graphs and Complexity
J. Nešetřil and M. Fiedler (Editors)
© 1992 Elsevier Science Publishers B.V. All rights reserved.

On the Kauffman Polynomial of Planar Matroids

FRANÇOIS JAEGER

We present a matroid version of the Kauffman polynomial of links which is defined for planar matroids and we discuss the possibility of its generalization to a larger class of matroids.

1. The Kauffman polynomial

1.1. The Kauffman polynomial of link diagrams

For more details on the definitions to follow, the reader can refer to [2], [3], [5].

A *link* consists of a finite set of disjoint simple closed curves in 3-dimensional space. Two links are *ambient isotopic* if there exists a continuous deformation of the ambient space which carries one onto the other. We shall restrict our attention to *tame links*, which are ambient isotopic to links made up of polygonal curves. Such links can be represented by *diagrams*. A diagram can be considered as a finite 4-regular plane graph (the image of the link under a suitable plane projection) together with an indication at each vertex of the 3-dimensional structure of the corresponding crossing (this is done with an obvious pictorial convention, see for instance the typical diagram shown on Figure 1). Note that we must adopt a

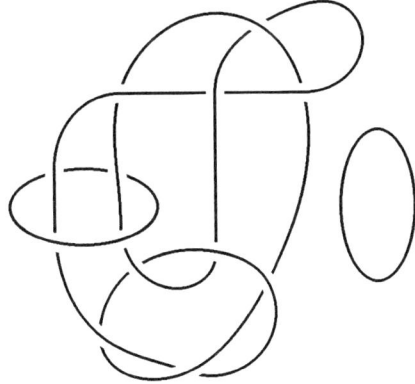

FIGURE 1

slightly generalized definition of 4-regular plane graphs by allowing "free loops" which are edges without end-vertices represented by simple closed curves disjoint from the rest of the graph. Two diagrams will be considered the same if one can be obtained from the other by an isotopy of the plane.

The ambient isotopy of links is described combinatorially in terms of diagrams by Reidemeister's Theorem: two diagrams represent ambient isotopic links if and only if one can be obtained from the other by a finite sequence of simple operations called *Reidemeister moves*. Such a move consists in the selection of a disk, the interior of which intersects the diagram in one of the configurations shown in Figure 2, and the replacement of this configuration by an equivalent one without altering the exterior of the disk. Reidemeister's Theorem allows the combinatorial construction of invariants of links under ambient isotopy as valuations of diagrams which are invariant under Reidemeister moves. A similar approach is valid for *oriented links* (each curve has received an orientation), using "oriented" Reidemeister moves for oriented diagrams.

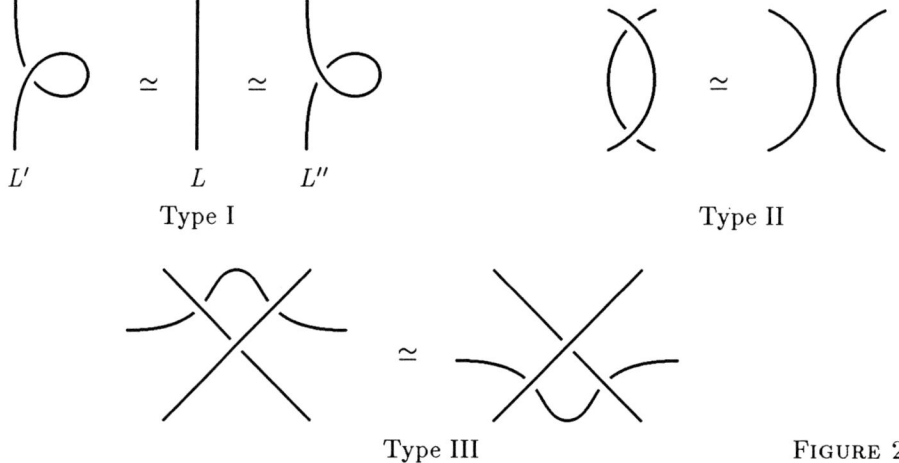

FIGURE 2

We shall denote by $\tilde{L}$ the *mirror image* of the diagram L (obtained from L by changing the crossing structure at every vertex). *The connected sum* of two diagrams L' and L'' is obtained by opening one edge of each and gluing the free ends as shown on Figure 3. A *mutation* is an operation on diagrams which is

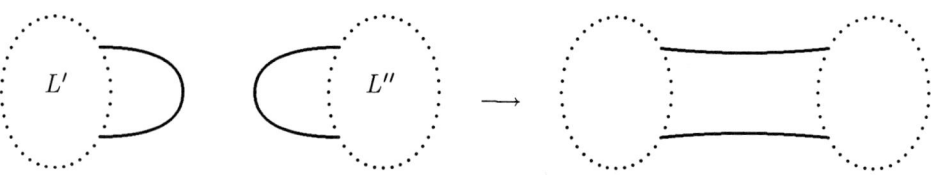

FIGURE 3

performed as follows. Choose coordinate axes Ox, Oy, Oz such that the diagram lies in the (x,y) plane. Find a simple closed curve C in this plane which intersects the diagram in the interior of four edges. Using an appropriate plane deformation we may assume that the curve C is the circle $\{(x,y,0) \mid x^2 + y^2 = 2\}$ and that it intersects the diagram in the points $(1,1,0)$, $(1,-1,0)$, $(-1,1,0)$, $(-1,-1,0)$. We shall consider that the portion of diagram situated inside the disk bounded by C represents a portion of link which lies in a small neighbourhood of this disk. Now rotate the disk by one half-turn around one of the coordinate axes Ox, Oy, Oz. The portion of diagram situated inside the disk will be modified in such a way as to represent the effect of the rotation on the corresponding portion of link. See Figure 4 for an example.

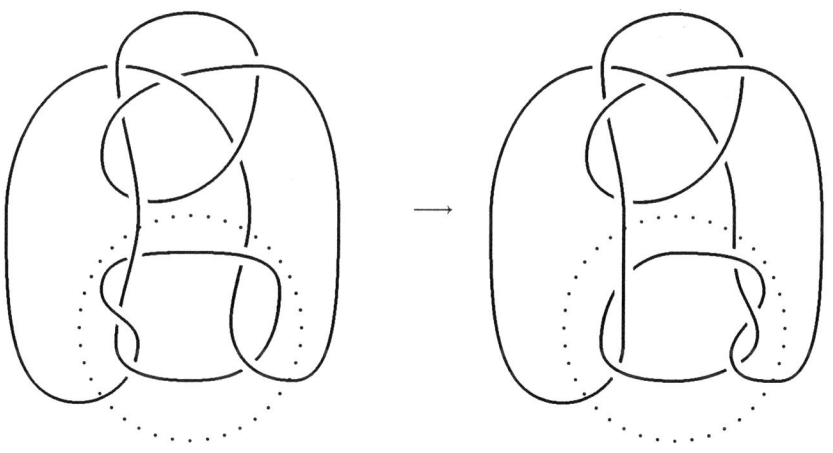

FIGURE 4

The Kauffman polynomial in its "Dubrovnik form" [7] is a mapping D from the class of diagrams to the ring $Z[a, a^{-1}, z, z^{-1}]$ of Laurent polynomials with integer coefficients in the two variables a, z. The mapping D is uniquely defined by properties (i)–(iv) of the following Proposition. We shall also need the other properties (v)–(viii) which can be easily derived from the previous ones (see [7]).

Proposition 1.

(i) If the diagrams L, L', L'' are related by Reidemeister moves of type I as shown on Figure 2, $D(L') = aD(L)$ and $D(L'') = a^{-1}D(L)$.

(ii) D is invariant under Reidemeister moves of types II and III (see Figure 2).

(iii) D takes the value 1 on the diagram consisting only of one free loop.

(iv) $D(L^+) - D(L^-) = z(D(L^0) - D(L^\infty))$ whenever the diagrams L^+, L^-, L^0, L^∞ are identical outside some disk and behave as indicated on Figure 5 inside this disk.

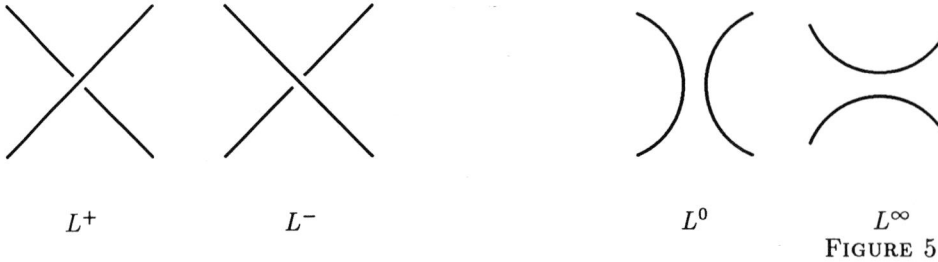

$$L^+ \qquad L^- \qquad\qquad L^0 \qquad L^\infty$$

FIGURE 5

(v) $D(\tilde{L})(a, z) = D(L)(a^{-1}, -z)$.

(vi) If the diagram L is the disjoint union of two diagrams L' and L'', then $D(L) = (z^{-1}(a - a^{-1}) + 1)D(L')D(L'')$.

(vii) If the diagram L is the connected sum of two diagrams L' and L'', then $D(L) = D(L')D(L'')$.

(viii) D is invariant under mutations.

The mapping D is not itself a link invariant but the form of (i) allows one to easily derive from D an invariant of oriented links [7]. The *writhe* $w(L)$ of an oriented link diagram L is the sum of the signs of its crossings as defined on Figure 6. The writhe is invariant under the oriented versions of Reidemeister moves II and III and it is not difficult to prove that $a^{-w(L)}D(L)$ (where $D(L)$ is in fact the value of D on the unoriented diagram obtained from L by forgetting its orientation) defines an invariant of oriented links. We deduce from this fact the following useful criterion for the invariance of D.

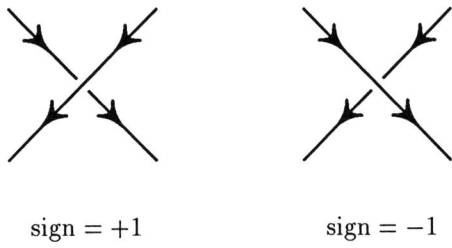

$$\text{sign} = +1 \qquad\qquad \text{sign} = -1$$

FIGURE 6

Invariance Lemma. *If two oriented diagrams represent ambient isotopic oriented links and have the same writhe, D takes the same value on the corresponding unoriented diagrams.*

1.2. The Kauffman polynomial of signed plane graphs

A *signed plane graph* is a pair (G, s) where $G = (V, E)$ is a plane graph and s is a mapping from E to $\{-1, +1\}$ which associates to every edge e its *sign* $s(e)$. We now describe a classical construction which associates to every signed plane graph (G, s) a diagram $K(G, s)$. Recall that the *medial graph* of G, denoted here by

$K(G)$, is a plane 4-regular graph obtained as follows. First one selects an interior point $m(e)$ on each edge e of G (these will be the vertices of $K(G)$). Then for any cyclic sequence of edges $e_1 \ldots e_k$ corresponding to a walk around a component of the boundary of a face f of G, and for $i = 1, \ldots, k$ (indices being read modulo k), the points $m(e_i)$, $m(e_{i+1})$ are joined by an edge drawn as a simple curve, the interior of which lies inside f and close to its boundary. We complete this usual definition by adding to $K(G)$ for every isolated vertex v of G a free loop drawn as a small simple closed curve around v. Then the diagram $K(G, s)$ is defined on $K(G)$ by choosing the crossing structure at every vertex according to the sign of the corresponding edge of G as described on Figure 7. An example of the construction of $K(G, s)$ is given on Figure 8. One may imagine the link represented by $K(G, s)$ as the boundary of a surface constructed by joining disks (corresponding to the vertices of G) by bands (corresponding to the edges of G), each band carrying a half-twist, the sign of which is indicated by the sign of the corresponding edge.

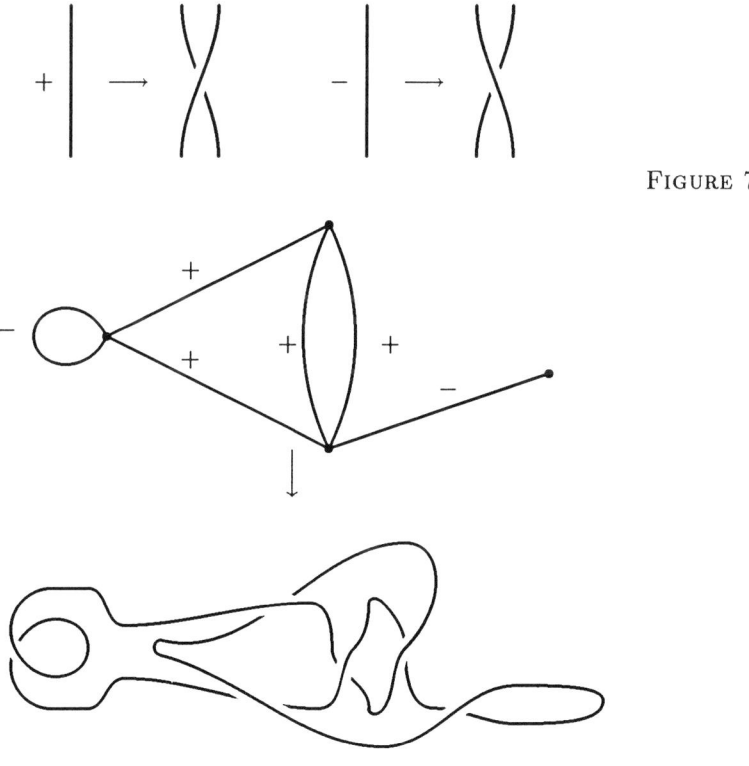

FIGURE 7

FIGURE 8

We shall also need the following definitions and notations. Two edges of the graph G will be said *parallel* if they have the same pair of distinct ends. Two edges of G will be said *coparallel* if they are not bridges and form a cocycle. When

G is connected, G^* will denote a geometric dual of G. Then two edges of G are coparallel iff the corresponding edges in G^* are parallel.

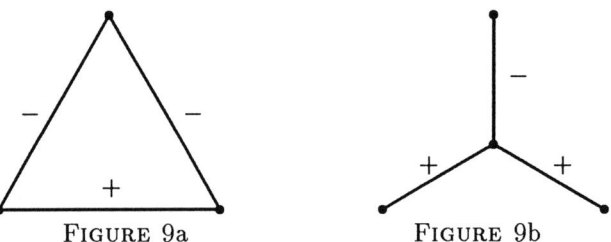

FIGURE 9a   FIGURE 9b

Assume now that the signed plane graph (G, s) has a face bounded by a triangle t with edge-signs as shown on Figure 9a. Let (G', s') be obtained from (G, s) by adding a new vertex inside t joined by signed edges to the vertices of t and deleting the edges of t, as shown on Figure 9b. Then we shall say that (G', s') is obtained from (G, s) by a *star-triangle transformation*.

For a subset F of E and a sign mapping s, we denote by $G - F$ (respectively: $G.F$) the graph obtained from G by deleting (respectively: contracting) the edges of F, and by $s \setminus F$ the restriction of s to $E - F$. When $F = \{e\}$ we shall use the simpler notations $G - e$, $G.e$ and $s \setminus e$.

Now, motivated by a result of Lipson [10], we define the *Kauffman polynomial of the signed plane graph* (G, s) as $D(K(G, s))$ and denote it simply by $D(G, s)$.

Proposition 2. *The Kauffman polynomial of signed plane graphs satisfies the following properties:*

(i) *If e is a loop (respectively: bridge) of G, $D(G, s) = a^{s(e)} D(G - e, s \setminus e)$ (respectively: $D(G, s) = a^{-s(e)} D(G.e, s \setminus e)$).*

(ii) *If the two edges e, e' are parallel (respectively: coparallel and not parallel) and $s(e) + s(e') = 0$, $D(G, s) = D(G - \{e, e'\}, s \setminus \{e, e'\})$ (respectively: $D(G, s) = D(G.\{e, e'\}, s \setminus \{e, e'\})$).*

(ii') *D is invariant under star-triangle transformations.*

(iii) *If G consists only of one isolated vertex and s is the empty mapping, $D(G, s) = 1$.*

(iv) *Let e be an edge of (G, s) with $s(e) = 1$ which is not a loop, and let the sign mapping s' be defined by $s' \setminus e = s \setminus e$, $s'(e) = -1$. Then*

$$D(G, s) - D(G, s') = z(D(G.e, s \setminus e) - D(G - e, s \setminus e)).$$

(v) *$D(G, -s)(a, z) = D(G, s)(a^{-1}, -z)$. When G is connected and G^* is a geometric dual of G, $D(G^*, s)(a, z) = D(G, s)(a^{-1}, -z)$.*

(vi) *If (G_i, s_i), $i = 1, \ldots, k$, are the connected components of (G, s), then*

$$D(G, s) = (z^{-1}(a - a^{-1}) + 1)^{k-1} \prod_{i=1,\ldots,k} D(G_i, s_i).$$

(vii) If G is connected and (G_i, s_i), $i = 1, \ldots, k$, are the blocks of (G, s), then

$$D(G, s) = \prod_{i=1,\ldots,k} D(G_i, s_i).$$

(viii) Let G' be a subgraph of G and let G'' be the subgraph of G formed by the edges not in G'. Assume that G' and G'' meet in at most two vertices. Then $D(G, s)$ is invariant under plane motions performed on G' (preserving the signs of edges) which leave invariant the set of vertices common to G' and G''.

Sketch of proof. (iii) This is an immediate consequence of Proposition 1 (iii).

(iv) If e is not a loop, $K(G, s)$, $K(G, s')$, $K(G.e, s \setminus e)$ and $K(G - e, s \setminus e)$ will play the role of L^+, L^-, L^0, and L^∞ respectively in Proposition 1 (iv).

(v) $K(G, -s)$ is the mirror image of $K(G, s)$ and hence $D(G, -s)(a, z) = D(G, s)(a^{-1}, -z)$ by Proposition 1(v). When G is connected it is easy to see that $K(G)$ and $K(G^*)$ are the same when considered as embedded on the sphere. Then corresponding orientations of $K(G^*, s)$ and $K(G, -s)$ have the same writhe and represent ambient isotopic oriented links. The result follows by the Invariance Lemma.

(vi) This is immediate by induction on k using Proposition 1 (vi).

(vii) If $k \geq 2$, we find a block G_i which has only one vertex in common with the rest of the graph (let us denote this remaining part by (G', s')) and which can be separated from it by a simple closed curve through that vertex. Then $K(G, s)$ is easily seen to be a connected sum of $K(G_i, s_i)$ and $K(G', s')$. The result follows by induction on k using Proposition 1 (vii).

(viii) If G' is disjoint from G'' the only non trivial case is the effect of a plane reflection on G'. Denoting by s' and s'' the restrictions of s to G' and G'', it follows from (vi) that $D(G, s) = (z^{-1}(a - a^{-1}) + 1)D(G', s')D(G'', s'')$. Moreover one easily shows using the Invariance Lemma that $D(G', s')$ is not affected by a plane reflection.

If G' has exactly one vertex in common with G'', it follows from (vi) and (vii) that $D(G, s) = D(G', s')D(G'', s'')$ and the proof is similar to that of the previous case.

Finally if G' and G'' have exactly two vertices in common, it is easy to see that the effect of a plane motion on G' which fixes or exchanges these two vertices corresponds to a mutation of the diagram $K(G, s)$. The result now follows from Proposition 1 (viii).

(i) By (vi) and (vii) we may restrict our attention to the case of graphs with one edge, which is easy by Proposition 1 (i) and (iii).

(ii) By (vi) and (vii) we may assume that G is 2-connected. If e, e' are parallel, we may assume that the digon $\{e, e'\}$ bounds a finite face (otherwise we apply (viii) to an appropriate plane reflection performed on the subgraph defined by e and the interior of the digon $\{e, e'\}$). Then it is easy to see that $K(G, s)$ and $K(G - \{e, e'\}, s \setminus \{e, e'\})$ are related by a Reidemeister move of Type II and the result follows from Proposition 1 (ii). Finally if e, e' are coparallel and not parallel, we apply the previous result to G^*, obtaining $D(G^*, s) = D(G^* - \{e, e'\}, s \setminus \{e, e'\})$. Since $G^* - \{e, e'\}$ is connected and has $G.\{e, e'\}$ as its dual, the result follows from (v).

(ii′) It is easy to see that $K(G, s)$ and $K(G', s')$ are related by a Reidemeister move of type III and the result follows from Proposition 1 (ii). □

Lipson, using the famous result by Whitney [14] that a 3-connected planar graph has an essentially unique plane embedding, has shown ([10], Corollary 6) that any two plane embeddings of the same planar graph can be obtained from one another by a finite sequence of modifications of the type described in Proposition 2 (viii). Consequently, $D(G, s)$ depends only on the abstract graph structure of (G, s) and not on its particular plane embedding.

1.3. The Kauffman polynomial of signed planar matroids

We may now go one step further and remark that properties (vi), (vii), (viii) of Proposition 2 are matroidal in nature. To exploit this fact we need the following definitions (see [12], [13] for the standard matroid terminology).

A matroid will be called *planar* if it is the cycle matroid of some planar graph. We shall call *signed matroid* a pair (M, s), where M is a matroid on the ground set E and s is a mapping from E to $\{-1, +1\}$. We extend in the obvious way to signed planar matroids the definitions and notations introduced for signed plane graphs in Section 1.2. In particular two elements of M will be said *parallel* (respectively: *coparallel*) if they form a circuit (respectively: cocircuit). We shall say that (M, s) and (M', s') are related by a *star-triangle transformation* iff they can be represented by plane graphs which are related by such a transformation in the previous sense (actually, this concept can be defined in purely matroid theoretical terms, but this is rather unnatural and too technical to be presented here).

Now given a signed planar matroid (M, s), we choose any connected plane graph G whose cycle matroid $M(G)$ is isomorphic to M, we identify s with a sign mapping for G and *we define the Kauffman polynomial of (M, s), denoted by $D(M, s)$, as equal to $D(G, s)$*.

Proposition 3. *D is a well defined invariant of signed planar matroids which satisfies the following properties:*

(i) *If e is a loop (respectively: coloop) of M, $D(M, s) = a^{s(e)} D(M - e, s \setminus e)$ (respectively: $D(M, s) = a^{-s(e)} D(M.e, s \setminus e)$).*

(ii) *If the two elements e, e' are parallel and not coparallel (respectively: coparallel and not parallel) and $s(e) + s(e') = 0$, $D(M, s) = D(M - \{e, e'\}, s \setminus \{e, e'\})$ (respectively: $D(M, s) = D(M.\{e, e'\}, s \setminus \{e, e'\})$). D takes the value $(z^{-1}(a - a^{-1}) + 1)$ on the matroid on two elements with opposite signs which are parallel and coparallel.*

(iii) *D is invariant under star-triangle transformations.*

(iv) *If M is the empty matroid and s is the empty mapping, $D(M, s) = 1$.*

(v) *Let e be an element of (M, s) with $s(e) = 1$ which is neither a loop nor a coloop, and let the sign mapping s' be defined by $s' \setminus e = s \setminus e$, $s'(e) = -1$. Then $D(M, s) - D(M, s') = z(D(M.e, s \setminus e) - D(M - e, s \setminus e))$.*

(vi) If (M_i, s_i), $i = 1, \ldots, k$, are the components of (M, s), then

$$D(M, s) = \prod_{i=1,\ldots,k} D(M_i, s_i).$$

(vii) $D(M, -s)(a, z) = D(M, s)(a^{-1}, -z) = D(M^*, s)(a, z)$.

Sketch of Proof. D is well defined for matroids because it takes the same value on two signed connected plane graphs which have isomorphic signed cycle matroids. This follows from Proposition 2 (vii) and (viii), together with Whitney's 2-isomorphism Theorem ([15], see also [13], Chapter 6) and the above-mentioned result of Lipson [10]. Then properties (i)-(vii) easily follow from Proposition 2, choosing appropriate graph representations of the matroids involved. □

2. Questions and perspectives

It is not difficult to show that the existence of the Kauffman polynomial of links is in fact equivalent to the existence of the invariant of signed planar matroids described in Proposition 3. It would be nice to be able to describe this invariant in the matroid context, without reference to representations by signed plane graphs. This might give new insight on the topological meaning of the Kauffman polynomial and its relationship with the Goeritz matrix of links (see [10]). This could also lead to the generalization of the Kauffman invariant to a larger class of matroids. For instance a special case of the Kauffman polynomial, the *bracket polynomial* [6], is closely related to the Tutte polynomial $t(x, y)$ evaluated along the curve $xy = 1$ (see for instance [8], [11]), and hence can easily be extended to arbitrary matroids. Two approaches could be attempted for a generalization of the matroid version of the Kauffman polynomial. The first one would be to find appropriate "state models" (see [6]) in terms of representations of the matroids by chain groups rather than by graphs. It would be natural in this case to restrict one's attention to regular matroids, which form an autodual class closed under minors and k-sums for $k \leqslant 3$ (note that a star-triangle transformation is a special kind of 3-sum). The second approach would be restricted to matroids which can be transformed into the empty one by reductions associated as in Proposition 3 with loops, coloops, parallel or coparallel elements and by a suitable generalization of the star-triangle transformation, and would be based as in the planar case [7] on the proof of the consistency of the defining axioms. In particular it would be already interesting to examine this possibility in the case of graphic matroids. The corresponding class of "reducible" graphs strictly contains the planar graphs since replacing a "star" of the Kuratowski complete bipartite graph $K_{3,3}$ by a triangle yields a planar graph.

When working along these lines, one can take advantage (as in [4], [9]) of the following observations. The "exchange rule" (v) of Proposition 3 allows one to reformulate, without any loss of information, the Kauffman invariant as an invariant of ordinary, unsigned, planar matroids, at the cost of more complicated formulas for the rules (ii) and (iii). On the other hand, by combining with appropriate weights the Kauffman invariants associated with the various minors of a matroid, one can obtain more general, many-variable invariants. Motivated by these considerations, we have carried out an axiomatic study of this type of invariants in the particularly simple case of series parallel matroids, for which the star-triangle reductions are unnecessary (see for instance [1], or [13], Chapter 6). One obtains the following result.

Proposition 4. *Let R be a commutative ring with unit 1 and special elements x, y, z, a, b, c, a', b', c' such that $bb' - cc' = by^2 + (ac' - a'b)y - c'z = b'x^2 + (a'c - ab')x - cz$. There exists a unique R-valued invariant f of series parallel matroids which satisfies the following properties:*

(i) *If e is a loop (respectively: coloop) of M, $f(M) = yf(M - e)$ (respectively: $f(M) = xf(M.e)$).*

(ii) *If the two elements e, e' are parallel and not coparallel (respectively: coparallel and not parallel), $f(M) = a'f(M - e) + b'f(M - \{e, e'\}) + c'f(M - e.e')$ (respectively: $f(M) = af(M.e) + bf(M.\{e, e'\}) + cf(M.e - e')$). D takes the value z on the matroid on two elements which are parallel and coparallel.*

(iii) *If M is the empty matroid, $f(M) = 1$.*

(iv) *If M_i, $i = 1, \ldots, k$, are the components of M, then $f(M) = \prod_{i=1,\ldots,k} f(M_i)$.*

(v) $f(M^*)(x, y, z, a, b, c, a', b', c') = f(M)(y, x, z, a', b', c', a, b, c)$.

The invariant f apparently cannot be derived from the Kauffman polynomial alone and contains it as a special case. Full details will appear elsewhere.

References

[1] T. Brylawski, *A combinatorial model for series-parallel networks*, Trans. Amer. Math. Soc. 154 (1971), 1–22.

[2] G. Burde, H. Zieschang, *Knots*, de Gruyter, Berlin, New York, 1985.

[3] R. H. Crowell, R. H. Fox, *Introduction to Knot Theory*, Springer-Verlag, New York, Heidelberg, Berlin, 1963.

[4] D. Jonish, K. C. Millett, *Isotopy invariants of graphs*, preprint.

[5] L. H. Kauffman, *Formal knot theory*, Mathematical notes 30, Princeton university press, 1983.

[6] ———, *State models and the Jones polynomial*, Topology 26 (1987), 395–407.

[7] ———, *An invariant of regular isotopy*, Trans. AMS **318** (2) (1990), 417–471.

[8] ———, *A Tutte polynomial for signed graphs*, Discrete Applied Math. 25 (1989), 105–127.

[9] L. H. Kauffman, P. Vogel, *Link polynomials and a graphical calculus*, preprint.

[10] A. S. Lipson, *Link signature, Goeritz matrices and polynomial invariants*, L'Enseignement Mathématique 36 (1990), 93–114.

[11] M. B. Thistlethwaite, *A spanning tree expansion of the Jones polynomial*, Topology 26 (1987), 297–309.

[12] D. J. A. Welsh, *Matroid Theory*, Academic Press, London, 1976.

[13] *Theory of matroids*, Encyclopedia of Mathematics and its Applications 26 (N. White, ed.), Cambridge University Press, 1986.
[14] H. Whitney, *Congruent graphs and the connectivity of graphs*, Amer. J. Math. 54 (1932), 150–168.
[15] _____, *2-isomorphic graphs*, Amer. J. Math. 55 (1933), 245–254.

François Jaeger
LSD, IMAG,
Grenoble, France

On Symmetry Groups of Selfdual Convex Polyhedra

Stanislav Jendroľ

Two polyhedra P_1 and P_2 are said to be duals of each other provided there exists a bijection δ from the family of vertices and faces of P_1 to the family of faces and vertices of P_2 which reverses inclusion. If there exists a duality map from the polyhedron P to itself we say that P is selfdual (see e.g. [2, 5, 7, 9]).

Results concerning selfduality of polyhedra have occured since the second half of the 19th century, see e.g. [3, 7, 8, 10, 11, 12]. Despite this familiarity one aspect of selfduality escaped attention until very recently. Its publication by Grünbaum and Shephard [7] spurred interest in the area and led to a number of new results, see e.g. [1, 2, 9, 14, 15]. Grünbaum and Shephard noticed that for a selfduality map δ of a selfdual polyhedron P the map δ^2 need not be the identity map. They defined the rank $r(P)$ of a selfdual polyhedron P as the minimum order of its selfduality maps. They asked if every selfdual polyhedron was of rank 2. This question was negatively answered in [9] and the problem was solved completely independently by several authors [2, 14, 15].

Ashley et al. [2] have considered selfdual harmonious tilings and polyhedra. Among other things they have determined all wallpaper groups which can appear as symmetry groups of harmonious selfdual tilings and have shown that the rank of a selfdual tiling of the plane is either 2 or 4 or ∞.

In this note we are concerned with the question which groups can appear as symmetry groups of selfdual polyhedra. We also discuss the ranks of selfdual polyhedra with a given symmetry group. The notations introduced in [4, 5, 6] are used throughout.

Theorem 1. *The group G is a symmetry group $S(P)$ of a selfdual convex polyhedron P if and only if $G \in \{[q], q \geq 1; [q]^+, q \geq 1; [2,2], [2,2]^+, [2,2^+], [2^+,4], [2^+,2^+], [2^+,4^+], [3,3], [3,3]^+\}$.*

Proof. Mani [13] has shown that one can find a convex polyhedron P to every finite 3-connected planar graph H such that the graph of P is isomorphic to H and such that the symmetry group of P is isomorphic to the automorphism group of H. To prove the positive part of Theorem 1 it is therefore sufficient to find examples of 3-connected selfdual planar graphs with suitable groups of automorphisms. An Archimedian q-sided pyramid can be taken as an example of a selfdual polyhedron with the symmetry group $[q]$, $q \geq 3$. A suitably modified 4-sided pyramid can serve

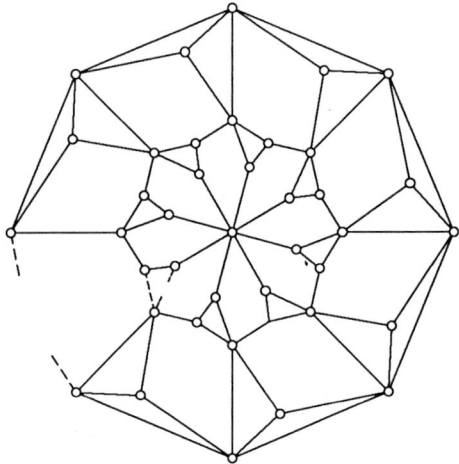

FIGURE 1

as an example of a polyhedron with the group [1] or [2] respectively. In Fig. 1 there is a graph of a polyhedron P_q of Ashley et al. [2, Theorem 3] with the symmetry group $[q]^+$, $q \geqslant 2$ and rank $r(P_q) = 2^k$ where k is the highest power of 2 dividing q. The platonic tetrahedron has the symmetry group isomorphic to [3, 3] and is itself selfdual. In Fig. 2 there is the graph of a selfdual polyhedron of Oudaise [15] with the symmetry group $[3,3]^+$. For examples of selfdual convex polyhedra with symmetry groups [2, 2], $[2,2]^+$, $[2,2^+]$, $[2^+,4]$, $[2^+,2^+]$ or $[2^+,4^+]$ see Figures 3-8 respectively, where the polyhedra are drawn in perspective views. Dashed lines in the Figures 3-8 denote the axes of 2-fold rotations of the polyhedra. The verification of the claims concerning groups of symmetries (and so, automorphism groups) and the selfduality of these polyhedra is routine, and is omitted.

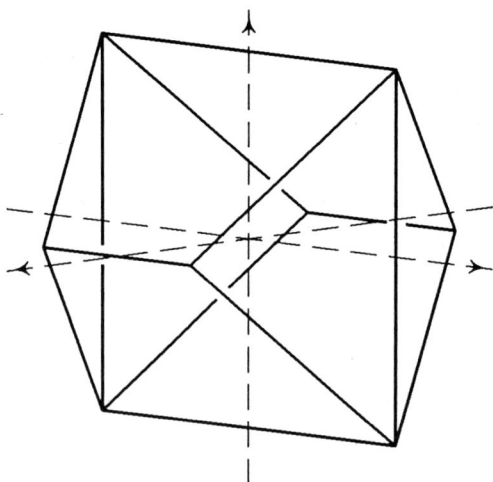

FIGURE 3

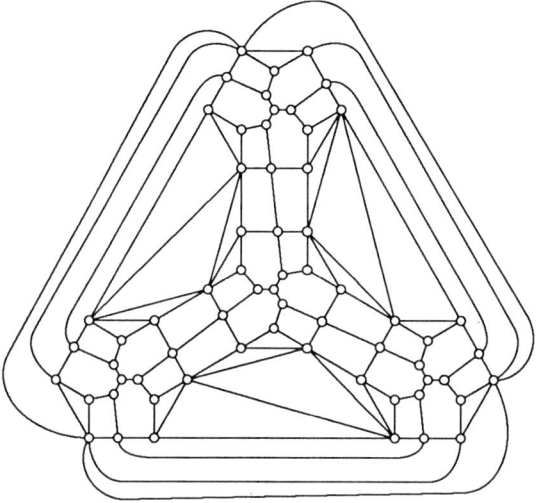

FIGURE 2

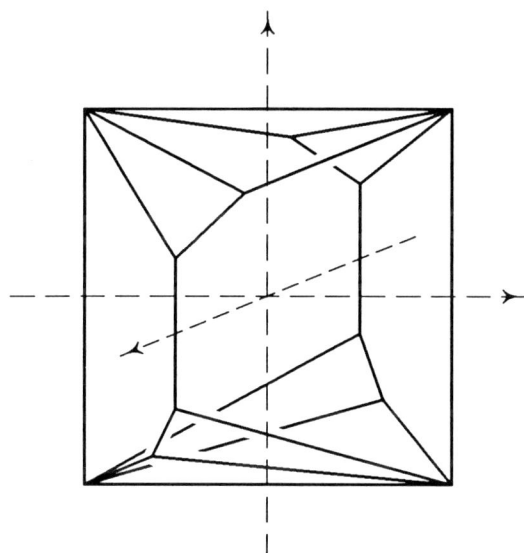

FIGURE 4

In the proof of the non-existence part of Theorem 1 the following property of selfdual polyhedra will be used

$$p_k(P) = v_k(P) \tag{A}$$

where $p_k(P)$ and $v_k(P)$ denote the number of k-gonal faces and k-valent vertices of the polyhedron P, respectively. Suppose P is a selfdual polyhedron with symmetry group $S(P)$ isomorphic to the group $[2, q]$, $[2, q]^+$, $[2, q^+]$, $[2^+, 2q]$ or $[2^+, 2q]$, $q \geqslant 3$, respectively. Because of properties of $S(P)$ (compare with [6]) an axis of the q-fold

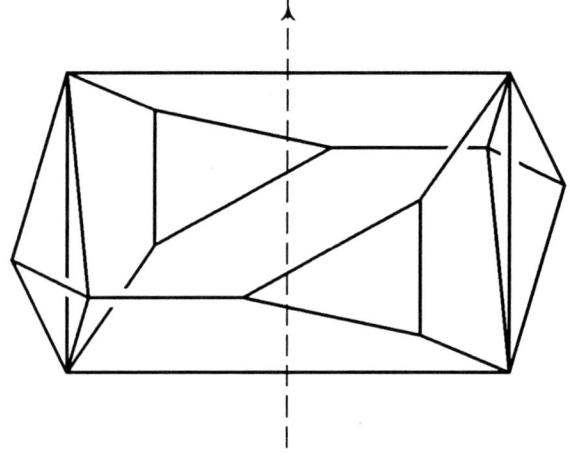

FIGURE 5

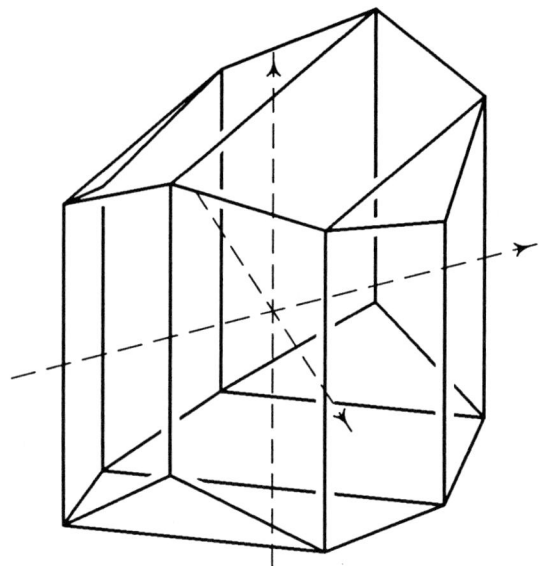

FIGURE 6

rotation of P passes through two congruent t-gonal faces or two t-valent vertices, $t \equiv 0 \pmod{q}$. The number of images of any other face or vertex by the symmetries of $S(P)$ is a multiple of q. Therefore for all $k \geqslant 3$, $k \neq t$ we have

$$v_k(P) \equiv 0 \pmod{q} \quad \text{and} \quad p_k(P) \equiv 0 \pmod{q}.$$

For the value t there is

$$v_t(P) \equiv a \pmod{q} \quad \text{and} \quad p_t(P) \equiv (2-a) \pmod{q}, \ a \in \{0,2\}.$$

Both these possibilities yield a contradiction with (A) which proves the nonexistence of a polyhedron P with properties required.

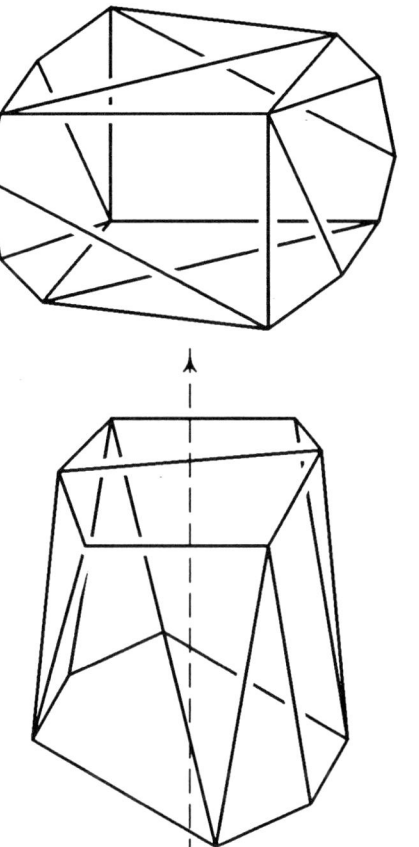

FIGURE 7

FIGURE 8

Let P be a selfdual polyhedron with the symmetry group $S(P)$ isomorphic to $[3,4]$, $[3,4]^+$ or $[3^+,4]$ respectively. Because of three axes of a 4-fold rotation and four axes of a 3-fold rotation of $S(P)$ there are six mutually congruent s-gonal faces or s-valent vertices and eight mutually congruent t-gonal faces or t-valent vertices with $s \equiv 0 \pmod 4$ and $t \equiv 0 \pmod 3$. The number of images of any other face or vertex by the symmetries of $S(P)$ is a multiple of 12 (see [6]); therefore there is

$$p_k(P) \equiv 0 \pmod{12}, \ v_k \equiv 0 \pmod{12} \text{ for all } k \geqslant 3, \ k \neq s, t$$

and for $s \neq t$,

$$p_s(P) \equiv a \pmod{12}, \ v_s \equiv (6-a) \pmod{12}, \ a \in \{0,6\},$$
$$p_t(P) \equiv b \pmod{12}, \ v_t \equiv (8-b) \pmod{12}, \ b \in \{0,8\}$$

or for $s = t$

$$p_s(P) \equiv c \pmod{12} \text{ and } v_s \equiv (2-e) \pmod{12}, \ c \in \{0,2\}.$$

Each of these two cases yields a contradiction with (A). Analogous arguments as those above show the nonexistence of selfdual polyhedra with symmetry groups $[3,5]$ or $[3,5]^+$. By Grünbaum and Shephard [6] no other groups are symmetry groups of convex polyhedra. The proof of Theorem 1 is finished. □

The interesting question in this connection is, which ranks can have selfdual polyhedra with a given symmetry group? The proof of Theorem 1 and [2] provides

Theorem 2. *For every group G from the list of Theorem 1 there exists a selfdual polyhedron P with symmetry group $S(P)$ isomorphic to G and rank $r(P) = 2$.*

There are known two examples of selfdual polyhedra with symmetry group $[2,2^+]$ and rank 4 (see [2] and [9]). Every other known example of a selfdual convex polyhedron with rank other than 2 has symmetry group $S(P)$ isomorphic to $[q]^+$ for a suitable q, $q \geqslant 1$. See [1, 2, 14]. If we restrict our attention to centrally symmetric polyhedra, we have

Theorem 3 ([2]). *If P is a selfdual polyhedron with a center of symmetry, then its rank $r(P)$ is either 2 or 4.*

Corollary. *If P is a selfdual convex polyhedron with the symmetry group $S(P)$ isomorphic to $[2,2]$, $[2,2^+]$ or $[2^+,2^+]$, then $r(P) = 2$ or 4.*

We conjecture that the analogue of Theorem 3 is true if we consider a plane of reflexion instead of a center of symmetry.

A selfdual polyhedron P is said to be *harmonious* if its symmetry group $S(P)$ is isomorphic to its graph automorphism group $A(P)$ (see [2]).

Theorem 4. *If P is a harmonious selfdual polyhedron with the symmetry group $S(P)$ then for the rank, $r(P) \in \mathcal{R}$ where*
(i) if $S(P) \in \{[2,2]^+, [2^+,4^+]\}$ then $\mathcal{R} = \{2,4,8\}$,
(ii) if $S(P) = [3,3]$ then $\mathcal{R} = \{2,4,8,16,32\}$ and
(iii) if $S(P) = [3,3]^+$ then $\mathcal{R} = \{2,4,8,16\}$.

Proof. Because of harmonicity of P the rank of any selfduality map of P cannot be larger than double the order of the group of automorphisms of the net of P and therefore of the symmetry groups $S(P)$ of P (by Mani's result [13] that was mentioned above). For the rest of the proof see the properties of the symmetry group. □

Acknowledgement. The autor wishes to thank Professor Branko Grünbaum for stimulating discussions on selfdual polyhedra which initiated the present work.

References

[1] D. Archdeacon, R. B. Richter, *The construction and classification of selfdual polyhedra*, to appear.
[2] J. Ashley, B. Grünbaum, G. C. Shephard, W. Stromquist, *Selfduality groups and ranks of selfdualities*, to appear.
[3] M. Brückner, *Vielecke und Vielflache*, Teubner, Leipzig, 1900.

[4] H. S. M. Coxeter, W. O. J. Moser, *Generators and Relations for Discrete Groups*, 3rd ed., Springer-Verlag, Berlin, 1972.

[5] B. Grünbaum, *Convex polytopes*, Interscience, London, 1967.

[6] B. Grünbaum, G. C. Shephard, *Patterns on the 2-sphere*, Mathematika **28** (1981), 1–35.

[7] _____, *Is selfduality involutory?*, Amer. Math. Monthly **95** (1988), 729–733.

[8] O. Hermes, *Die Formen der Vielflache*, J. reine angew. Math. **122** (1900), 124–154.

[9] S. Jendroľ, *A non-involutory selfduality*, Discrete Math. **74** (1989), 325–326.

[10] E. Jucovič, *Selfconjugate K-polyhedra*, Mat. Fyz. Časopis SAV **12** (1962), 1–22. (Russian, with German summary)

[11] _____, *Characterization of the p-vector of a selfdual 3-polytope*, Combinatorial Structures and their Applications (R. Guy et al., eds.), Gordon and Breach, New York, 1970, pp. 185–187.

[12] T. P. Kirkman, *On autopolar polyhedra*, Philos. Trans. Roy. Soc. London **147** (1857), 183–215.

[13] P. Mani, *Automorphismen von polyedrischen Graphen*, Math. Annalen **192** (1971), 279–303.

[14] J. McCanna, *Is self-duality always involutory?*, Congressus Numerantium **72** (1990), 175–178.

[15] A. Oudaise, personal communication.

Stanislav Jendroľ
P. J. Šafárik University,
041 54 Košice, Jesenná 5, Czechoslovakia

A Remark on 2-(v, k, λ) Designs

Věroslav Jurák

Let $v > k \geqslant t$ and λ be positive integers. A $t-(v,k,\lambda)$ design is a pair (X, B) such that
(i) X is a v-set,
(ii) B is a collection of distinct k-subsets (blocks) of X,
(iii) each t-subset of X is contained in exactly λ members of B.

A $t-(v,k,\lambda)$ design is symmetric if it is incomplete (i.e. $|B| \neq \binom{v}{2}$), $t \geqslant 2$ and $|B| = v$.

If a $t-(v,k,\lambda)$ design is symmetric then $t = 2$.

An isomorphism of designs (X, B) and (X, B') is a permutation p of X such that $b \in B$ implies $p(b) \in B'$. If $B' = B$ then an isomorphism is called an automorphism of the design (X, B).

For all foregoing definitions and theorems see [1].

In this paper we shall be concerned with the above mentioned symmetric 2-(v, k, λ) designs where $X = \{0, 1, \ldots, v-1\}$. Further, for each $a \in X$ which has no common divisor with v, let us write $ba = \{b_i a \pmod{v} : b_i \in b\}$ and $Ba = \{ba : b \in B\}$. Finally, a symmetric 2-(v, k, λ) design is cyclic if, for each non-zero $a \in X$, there is an ordering $^0 b$, $^1 b$, ..., $^{v-1} b$ of the elements of B such that $^{i+1} b = {^i b} + a = \{{^i b_j} + a \pmod{v} : {^i b_j} \in {^i b}\}$.

Now, here we have the **main propositions**:

For each symmetric 2-(v, k, λ) design (X, B) which is cyclic, there is an automorphisms group G of (X, B) ($r \in G$ iff $rB = B$) and the set M of isomorphisms of (X, B) and the conjugated design $(X, (-1)B)$ ($s \in M = (-1)G$ iff $sB = (-1)B$). Any element of G or M has no common divisor with v. Actually we write briefly only r or s for $p(B) = Br$ or Bs. Clearly, the design $(X, (-1)B)$ is also a symmetric 2-(v, k, λ) design which is cyclic.

All symmetric 2-(v, k, λ) designs which are not cyclic form also pairs of conjugated designs (X, B_i) and $(X, (-1)B_i)$. For each pair there is also the automorphisms group G of (X, B_i) and the set M of isomorphisms of (X, B_i) and $(X, (-1)B_i)$ as above.

Example: Collections B and B', where

$$B = \begin{matrix} ^0b & ^1b & ^2b & ^3b & ^4b & ^5b & ^6b \\ 1 & 2 & 3 & 4 & 5 & 6 & 0 \\ 2 & 3 & 4 & 5 & 6 & 0 & 1 \\ 4 & 5 & 6 & 0 & 1 & 2 & 3 \end{matrix}, \quad B' = \begin{matrix} 6 & 5 & 4 & 3 & 2 & 1 & 0 \\ 5 & 4 & 3 & 2 & 1 & 0 & 6 \\ 3 & 2 & 1 & 0 & 6 & 5 & 4 \end{matrix}$$

of distinct 3-subsets of $X = \{0, 1, \ldots, 6\}$ are the only mutually conjugated symmetric 2-$(7, 3, 1)$ designs which are cyclic. The automorphisms group of (X, B) or (X, B') is $G = \{1, 2, 4\}$ and the set of isomorphisms of (X, B) and (X, B') is $M = \{3, 5, 6\}$.

Further, the collections B_1 and B_1', B_2 and B_2', where

$$B_1 = \begin{matrix} 4 & 6 & 0 & 2 & 1 & 4 & 5 \\ 5 & 0 & 1 & 3 & 2 & 6 & 6 \\ 0 & 2 & 3 & 4 & 5 & 1 & 3 \end{matrix}, \quad B_1' = \begin{matrix} 3 & 1 & 0 & 5 & 6 & 3 & 2 \\ 2 & 0 & 6 & 4 & 5 & 1 & 1 \\ 0 & 5 & 4 & 3 & 2 & 6 & 4 \end{matrix}$$

$$B_2 = \begin{matrix} 6 & 3 & 5 & 4 & 2 & 1 & 1 \\ 0 & 4 & 0 & 5 & 3 & 3 & 2 \\ 1 & 0 & 2 & 6 & 6 & 5 & 4 \end{matrix}, \quad B_2' = \begin{matrix} 1 & 4 & 2 & 3 & 5 & 6 & 6 \\ 0 & 3 & 0 & 2 & 4 & 4 & 5 \\ 6 & 0 & 5 & 1 & 1 & 2 & 3 \end{matrix}$$

of distinct 3-subsets of $X = \{0, 1, \ldots, 6\}$ are mutally conjugated symmetric 2-$(7, 3, 1)$ designs which are not cyclic. The automorphisms group of (X, B_i) or (X, B_i') is $M = \{3, 5, 6\}$, for $i = 1, 2$.

Note that the isomorphisms $B_i \mapsto B_i + a = \{^j b + a : {}^j b \in B_i\}$ for all $a \in X$ and $i = 1, 2$ lead to all symmetric 2-$(7, 3, 1)$ designs which are not cyclic. The number of them is 28.

If $X = \{0, 1, \ldots, 6\}$ is the set of vertices of the complete graph K_7 then our 30 distinct symmetric 2-$(7, 3, 1)$ designs give all decompositions of the complete graph K_7 into seven complete graphs K_3.

For the construction of pairs of mutually conjugated symmetric 2-(v, k, λ) designs (X, B_j) and (X, B_j') which are not cyclic but with the automorphisms group G of (X, B_j) and the set M of isomorphisms of (X, B_j) and (X, B_j'), we further consider only sums of elements of all combinations of distinct integers $0, 1, \ldots, v-1$ taken k at a time with the correction modulo v.

In our example the collection B_1 has sums of elements of blocks equal to 2, 1, 4, 2, 1, 4, 0 respectively. The first three blocks of B_1 are taken from the collection B. The next three blocks are the only possible triples with sums of elements of blocks equal to 2, 1, 4 which are not contained in collections B and B' and which can be joined to the first three blocks. The last block is the only possible triple which can be joined to our six blocks.

Further, the collection B_2 has sums of elements of blocks equal to 0, 0, 0, 1, 4, 2, 0 respectively. The first three blocks are the only possible blocks with sums of elements equal to 0 which are not from the collections B and B'. The next three blocks are the only possible triples with sums of elements of blocks equal to 1, 4, 2 respectively which are not from the collections B and B' and which we can join to the first blocks. The last block is the only possible block which we can join to our blocks.

The sequences 2, 1, 4, 2, 1, 4, 0 and 0, 0, 0, 1, 4, 2, 0 of sums of elements of blocks of collections B_1 and B_2 fix the action of the group G and the set M on designs (X, B_1) and (X, B_2).

References

[1] Norman L. Biggs, Arthur T. White, *Permutation Groups and Combinatorial Structures*, London Mathematical Society Lecture Note Series 33, Cambridge University Press, 1979.

Věroslav Jurák
Technická 2,
Praha 6 – Dejvice,
160 00 Czechoslovakia

On a New Class of Intersection Graphs

MANFRED KOEBE

> We investigate spider graphs, a common generalization of interval, permutation, and circle graphs. A characterization of this new class of intersection graphs is given and a polynomial-time recognition algorithm is sketched.

1. Introduction

Intersection graphs are very interesting from the viewpoint of algorithmic graph theory because they can give hints with respect to the "borderline" between P and NP—if there is such a line (see [2] for classical results.) On the other hand, they have a wide range of direct applications, e.g. in VLSI layout (see e.g. [4], [1]).

In this paper we want to investigate a new class of the so-called *spider graphs*.

After giving some useful notation and examples we characterize spider graphs by forbidden substructures and in section 4 we sketch a polynomial-time recognition algorithm for spider graphs.

2. Notation and examples

Throughout this paper, $G = (V, E)$ always denotes a finite, undirected graph without loops and multiple edges. Let K be a fixed circle in the plane with an orientation fixed once and for all. On K there are positioned points, called *terminals*, with a numbering related to the orientation on K and an arbitrarily fixed first point. Furthermore we are given a finite system $\{T_i : i = 1, \ldots, n\}$ of pairwise disjoint sets of terminals on K, each set having at least two elements. T_i and T_j are said to *alternate* if there is no chord in K which separates the terminals of T_i and T_j. $G = (V, E)$ is called a *spider graph* if there is a collection $\{T_i\}$ on K and a bijection between V and $\{T_i\}$ such that the vertices corresponding to T_i and T_j in this bijection are adjacent iff T_i and T_j alternate on K, or equivalently, the minimal Steiner trees for the point sets T_i and T_j in the plane intersect. K together with $\{T_i : i = 1, \ldots, n\}$ is called a *spider representation* $S(G)$ of G. It is clear that the representation of a given graph is not unique. Let $T(P)$ denote the set of terminals representing $P \in V$ in a given representation.

The *domain* $D(P)$ of the vertex P is the segment of K between the first and the last element of $T(P)$—with respect to the terminal ordering on K mentioned above.

In comparison with other intersection graphs, spider graphs are a rather extensive class. For example, all trees, cycles, interval, permutation, trapezoid, and of course circle graphs have a spider representation. In all cases it can be constructed in a straightforward way (for details see [3]).

Theorem 1 ([3]). *Every chordal graph is a spider graph.*

3. Characterization of spider graphs

Because all chordal graphs are spider graphs, to get a characterization we have to inspect the chordless cycles with at least four vertices and its connections to the rest of the graph. Let c denote such a cycle in a graph G with a fixed ordering in the set $V(c)$ of its vertices, $G - c$ the subgraph of G after removing all vertices of c and all edges adjacent to a vertex of c. If Q is an element of $V(c)$, Q^- denotes the predecessor and Q^+ the successor of Q on c.

Let be $Q_i, Q_j \in V(c)$ and $R \notin V(c)$ a vertex of G with $t_1, t_2 \in T(R)$, $t_1 \in D(Q_i)$, $t_2 \in D(Q_i)$. Here $D(Q_i)$ denotes the smallest segment of K containing $T(Q_i)$, but no terminal of $\bigcup_{k=1}^{s} T(Q_k) \setminus (T(Q_i^-) \cup T(Q_i) \cup T(Q_i^+))$.

c is called a *cycle with empty centre*, if for every such R one of the following conditions holds:
(i) For every k, $i < k < j$, there is a $t'' \in T(R)$ with $t'' \in D(Q_k)$.
(ii) For every k, $k < i$ or $k > j$, there is a $t'' \in T(R)$ with $t'' \in D(Q)$.

Note that the existence of a chordless cycle of length greater than 3 always implies the existence of a chordless cycle with empty centre.

Theorem 2. *Let $G = (V, E)$ be a spider graph, $c = [Q, \ldots, Q]$ a chordless cycle of length at least 4 in G with empty centre. Then there are vertices $Q_{i_1}, \ldots, Q_{i_k}$ of c and induced subgraphs $G_{i_k}, \ldots, G_{i_k}$ of G such that:*
(i) Every vertex of G which is adjacent to a vertex of $G - G_{i_j} - (Q_{i_j}^-, Q_{i_j}^+)$ is adjacent to Q as well for $j = 1, \ldots, k$.
(ii) Every edge of $G - c$ belongs to at least one subgraph G_{i_j}, i.e. $\bigcup_{j=1}^{k} G = G_{i_j} - c$.

It is not complicated to see that for every Q_i of c the corresponding G_i is induced by all vertices which have at least one terminal in the domain $D(Q_i)$.

Now we can show that the condition of theorem 2 is also sufficient.

Theorem 3. *Every graph $G = (V, E)$ satisfying the condition of theorem 2 is a spider graph.*

Sketch of Proof. We use mathematical induction on $|V|$. Let $P \in V$ be a vertex such that $G' = G - P$ contains a chordless cycle of length at least 4 with empty centre which is not expandable at a neighbour of P. If at most one vertex of c is adjacent to P there is a $Q \in V(c)$ and a spider representation of G' such that for all vertices of the neighbourhood of P at least one terminal is contained in $D(Q)$. We repeat the considerations with the proper induced subgraph $G'(Q)$ of G' containing all vertices with at least one terminal in $D(Q)$. Now let $Q_{i_1}, \ldots, Q_{i_r}$, $r < 1$, be a subset of $V(c) \cap \text{Adj}(P)$ such that every neighbour of P is contained in at least one of the subgraphs $G'(Q_{i_j})$, $j = 1, \ldots, r$. We can deal with every subgraph separately and determine in every case a terminal set for P lying in $D(Q_{i_j})$. Because of the properties of c the union of these terminals models the vertex P in a correct way. If we get chordal subgraphs we know by theorem 1 that they have spider representations (for details see [3]). □

4. Recognition of spider graphs

Because the proof of theorem 3 is constructive we can use it to get an efficient recognition algorithm for spider graphs (for details see [3]):

Theorem 4. *The recognition problem for spider graphs G is polynomially solvable.*

References

[1] I. Dagan, M. C. Golumbic, R. Y. Pinter, *Trapezoid graphs and their coloring*, Discrete Appl. Math. **21** (1988), 35–46.

[2] M. C. Golumbic, *Algorithmic graph theory and perfect graphs*, Academic Press, New York, 1980.

[3] M. Koebe, *Spider graphs—a new class of intersection graphs*, submitted for publication.

[4] M. Koebe, *Intersection graphs and routing in VLSI layout*, Combinatorial Methods and Applications II, Preprint 25, Fachbereich Mathematik, EMAU Greifswald, 1990, pp. 50–53.

Manfred Koebe
Ernst-Moritz-Arndt-Universität,
Fachbereich Mathematik,
Jahnstr. 15a, 2200 Greifswald, Germany

Fourth Czechoslovakian Symposium on
Combinatorics, Graphs and Complexity
J. Nešetřil and M. Fiedler (Editors)
© 1992 Elsevier Science Publishers B.V. All rights reserved.

Asymptotic Normality of Isolated Edges in Random Subgraphs of the n-Cube

URSZULA KONIECZNA

We consider two types of random subgraphs of the n-cube. For these models we study the asymptotic behaviour of the number of isolated edges.

1. Introduction

The n-cube Q_n is the graph with 2^n vertices labelled by sequences $(a_1, \ldots, a_n)$, $a_i \in \{0, 1\}$ and $n2^{n-1}$ edges between vertices differing in exactly one coordinate.

In our paper we will consider two types of subgraphs of the n-cube: induced subgraphs and spanning subgraphs.

Choosing vertices of Q_n at random and independently of each other with the same probability p_v, we obtain a random induced subgraph f. Choosing edges of Q_n at random and independently of each other with the same probability p_e, we obtain a random spanning subgraph g. In this paper we shall investigate the asymptotic behaviour of the random variables $I(f)$ and $I(g)$ equal to the number of isolated edges in f and in g respectively.

K. Weber showed in his paper [5] that these random variables have asymptotically a Poisson and a normal distribution for same values of probability p_v and p_e but some "gaps" in the intervals of p_v and p_e were left to investigate. We want to prove that there is convergence in distribution to the normal distribution for p_e and p_v from these "gaps" too. Moreover we will show this convergence in the metric of Stein.

In 1970 C. Stein introduced a new technique to obtain estimates of the rate of convergence to the standard normal distribution. He used the estimate as follows:

$$\left| \int h(x) \, dF_n(x) - \int h(x) \, d\Phi(x) \right| \leqslant \kappa_n \|h\| \qquad (1.1)$$

for all bounded test functions h with bounded derivative, where $\|h\| = \sup_x |h(x)| + \sup_x |h'(x)|$, $F_n(x)$ is the distribution function being approximated and $\Phi(x)$ is that of the standard normal distribution. The quantity κ_n in (1.1) provides an upper estimate of the distance between F_n and Φ in a metric which we denote d_1. We shall use here the estimate for the so called decomposable random variables given by A. D. Barbour, M. Karoński, A. Ruciński in the paper [2].

For convenience we denote a random variable with the standard normal distribution by $N(0,1)$. Moreover $\mathbf{E}X$ stands for the expectation of the random variable X, $\operatorname{var}X$ for the variance and $\tilde{X} = (X - \mathbf{E}X)/\sqrt{\operatorname{var}X}$. Finally $\varrho(v, v')$, v, $v' \in V(Q_n)$ is the Hamming distance between v and v' i.e. the number of coordinates in which v and v' differ. The distance between two edges e and e' of $E(Q_n)$ is defined as $\varrho(e, e') = \min \varrho(v, v')$ where the minimum is taken over all vertices v from e and v' from e'.

2. Main results

Now we formulate theorems concerning the convergence to the standard normal distribution for the random variables $I(g)$ and $I(f)$.

Theorem 1. Let $I(g)$ count the isolated edges in g. Then

$$d_1\big(\tilde{I}(g), N(0,1)\big) = 0\big\{(\mathbf{E}I(g))^{-\frac{1}{2}}\big\}.$$

In particular, $\tilde{I}(g) \to N(0,1)$ if $\mathbf{E}I(g) \to \infty$ as $n \to \infty$ i.e. when $n2^n p_e \to \infty$ and $2\sqrt{2}np_e - 2n(\sqrt{2}-1) - \ln n \to -\infty$.

Proof. Let $E = E(G)$ be the set of edges of the graph G. For every $\alpha, \beta, \gamma \in E$ we define:

$$Y_\alpha = \begin{cases} 1 & \text{when } \alpha \text{ is an isolated edge in } G \\ 0 & \text{otherwise} \end{cases}$$

$$Y_\beta^{(\alpha)} = \begin{cases} 1 & \text{when } \beta \text{ is an isolated edge in } G - \alpha \\ 0 & \text{otherwise} \end{cases}$$

$$S = \sum_{\alpha \in E} Y_\alpha, \quad m_\alpha = \mathbf{E}Y_\alpha, \quad \sigma^2 = \operatorname{var} S.$$

Hence S is the number of all isolated edges in G.

Let $X_\alpha = (Y_\alpha - m_\alpha)/\sigma$ and

$$Z_\alpha = \sigma^{-1}\Big\{\sum_{\beta \cap \alpha \neq \emptyset} Y_\beta + \sum_{\beta \cap \alpha = \emptyset}\big(Y_\beta - Y_\beta^{(\alpha)}\big)\Big\} = \sum_{\beta \in E} Z_{\alpha\beta},$$

$$V_{\alpha\beta} = \sigma^{-1}\Big\{\sum_{\substack{\gamma \cap \beta \neq \emptyset \\ \gamma \cap \alpha = \emptyset}} Y_\gamma^{(\alpha)} + \sum_{\substack{\gamma \cap \beta = \emptyset \\ \gamma \cap \alpha = \emptyset}}\big(Y_\gamma^{(\alpha)} - Y_\gamma^{(\alpha \cup \beta)}\big)\Big\}.$$

In [2] it was proved that $d_1\big(\tilde{S}, N(0,1)\big) \leqslant K\kappa$ where

$$\kappa = \frac{1}{2}\sum_{\alpha \in E} \mathbf{E}\big(|X_\alpha|Z_\alpha^2\big) + \sum_{\alpha \in E}\sum_{\beta \in E}\big(\mathbf{E}|X_\alpha Z_{\alpha\beta}V_{\alpha\beta}| + \mathbf{E}|X_\alpha Z_{\alpha\beta}|\,\mathbf{E}|Z_\alpha + V_{\alpha\beta}|\big) \quad (2.1)$$

and K is a universal constant.

In particular if $\kappa \to 0$ as $n \to \infty$ then $\tilde{S} \xrightarrow{D} N(0,1)$.

Let $G = g$. Notice that $P(Y_\alpha = 1) = p_e(1-p_e)^{2(n-1)} = m_\alpha$. Hence $\mathbf{E}S = n2^{n-1}m_\alpha$. The variance σ^2 of the random variable S is given by

$$\sigma^2 = \sum_{\alpha,\beta \in E} \text{cov}(Y_\alpha, Y_\beta),$$

$$\text{cov}(Y_\alpha, Y_\beta) = \begin{cases} A & \text{when } \alpha = \beta \\ B & \text{when } \alpha \neq \beta, \alpha \cap \beta \neq \emptyset \\ C & \text{when } \varrho(\alpha,\beta) = 1 \\ 0 & \text{when } \varrho(\alpha,\beta) \geq 2 \end{cases}$$

where

$$A = \mathbf{E}Y_\alpha^2 - (\mathbf{E}Y_\alpha)^2 = m_\alpha - m_\alpha^2; \quad B = -m_\alpha^2;$$
$$C = p_e^2(1-p_e)^{4n-6} + p_e^2(1-p_e)^{4n-5} - m_\alpha^2.$$

Therefore

$$\sigma^2 = \mathbf{E}S\left[1 - m_\alpha - 2(n-1)m_\alpha + (n-1)m_\alpha(1-p_e)^{-2} \right.$$
$$\left. + (n-1)(2n-3)m_\alpha(1-p_e)^{-1} - 2(n-1)^2 m_\alpha\right].$$

One can easily check that $\sigma^2 > c\mathbf{E}S$ where c is a positive constant. To complete the proof we will show that the first and the third term in the formula (2.1) are equal to $O(\sigma^{-3}\mathbf{E}S)$ and the second one is equal to 0.

The first term can be estimated as follows

$$\sum_{\alpha \in E} \mathbf{E}\left(|X_\alpha|Z_\alpha^2\right) \leq \sigma^{-1} \sum_{\alpha \in E} \left[\mathbf{E}\left(|Y_\alpha|Z_\alpha^2\right) + m_\alpha \mathbf{E}Z_\alpha^2\right]$$

where

$$\mathbf{E}Z_\alpha^2 = \sigma^{-2}\left\{m_\alpha + 2\sum_{\beta \in E}\left[Y_\alpha\left(Y_\beta - Y_\beta^{(\alpha)}\right)\right] + \sum_{\beta \in E}\sum_{\gamma \in E}\left[\left(Y_\beta - Y_\beta^{(\alpha)}\right)\left(Y_\gamma - Y_\gamma^{(\alpha)}\right)\right]\right\}$$

and

$$\mathbf{E}\left(|Y_\alpha|Z_\alpha^2\right) = \sigma^{-2}\left\{m_\alpha + 2\sum_{\beta \in E}\mathbf{E}\left[Y_\alpha^2\left(Y_\beta - Y_\beta^{(\alpha)}\right)\right] \right.$$
$$\left. + \sum_{\beta \in E}\sum_{\gamma \in E}\mathbf{E}\left[Y_\alpha\left(Y_\beta - Y_\beta^{(\alpha)}\right)\left(Y_\gamma - Y_\gamma^{(\alpha)}\right)\right]\right\}.$$

Let β and γ be such that $\varrho(\alpha,\beta) \leq 1$ and $\varrho(\alpha,\gamma) \leq 1$ for a fixed α. Then

$$\sum_{\beta \in E}\mathbf{E}\left[Y_\alpha\left(Y_\beta - Y_\beta^{(\alpha)}\right)\right] = \sum_{\beta \in E}\mathbf{E}\left[Y_\alpha^2\left(Y_\beta - Y_\beta^2\right)\right]$$
$$= \sum_{\beta \in E}\sum_{\gamma \in E}\left[Y_\alpha\left(Y_\beta - Y_\beta^{(\alpha)}\right)\left(Y_\gamma - Y_\gamma^{(\alpha)}\right)\right] = 0;$$

$$\sum_{\beta \in E} \sum_{\gamma \in E} \left[\left(Y_\beta - Y_\beta^{(\alpha)} \right) \left(Y_\gamma - Y_\gamma^{(\alpha)} \right) \right] = O(\mathbf{E}S).$$

In this way the first term of (2.1) is equal to $O(\sigma^{-3}\mathbf{E}S)$. For the second term we have

$$\sum_{\alpha \in E} \sum_{\beta \in E} \mathbf{E}|X_\alpha Z_{\alpha\beta} V_{\alpha\beta}| = \sigma^{-1} \sum_{\beta \cap \alpha \neq \emptyset} \mathbf{E}|X_\alpha Y_\beta V_{\alpha\beta}| + \sigma^{-1} \sum_{\beta \cap \alpha = \emptyset} \mathbf{E}|X_\alpha (Y_\beta - Y_\beta^{(\alpha)}) V_{\alpha\beta}|.$$

Moreover $\sum_{\beta \cap \alpha \neq \emptyset} \mathbf{E}|X_\alpha Y_\beta V_{\alpha\beta}| = 0$ because

(i) $Y_\alpha Y_\beta = 0$ if $\alpha \neq \beta$,
(ii) $V_{\alpha\beta} = 0$ if $\alpha = \beta$,
(iii) $Y_\beta Y_\gamma^{(\alpha)} = 0$ if $\gamma \cap \beta \neq \emptyset$ and $\gamma \cap \alpha = \emptyset$,
(iv) $Y_\beta \left(Y_\gamma^{(\alpha)} - Y_\gamma^{(\alpha \cup \beta)} \right) = 0$ if $\gamma \cap \beta = \emptyset$ and $\gamma \cap \alpha = \emptyset$.

Similarly taking the terms with $\beta \cap \alpha = 0$, we obtain

$$\sum_{\beta \cap \alpha = 0} \mathbf{E}|X_\alpha (Y_\beta - Y_\beta^{(\alpha)}) V_{\alpha\beta}| = 0.$$

Finally,

$$\sum_{\alpha \in E} \sum_{\beta \in E} \mathbf{E}|X_\alpha Z_{\alpha\beta}| \mathbf{E}|Z_\alpha + V_{\alpha\beta}| \leqslant \sigma^{-1} \left\{ \sum_{\alpha \in E} \sum_{\beta \in E} \mathbf{E}(Y_\alpha |Z_{\alpha\beta}|)(\mathbf{E}|Z_\alpha| + \mathbf{E}|V_{\alpha\beta}|) \right.$$

$$\left. + \sum_{\alpha \in E} \sum_{\beta \in E} m_\alpha \mathbf{E}|Z_{\alpha\beta}|(\mathbf{E}|Z_\alpha| + \mathbf{E}|V_{\alpha\beta}|) \right\}$$

where

$$\mathbf{E}|Z_\alpha| \leqslant \sum_{\beta \in E} \mathbf{E}|Z_{\alpha\beta}| = O(\sigma^{-1}); \qquad \mathbf{E}|V_{\alpha\beta}| = O(\sigma^{-1})$$

and

$$\sum_{\beta \in E} Y_\alpha |Z_{\alpha\beta}| = \sigma^{-1} Y_\alpha.$$

So the third part of κ is equal to $O(\sigma^{-3} \mathbf{E}S)$ also and this completes the proof. $\square$

Theorem 2. Let $I(f)$ count the isolated edges in f. Then $d_1(\tilde{I}(f), N(0,1)) = O\left\{ (\mathbf{E}I(f))^{-\frac{1}{2}} \right\}$. In particular $\tilde{I}(f) \xrightarrow{D} N(0,1)$ if $\mathbf{E}I(f) \to \infty$ as $n \to \infty$ i.e. when $n 2^n p_v^2 \to \infty$ and $2\sqrt{2} n p_v - 2n(\sqrt{2} - 1) - \ln n \to -\infty$.

Since the proof of Theorem 2 follows the same line as that of the preceding theorem we omit it.

References

[1] A. D. Barbour, *Poisson convergence and random graphs*, Math. Proc. Comb. Soc. **92** (1982), 319–359.

[2] A. D. Barbour, M. Karoński, A. Ruciński, *Central limit theorem for decomposable random variables with application to random graphs*, J. Comb. Theory B **47** (1989), 125–145.

[3] U. Konieczna, *Asymptotic normality of the vertex degree in random subgraphs of the n-cube*, to appear.

[4] C. Stein, *A bound for the error in the normal approximaton to the distribution of a sum of dependent random variables*, Proc. VIth Berk. Symp. Math. Stat. Prob. **2** (1970), 583–602.

[5] K. Weber, *Poisson convergence in the n-cube*, Math. Nachr. **131** (1987), 49–57.

Urszula Konieczna
Department of Applied Mathematics,
Academy of Technology and Agriculture,
Al. Prof. Kaliskiego 7,
85-790 Bydgoszcz,
Poland

Fourth Czechoslovakian Symposium on
Combinatorics, Graphs and Complexity
J. Nešetřil and M. Fiedler (Editors)
© 1992 Elsevier Science Publishers B.V. All rights reserved.

On Bounds of the Bisection Width of Cubic Graphs

A. V. KOSTOCHKA, L. S. MEL'NIKOV

1. Introduction

Let G be a finite graph. The *bisection width* $bw(G)$ of graph G is the minimal number of edges between vertex sets A and $\bar{A}$ of almost equal size, i.e. $A \cup \bar{A} = V(G)$ and $||A| - |\bar{A}|| \leq 1$. If $A \subseteq V(G)$, then $E(A, \bar{A})$ denotes the set of edges of G having one end in A and the other end in $V(G) \setminus A = \bar{A}$. There is another important characteristics—the *isoperimetric number* $i(G)$ of graph G, which equals the minimum of the ratio $|E(A, \bar{A})|/|A|$ for all $A \subseteq V(G)$ such that $2|A| \leq |V(G)| = n$.

From the definitions the following inequality for these characteristics follows:

$$i(G) \leq \frac{2}{n} bw(G). \tag{1}$$

The importance of investigation of the bisection width $bw(G)$ and the isoperimetric number $i(G)$ lies in various interesting interpretations of these numbers (see [6], [7]). The problems of finding the bisection width and the isoperimetric number of a given graph are known to be NP-hard [5]. Moreover, it was shown in [2], [6] that the problem of finding the bisection width is NP-hard even for the special class of graphs with maximum degree 3. Generalizations have been considered in [1], [2].

In [2] a method was given for transforming a regular graph G with n vertices into a cubic graph G^* with $O(n^6)$ vertices such that any minimum bisection of G^* uses only edges of G. Therefore we are interested in the examination of cubic graphs.

Buser [3] showed the existence of such cubic graphs G that $i(G) \geq 1/128$. Clark and Entringer [4] obtained in their latest work the upper bound $bw(G) \leq \frac{1}{3}(n+138)$ for cubic graphs.

In the present work the bounds for $i(G)$ and $bw(G)$ are improved. We show the existence of such cubic graphs G_0 that $i(G_0) \geq 1/4.95$ and for every cubic graph we prove $bw(G) \leq n/4 + O(\sqrt{n} \log n)$. The last result is a corollary of a general result about q-regular graphs.

2. An upper bound for the bisection width of regular graphs

First we need some preliminary results.

Lemma 1. Let T be a tree with maximal degree q. Then for any k ($k \leqslant |V(T)|$) the set $V(T)$ can be divided in two parts V_1 and V_2 such that
(a) $|V_1| = k$;
(b) $T \langle V_2 \rangle$ is a connected subgraph;
(c) the number of components of $T \langle V_1 \rangle$ is not more than $1 + \log_{\frac{q-1}{q-2}} k$.

Proof. By induction on k. For $k = 1$ the statement holds. Let $k > 1$. Let v_0 be an arbitrary pending vertex of the tree T. Consider T as a rooted (directed) tree $\vec{T}$ with root v_0. For any vertex v of T denote by $\vec{T}_v$ the rooted subtree of $\vec{T}$ with root v_0. All vertices are in a natural way partitioned into levels, level 0 consisting of the only vertex v_0. From among those vertices v for which the number of vertices in $\vec{T}_v$ is greater than k, choose a vertex of the highest level and call it v_1. Denote the immediate successors of v_1 by $x_1, \ldots, x_s$. As the maximal degree is $\Delta(T) = q$ and $\deg(v_0) = 1$, $s \leqslant q - 1$. Denote by n_i ($1 \leqslant i \leqslant s$) the number of vertices of $\vec{T}_{x_i}$. By how v_1 was chosen, we have $n_1 + \ldots + n_s + 1 > k$. But then, if $n_1 = \max\{n_i : 1 \leqslant i \leqslant s\}$,

$$n_1 \geqslant \frac{k}{s} \geqslant \frac{k}{q-1} \qquad (2)$$

Consider $T' = T - V(\vec{T}_{x_i})$, $k' = k - n_1$. From (2) we have

$$k' \leqslant \frac{q-2}{q-1} k. \qquad (3)$$

By the induction hypothesis, T' may be split into a tree of $(|V(T')| - k')$ vertices and a graph of k' vertices and not more than

$$1 + \log_{\frac{q-1}{q-2}} k'$$

components. By (3), this quantity is less or equal $\log_{\frac{q-1}{q-2}} k$. Adding $V(T_{x_1})$, we get the required splitting. $\square$

Lemma 2. Let G be a multigraph of $2n$ vertices and m edges. Then there exists a splitting of $V(G)$ into V_1 and V_2 of equal size such that

$$|E(V_1, V_2)| \leqslant \frac{mn}{2n-1}.$$

Proof. The number of all possible splittings of $V(G)$ into halves is $\frac{1}{2}\binom{2n}{n}$. Each edge connects vertices of different halves in $\binom{2n-2}{n-1}$ splittings. Hence the mean number of edges in a splitting is

$$\frac{m\binom{2n-2}{n-1}}{\frac{1}{2}\binom{2n}{n}} = \frac{mn}{2n-1}$$

and there must be a splitting (V_1, V_2) with $|E(V_1, V_2)|$ not more than this mean value. $\square$

Theorem 1. *For any given natural number q ($q \geq 2$) and for any connected q-homogeneous graph G having n vertices, its bisection width fulfils*

$$bw(G) \leq \frac{q-2}{4} n + O(q\sqrt{n} \log n).$$

Proof. Let n be large and G a q-homogeneous graph on n vertices. Denote $s = \lfloor \sqrt{n} \rfloor$, $m = \lfloor \frac{n}{2s} \rfloor$, $t = n - 2sm$.

Choose in G an arbitrary spanning tree T. According to lemma 1, $V(G)$ may be partitioned into $2m + 1$ subsets $V_0, V_1, \ldots, V_{2m}$ such that
(a) $|V_i| = s$ for any i, $1 \leq i \leq 2m$
(b) the number of connected components of $G \langle V_i \rangle$ is not more than

$$1 + \log_{\frac{q-1}{q-2}} s$$

for any $1 \leq i \leq 2m$;
(c) the graph $G \langle V_0 \rangle$ is connected.

Construct now the multigraph $\tilde{G} = (\tilde{U}, \tilde{E})$ by the following rules. The vertex set $\tilde{V}$ is formed by the partition classes $\{V_1, V_2, \ldots, V_{2m}\}$ and the vertex V_i is connected with the vertex V_j by as many edges as are the sets V_i and V_j in G.

By the construction, $|E(G \langle V_i \rangle)| \geq s - 1 + \log_{\frac{q-1}{q-2}} s$ for $1 \leq i \leq 2m$ and $|E(G \langle V_0 \rangle)| \geq t - 1$. Hence,

$$|\tilde{E}| \leq |E(G)| - 2m(s - 1 - \log_{\frac{q-1}{q-2}} s) - t + 1 =$$
$$= |E(G)| - |V(G)| + 2m(1 + \log_{\frac{q-1}{q-2}} s) + 1 =$$
$$= \frac{q-2}{2} + O(q\sqrt{n} \log n).$$

By lemma 2, $\tilde{V}$ my by partitioned into $\tilde{V}_1$ and $\tilde{V}_2$ of equal size such that

$$|E_{\tilde{G}}(\tilde{V}_1, \tilde{V}_2)| \leq \frac{|\tilde{E}|m}{2m-1} \leq \frac{q-2}{4} n + O(q\sqrt{n} \log n).$$

Finally partition V_0 arbitrarily into W_1 and W_2 of size $\lfloor \frac{t}{2} \rfloor$ and $\lceil \frac{t}{2} \rceil$ respectively. Let now $X_1 = W_1 \cup \{v \in V_i : V_i \in \tilde{V}_1\}$, $X_2 = V(G) - X_1$. Then

$$|E_G(X_1, X_2)| = |E_{\tilde{G}}(\tilde{V}_1, \tilde{V}_2) \cup E_G(W_1, X_2) \cup E_G(W_2, X_1)| \leq$$
$$\leq \frac{q-2}{4} n + O(q\sqrt{n} \log n) + 2 \left\lceil \frac{t}{2} \right\rceil q \leq \frac{q-2}{4} n + O(q\sqrt{n} \log n).$$

□

Corollary 1. *For any cubical graph G,*

$$bw(G) \leq \frac{n}{4} + O(\sqrt{n} \log n).$$

Corollary 2. *There exist cubical graphs G_1 such that*

$$bw(G_1) \geqslant \frac{n}{9.9}.$$

Because of lack of space, we omit here the rather complicated proof of existence of a graph G_0 such that its isoperimetric number $i(G_0) \geqslant \frac{1}{4.95}$. We hope to be able to publish it later.

References

[1] E. R. Barnes, A. J. Hoffman, *Partitioning, spectra and linear programming*, Progress in Combinatorial Optimization (W. Pulleyblank, ed.), Academic Press, New York, 1984.

[2] T. H. Bui, S. Chaudhuri, T. Leighton, M. Sipser, *Graph bisection algorithms with good average case behavior*, FOCS (1984); Combinatorica **7** (1987), 171–191.

[3] P. Buser, *On the bipartition of graphs*, Discrete Appl. Math. **9** no. 1 (1984), 105–109.

[4] L. H. Clark, R. C. Entringer, *The bisection width of cubic graph*, Bull. Austral. Math. Soc. **39** no. 3 (1988), 389–396.

[5] M. R. Garey, D. S. Johnson, L. Stockmeyer, *Some simplified NP-complete graph problems*, Theor. Comput. Sci. **1** (1976), 237–267.

[6] R. M. MacGregor, *On partitioning a graph: a theoretical and empiric study*, Ph. D. Thesis, Univ. of California, Berkeley, 1978.

[7] B. Mohar, *Isoperimetric numbers of graphs*, J. Combin. Theory B **47** no. 3 (1989), 274–291.

A. V. Kostochka, L. S. Mel'nikov
Institute of Mathematics, Siberian Branch,
Academy of Sciences of the USSR,
Novosibirsk 630090, USSR

On Random Cubical Graphs

A. V. KOSTOCHKA, A. A. SAPOZHENKO, K. WEBER

The historical development and the state of the art concerning random subgraphs of the n-cube graph are summarized.

A graph G is called cubical if it is embeddable in an n-cube graph Q_n, i.e. G is a subgraph of some Q_n (cf. [6]). A method to generate all cubical graphs at random was introduced by Dyer and Frieze at the 2nd "Random Graphs" seminar, Poznań 1985 (cf. [4]). Let $h_n = (V_n, E_n)$ be the random subgraph of Q_n produced as follows: V_n is randomly sampled from the vertex set of Q_n so that $P(x \in V_n) = p_v$ independently for each vertex $x \in Q_n$. E_n is now randomly sampled from the set of edges induced by V_n in Q_n so that $P(xy \in E_n) = p_e$ independently for each induced edge xy. But already many years earlier the following two special models of random subgraphs of Q_n were investigated. For $p_e = 1$ all induced edges are chosen, and we get random induced subgraphs f_n of Q_n (they may be interpreted as random Boolean functions). If $p_v = 1$ then all 2^n vertices belong to the random subgraph, i.e. we get random spanning subgraphs g_n of Q_n.

One says that a random graph has a certain property if it has this property with probability tending to one as n tends to infinity. In all what follows let $p = p_v p_e$ and $q = 1 - p$.

1. Random induced subgraphs (random Boolean functions) f_n

Jablonski (1958): cube symmetry ($p = \frac{1}{2}$)
Zhuravljov (1962): subcube dimension $d \leqslant \lfloor \log_2 n \rfloor$ ($p = \frac{1}{2}$)
Glagoljev (1967): subcube coverings ($p = \frac{1}{2}$)
Sapozhenko (1967...): components, radius, diameter ($p = \frac{1}{2}$)
Sapozhenko (1967...): subcube coverings ($p = \frac{1}{2}$)
Toman (1979): components (fixed $p \neq \frac{1}{2}$)
Weber (1982...): subcube coverings
Korshunov, Sapozhenko (1983): $P(|E(f_n)| = 0) \sim 2\sqrt{e} 2^{-2^{n-1}}$ ($p = \frac{1}{2}$)
Weber (1983...): components, matchings, graph theoretic parameters
Škoviera (1986): subcube coverings

Mahrhold,
Weber (1987): planarity threshold

Up to now the theory of random cubical graphs is not so developed as the theory of usual random graphs, i.e. random spanning subgraphs of the complete graph. But the history starts about at the same time (around the year 1960), and it is with random Boolean functions for the case $p = \frac{1}{2}$. In this case a property of random Boolean functions is nothing else than a property of almost all or asymptotically all of the 2^{2^n} Boolean functions f_n. In 1958 Jablonski showed that almost all f_n are asymmetric concerning cube symmetry, i.e. they are invariant only concerning the trivial automorphism of Q_n. Zhuravljov proved that the dimension of subcubes of almost all f_n is upperbounded by $\lfloor \log_2 n \rfloor$. The latter consideration was considerably extended by Glagoljev by giving estimates for many parameters characterizing the complexity of the Boolean function minimatization process as well as the complexity of subcube coverings (coverings of the vertex set of Q_n by subcubes) for almost all f_n. The first results on graph theoretic parameters are due to Sapozhenko ([15]) who could also considerably improve many of the Glagoljev results. The case $p \neq \frac{1}{2}$ was first considered in the literature by Toman ([21]). In 1977 Weber started the investigation of subcube coverings of f_n for arbitrary probability p. The papers appeared since 1982 (for a survey see [25]). Since 1983 Weber published results concerning graph theoretic properties and parameters of f_n for arbitrary probability p. For example, in 1987 we determined together with Mahrhold the threshold probability for planarity ([13]). In 1983 Korshunov and Sapozhenko determined the asymptotics for the probability that set of induced edges of f_n is empty (f_n consists only of isolated vertices). This result yields the asymptotic number of independent sets in Q_n: $2\sqrt{e}\, 2^{2^{n-1}}$.

2. Random spanning subgraphs g_n

The history of random spanning subgraphs g_n started some years later with a paper of Burtin.

Burtin (1977): connectedness (fixed $p \neq \frac{1}{2}$)
Erdős,
Spencer (1979): components $(p = \frac{1}{2})$
Toman (1980): components $(p = \frac{1}{2})$
Ajtai,
Komlós,
Szemerédi (1982): components $\left(p = \frac{1+\varepsilon}{n}\right)$
Toman (1982): radius (fixed $p \geq \frac{1}{2}$)
Weber (1982...): $\Big\}$ see above
Weber (1983...):
Bollobás (1983): connectivity $\left(p = \frac{1}{2}\left(1 + \frac{c}{n}\right)\right)$
Mahrhold,
Weber (1987): planarity threshold
Kostochka (1989): existence of perfect matching for fixed $p > \frac{1}{2}$
Bollobás (1989): existence of perfect matching for fixed $p > \frac{1}{2}$

Further comments on this and the following section the reader may find in Sections 4 and 5.

3. Random cubical graphs h_n

Dyer, Frieze (1985):	thresholds for s-connectivity, $s = 1, 2, \ldots$
Konieczna, Weber (1987):	matchings, graph theoretic parameters
Mahrhold, Weber (1988):	radius (fixed $p \geq \frac{1}{2}$)
Weber (1989):	components (fixed p)
Sapozhenko, Weber (1989):	radius, diameter (fixed p)

4. Evolution

What is known about the component structure of the random graphs f_n, g_n resp. h_n? Let us answer this question with the following diagram and some comments. Note that we have for the cardinalities of the vertex sets $|V(f_n)| \sim p2^n$, $|V(g_n)| = 2^n$ and $|V(h_n)| \sim p_v 2^n$.

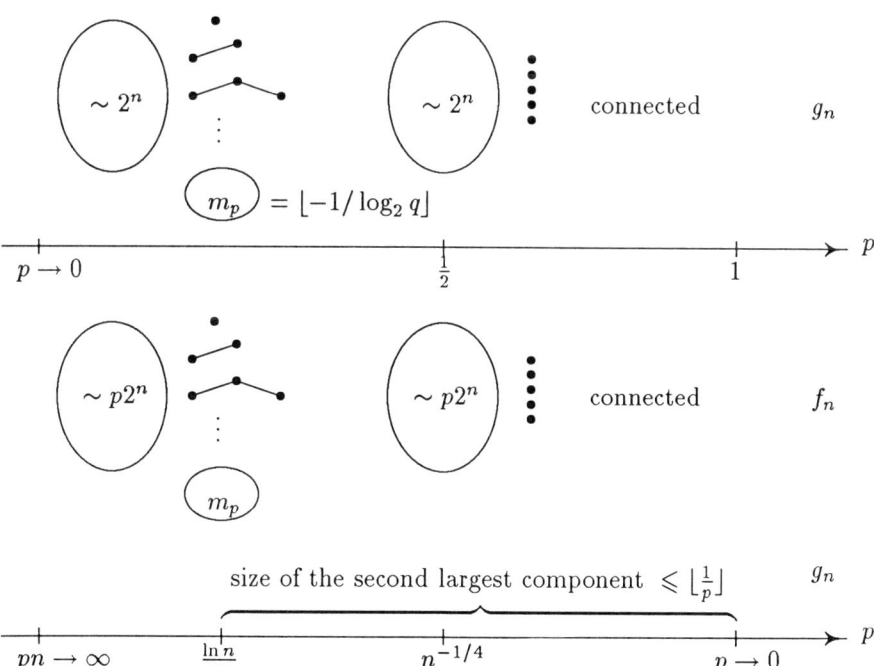

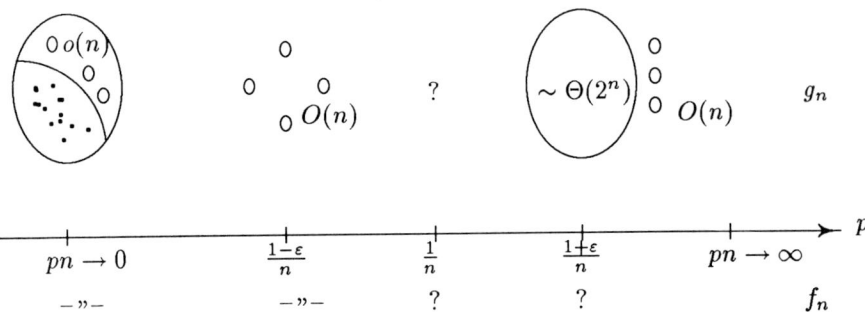

For fixed probabilities p we have exactly the same results as it is shown in the diagram for f_n and g_n for the random graphs h_n: For $p > \frac{1}{2}$ h_n is connected, for $p = \frac{1}{2}$ outside one large component there are only isolated vertices (their number is asymptotically Poisson distributed), for $p < \frac{1}{2}$ outside one large component with asymptotically all vertices there exist only components of sizes $1, 2, \ldots, m_p = \lfloor -1/\log_2 q \rfloor$ ([28]). I.e., in particular, the second largest component has the constant size m_p. Weber conjectured that the same is true for $p = p(n) \to 0$ as $n \to \infty$ as long as $pn \to \infty$. But in [26] he could prove only for g_n that the size of the second largest component is at most $\lfloor \frac{1}{p} \rfloor$ if $p \to 0$ s.t. $p \geqslant n^{-\frac{1}{4}+\varepsilon}$ ($m_p \sim \frac{1}{p}\log_2 e$ if $p \to 0$).

Kostochka ([11]) improved this up to $p \geqslant \frac{1}{n}\ln n$. For $p = \frac{1+\varepsilon}{n}$ ($\varepsilon > 0$) g_n consists of one large component of size $\Theta(2^n)$, and small components of size $O(n)$ (cf. [1]). We have no analogous result for f_n or h_n. If $p = \frac{1-\varepsilon}{n}$ then all components both of g_n and f_n are of order $O(n)$ (cf. [26]). We do not know the component structure for $p = \frac{1}{n}$. If $pn \to 0$ then asymptotically all vertices are isolated and the size of the largest component is $o(n)$ (whereas for $pn \to \infty$ asymptotically all vertices have a degree about pn).

If $2^{-o(n)} \leqslant p \leqslant n^{-\varphi}$, where $\varphi = \varphi(n) \to \infty$ arbitrarily slowly, as $n \to \infty$, then the largest component both of f_n and g_n contains asymptotically $n/\log_2 \frac{1}{p}$ vertices (cf. [24]).

5. Matchings and graph theoretic parameters

For $pn \to \infty$ h_n contains an asymptotically optimal maximum matching, i.e. a matching of cardinality $\text{mat}(h_n) \sim \frac{1}{2}|V(h_n)| \sim p_v 2^{n-1}$ (cf. [9]). Gallai's identities and König's theorem (all cubical graphs are bipartite) imply the same asymptotics $p_v 2^{n-1}$ for the independence number, the vertex covering number and the edge covering number.

The corresponding special results $\text{mat}(f_n) \sim p 2^{n-1}$ and $\text{mat}(g_n) \sim 2^{n-1}$ were shown by Weber already in 1982.

At the first "Random Graphs" seminar, Poznań 1983, Weber posed the conjecture that for fixed probabilities $p > \frac{1}{2}$ even $\text{mat}(g_n) = 2^{n-1}$ is true, i.e. the existence of a perfect matching. This conjecture was independently proved by Kostochka and

Bollobás in 1989 (cf. [12], [2]). But the existence of a Hamilton cycle in this latter case could not be proved up to now.

If $pn \to \infty$ such that $\log_2 \frac{1}{p} = o(n)$ then the maximum degree $\Delta(h_n) \sim n/\log_2 \frac{1}{p}$ (cf. [9]). This result is interesting because the probability $p = \frac{1}{n}$ which is important as a threshold probability for the component structure and many other properties and parameters has no influence on it.

Finally let us mention quite recent results of Sapozhenko and Weber concerning the diameter D of h_n for arbitrarily fixed probabilities p. It holds (cf. [19]) $D(h_n) = n$ if $p > \frac{1}{2}$, $D(h_n) = n + 1$ if $p = \frac{1}{2}$ and $n + m_p \leqslant D(h_n) \leqslant n + m_p + 8$ if $p < \frac{1}{2}$ (the same m_p as above).

Conjecture. $D(h_n) = n + m_p$ for arbitrary fixed probabilities p.

References

[1] M. Ajtai, J. Komlós, E. Szemerédi, *Largest component of a k-cube*, Combinatorica **2** (1982), 1–8.

[2] B. Bollobás, *Complete matchings in random subgraphs of the cube*, Random Structures & Algorithms **1** (1990), 95–104.

[3] Yu. D. Burtin, *On the probability of connectedness of random graphs in the n-cube*, Problemy Pered. Infor. **13** no. 2 (1977), 90–95. (Russian)

[4] M. E. Dyer, A. M. Frieze, L. R. Foulds, *On the strength of connectivity of random subgraphs of the n-cube*, Annals of Discrete Math. **33** (1987), 17–40.

[5] P. Erdős, J. Spencer, *Evolution of the n-cube*, Comput. Math. Appl. **5** (1979), 33–39.

[6] M. R. Garey, R. L. Graham, *On cubical graphs*, J. Combin. Theory Ser. B **18** (1975), 84–95.

[7] V. V. Glagoljev, *Some estimations for disjunctive normal forms of Boolean functions*, Problemy Kibernet. **19** (1967), 75–94. (Russian)

[8] S. V. Jablonski, *Functional constructions in k-valued logic*, Trudy Mat. Inst. V. A. Steklova **51** (1958). (Russian)

[9] U. Konieczna, K. Weber, *Graph theoretic parameters of random subgraph of the n-cube*, Dresdner Reihe zur Forschung no. 9 (1988), 25–28.

[10] A. D. Korshunov, A. A. Sapozhenko *On the number of binary codes with distance 2*, Problemy Kibernet. **40** (1983), 111–140. (Russian)

[11] A. V. Kostochka, *Estimations for the number of connected subgraphs with a small edge boundary. Preprint*, Novosibirsk, 1990. (Russian)

[12] _____, *Maximum matching and connected components of random spanning subgraphs of the n-dimensional unit cube*, Diskretny Analiz **48** (1989), 23–39. (Russian)

[13] K. Mahrhold, K. Weber, *Planarity thresholds for two types of random subgraphs of the n-cube*, Comment. Math. Univ. Carolinae **30** no. 1 (1989), 71–73.

[14] _____, *On the radius of random subgraphs of the n-cube*, to appear.

[15] A. A. Sapozhenko, *Metric properties of almost all Boolean functions*, Diskretny Analiz **10** (1967), 91–119. (Russian)

[16] _____, *Geometric structure of almost all Boolean functions*, Problemy Kibernet. **30** (1975), 227–261. (Russian)

[17] _____, *On the greatest length of an irredundant disjunctive normal form of almost all Boolean functions*, Mat. Zametki **4** no. 6 (1967), 649–658. (Russian)

[18] _____, *On the complexity of disjunctive normal forms obtaining by a greedy algorithm*, Diskretny Analiz **21** (1972), 62–71. (Russian)

[19] A. A. Sapozhenko, K. Weber, *Radius and diameter of random subgraphs of the hypercube*, submitted.

[20] M. Škoviera, *On the minimatization of random Boolean functions. Part 1+2*, Computers and Artificial Intell. **5** no. 4 (1986), 321–334; resp. no. 6, 493–509.

[21] E. Toman, *Geometric structure of random Boolean functions*, Problemy Kibernet. **35** (1979), 111–132. (Russian)

[22] _____, *On the probability of connectedness of random subgraphs of the n-cube*, Math. Slovaca **30** no. 3 (1980), 251–265. (Russian)

[23] K. Weber, *Random graphs in the n-cube*, Seminarberichte der HUB no. 56 (1984), 87–92.

[24] _____, *On the evolution of random graphs in the n-cube.*, Teubner-Texte zur Mathematik, vol. 73, Leipzig, 1985, pp. 203–206.

[25] _____, *Subcube coverings of random graphs in the n-cube*, Annals of Discr. Math. **28** (1985), 319–336.

[26] _____, *On components of random graphs in the n-cube*, Elektron. Inf. verarb. Kybern. EIK **22** no. 12 (1986), 601–613.

[27] _____, *On the independence number of random graphs in the n-cube*, Annals of Discr. Math. **33** (1987), 333–337.

[28] _____, *On components of random subgraphs of the n-cube*, submitted.

[29] Yu. I. Zhuravljov, *Set theoretic methods in the algebra of logic*, Problemy Kibernet. **8** (1962), 5–44. (Russian)

A. V. Kostochka
Institute of Mathematics,
Siberian Branch,
Academy of Sciences of the USSR,
630090 Novosibirsk,
USSR

K. Weber
Ingenieurhochs. f. Seefahrt,
Warnemünde-Wustrow,
Germany

On the Computational Complexity of Seidel's Switching

JAN KRATOCHVÍL, JAROSLAV NEŠETŘIL, ONDŘEJ ZÝKA

Given a graph G and a vertex v of G, Seidel's switching of v in G results in making v adjacent to exactly those vertices it was nonadjacent to in G, while the rest of G remains unchanged. Graphs G and H are called switching equivalent if G can be made isomorphic to H by a sequential application of Seidel's switching. When P is a graph property, we consider the following problem

$S(P)$ Instance: A graph G.
Question: Is G switching equivalent to a graph possessing the property P?

The purpose of this extended abstract is to present examples showing that in general, the complexity of $S(P)$ is independent on the complexity of P. On the other hand, we prove that deciding whether two given graphs are switching equivalent is isomorphism-complete.

1. Preliminaries

All graphs considered are finite, undirected and without loops or multiple edges. Given a graph G, the vertex set and the edge set of G are denoted by $V(G)$ and $E(G)$, respectively. The edge joining vertices u, v is denoted by uv.

Graphs G and H are isomorphic (denoted by $G \cong H$) if there is a bijection $f \colon V(G) \to V(H)$ such that $uv \in E(G)$ iff $f(u)f(v) \in E(H)$ for all $u, v \in V(G)$. A graph G is called an induced subgraph of H if there exists an injection f of the same property. A graph is called G-free if it does not contain G as an induced subgraph.

Given a graph $G = (V, E)$ and a set $A \subset V$, *Seidel's switching* of A in G is the graph

$$S(G, A) = (V, E \div \{uv \mid u \in A, v \in V - A\}),$$

i.e. $S(G, A)$ is the Boolean sum of G and the complete bipartite graph $K_{A, V-A}$ ('$\div$' designates the symmetric difference of sets). We say that G is *switching equivalent* to H (denoted by $G \sim H$) if there is a subset $A \subset V(G)$ such that $S(G, A) \cong H$. Note that '$\sim$' is really an equivalence relation on graphs. A class of mutually

---

Supported by Sonderforschungsbereich 303 (DFG).

switching equivalent graphs is called a *switching class*. Recall that there is a one-to-one correspondence between switching classes and certain systems of triples called two-graphs [S].

Let P be a graph property. We consider the following problem

$S(P)$ Instance: A graph G.
         Question: Is G switching equivalent to a graph possessing the property P?

Our aim is to study the computational complexity of $S(P)$ for several properties P. It turns out that in general, the complexity of $S(P)$ is independent of that of P.

As usual, we call a property P polynomial (resp. NP-complete) if recognizing graphs having P is polynomial (resp. NP-complete), and we call P trivial if it is possessed by all graphs, or by none. Similarly, we call property P *switching polynomial* or *switching NP-complete* or *switching trivial* if $S(P)$ is polynomial, NP-complete or trivial, respectively.

In this paper we deal only with properties P which are in NP. Then also the switching versions $S(P)$ are in NP, and we will not mention this fact explicitly in the proofs. In Section 2, we present switching polynomial properties. Examples of switching NP-complete properties are shown in Section 3. In the last section we prove that deciding whether two given graphs are switching equivalent is isomorphism-complete. Throughout the paper, n denotes the number of vertices of the graph being considered.

2. Switching polynomial properties

Let us first see a switching trivial property:

Proposition 2.1. *"Containing a hamiltonian path" is an NP-complete but switching trivial property.*

Proof.   The NP-completeness of the HAMILTONIAN PATH problem is well known [GJ].

We show that every graph is switching equivalent to a graph containing a hamiltonian path. Suppose a graph G with vertices $u_1, u_2, \ldots, u_n$ is given. Define $A \subset V(G)$ by recursion

1. $u_1 \in A$,
2. $u_{i+1} \in A$ iff $\bigl(u_i \in A$ and $u_i u_{i+1} \in E(G)\bigr)$ or $\bigl(u_i \notin A$ and $u_i u_{i+1} \notin E(G)\bigr)$.

Then $u_1 u_2 \ldots u_n$ is a hamiltonian path in $S(G, A)$.   □

Let us now see switching polynomial properties which are not switching trivial. We are going to show that there are both polynomial and NP-complete properties of this kind.

Theorem 2.2. *"Not containing an induced copy of P_2" is a property both polynomial and switching polynomial. (Here $P_2 = \circ\text{-}\circ\text{-}\circ$ is the path of length two.)*

Proof. A graph is P_2-free iff it is the disjoint union of complete graphs. Such graphs can be recognized in time $O(m)$, m being the number of edges.

To decide whether G is switching equivalent to a P_2-free graph, we need to decide whether there is a set $A \subset V(G)$ such that switching A destroys all induced P_2's in G without creating new copies. This can be done as follows:

Algorithm 2.3. Input: A graph G.
1. Choose $x \in V(G)$ arbitrarily and consider $G' = S(G, \{u \mid ux \in E(G)\})$ (thus x is an isolated vertex in G'),
2. Set $G'' := G' - x$,
3. **While** there are vertices $u, v \in V(G'')$ such that $uv \in E(G'')$ and $\{y \mid y \in V(G''), y \neq u, yv \in E(G'')\} = \{y \mid y \in V(G''), y \neq v, yu \in E(G'')\}$ **do** $G'' := G'' - v$,
(When Step 3 is finished, G'' is reduced, i.e. no two adjacent vertices have the same neighbourhood)
4. **If** G'' is a star or a discrete graph **then** G is switching equivalent to a P_2-free graph **else** it is not.
(A star is a complete bipartite graph $K_{1,r}$ with $r \geqslant 0$.)

Proof of the correctness of the algorithm. Let there exist a P_2-free graph H switching equivalent to G. Then also $H \sim G'$. Suppose $H = S(G', A)$ with $x \notin A$. Since H has no P_2, A induces a complete graph in G', $V(G) - (A \cup \{x\})$ induces a disjoint union of complete graphs and there is no edge in G' joining A and $V(G) - (A \cup \{x\})$. Thus G' has the following shape:

$$V(G') = \{x\} \cup A \cup A_1 \cup A_2 \cup \ldots \cup A_r,$$

$$E(G') = \binom{A}{2} \cup \binom{A_1}{2} \cup \binom{A_2}{2} \cup \ldots \cup \binom{A_r}{2} \cup$$
$$\cup \{uv \mid u \in A, v \in A_1 \cup A_2 \cup \ldots \cup A_r\}$$

for some $r \geqslant 0$ (here $\binom{X}{2}$ denotes the set of all $\binom{|X|}{2}$ edges on X). Note that r may equal 0 and/or A may be empty. Reducing G'' (Step 3 of the algorithm) yields a star (if $A \neq \emptyset$) or a discrete graph (if $A = \emptyset$ or $r = 1$). On the other hand, H is clearly P_2-free, provided G' has the above described shape.

Algorithm 2.3 has worst-case running time $O(n^3)$. □

Theorem 2.4. *"Containing a hamiltonian cycle" is an NP-complete and switching polynomial property.*

Proof. The NP-completeness of HAMILTONIAN CYCLE is well known [GJ].

Given a graph G, let H be such that $G \sim H$ and H has the maximum possible number of edges. Then all vertices of H have degrees $\geqslant \frac{n-1}{2}$, and using the Chvátal condition [Ch], one can show that either H contains a hamiltonian cycle, or n is odd and $H \cong K_{\frac{n+1}{2}, \frac{n-1}{2}}$. In the latter case, every graph switching equivalent to H

is complete bipartite and hence does not contain a hamiltonian cycle. This completes the proof, since one can test quite easily whether a given graph is complete bipartite. □

3. Switching NP-complete properties

In this section we are going to show switching NP-complete properties. We will see that there are both polynomial and NP-complete properties of this kind. (Actually, it took us some time to find the first switching NP-complete property.)

Theorem 3.1. *"Being a regular graph" is a polynomial and switching NP-complete property.*

Proof. A graph is k-regular if every vertex has valency k. It is regular if it is k-regular for some k. Thus one can test regularity in time $O(m)$, m being the number of edges.

Let us say that a graph is $(4,6)$-*biregular* if it is bipartite, all vertices in one part having valency 4 and all vertices in the other part having valency 6. A $(4,6)$-biregular graph is called *balanced* if there is a coloring of its vertices by two colors (say black and white) such that for every vertex, the numbers of its black and white neighbours are equal. We have the following two lemmas

Lemma 3.2. *Given a $(4,6)$-biregular graph G, it is NP-complete to decide whether G is balanced.*

Lemma 3.3. *A $(4,6)$-biregular graph G is switching equivalent to a regular graph H if and only if G is balanced. (In that case H is $\frac{n}{2}$-regular.)*

The proofs of Lemmas 3.2 and 3.3 are slightly technical and the lemmas will be proved elsewhere. Straightforwardly, Lemmas 3.2 and 3.3 together yield a proof of Theorem 3.1. □

Remark. It is easy to see that for any fixed k, one can decide in time $\leqslant O(n^{k+1})$ whether a given graph is switching equivalent to a k-regular graph.

Once we know a switching NP-complete property, we are sure that there are properties which are both NP-complete and switching NP-complete:

Proposition 3.4. *Let P be a property such that $S(P)$ is NP-complete. Then $S(S(P))$ is NP-complete as well.*

The question in $S(P)$ is: Given G, does there exist H so that $H \sim G$ and H has P?

The question in $S(S(P))$ is: Given G, does there exist H so that $H \sim G$ and there is H' such that $H' \sim H$ and H' has P?

Since '$\sim$' is transitive, $H' \sim G$ has to be satisfied in the latter question, too. Thus the two questions are equivalent and $S(P) = S(S(P))$.

Corollary 3.5. *There exists a property which is both NP-complete and switching NP-complete.*

Proof. Consider the property $S(P)$ where P is "being regular". □

Remark. There are also natural properties which are both *NP*-complete and switching *NP*-complete. Consider e.g. the following problem:

Given a graph G, can its vertices be represented by segments of the lines of a two-dimensional grid so that no two segments are overlapping and any two segments are crossing if and only if the corresponding vertices are adjacent?

The *NP*-completeness of this question is proved in [Kra], the *NP*-completeness of the switching version can be proved by an easy construction.

4. Switching equivalence is isomorphism-complete

In this section we consider the relationship between the following two problems

SWITCHING EQUIVALENCE   Instance: Graphs G, H.
    Question: Is $G \sim H$?

GRAPH ISOMORPHISM   Instance: Graphs G, H.
    Question: Is $G \cong H$?

Theorem 4.1. *SWITCHING EQUIVALENCE and GRAPH ISOMORPHISM are polynomially equivalent.*

Proof. SWITCHING EQUIVALENCE $\propto$ GRAPH ISOMORPHISM. Let be given graphs G and H with vertex sets $\{u_1, u_2, \ldots, u_n\}$ and $\{v_1, v_2, \ldots, v_n\}$, respectively. Without loss of generality we may assume that the vertex u_1 is isolated in G (otherwise we would consider $G' = S(G, \{x \mid u_1 x \in E(G)\})$). Then $G \sim H$ iff there exist a set $A \subset V(H)$ and a permutation $\pi \in \mathrm{Sym}\{1, 2, \ldots, n\}$ so that

$$f: u_i \to v_{\pi(i)}$$

is an isomorphism of G and $S(H, A)$. In that case $v_{\pi(1)}$ is isolated in $S(H, A)$. However, for every vertex v_j of H, there is a uniquely determined switching set $A_j \subset V(H)$ such that v_j is isolated in $S(H, A_j)$ (namely, $A_j = \{x \mid v_j x \in E(H)\}$). Thus $G \sim H$ iff there exists a $j \in \{1, 2, \ldots, n\}$ such that $G \cong S(H, A_j)$, i.e. $G - u_1 \cong S(H, A_j) - v_j$. Hence to test switching equivalence of the graphs G, H, it suffices to test graph isomorphism for n pairs of graphs. □

To prove that SWITCHING EQUIVALENCE is as difficult as GRAPH ISOMORPHISM, we use the following lemma.

Lemma 4.2. *Given graphs G and H, we denote by G' (resp. H') the graph obtained from G (resp. H) by making each vertex adjacent to a new extra vertex of degree one. Suppose $n > 4$ and both G and H are connected. Then $G' \sim H'$ if and only if $G \cong H$.*

The proof of the lemma is technical and it will appear elsewhere. It follows immediately that GRAPH ISOMORPHISM $\propto$ SWITCHING EQUIVALENCE.

References

[Ch] V. Chvátal, *On hamiltonian ideals*, J. Comb. Theory, Ser. B **12** (1972), 163–168.

[GJ] M. R. Garey, D. S. Johnson, *Computers and intractability. A guide to the theory of NP-completeness*, W. H. Freeman, San Francisco, CA, 1979.

[Kra] J. Kratochvíl, *A special planar satisfiability problem and some consequences of its NP-completeness*, Submitted.

[S] J. J. Seidel, *A survey of two-graphs*, Teorie combinatorie, International Colloq. 1973, Rome, Atti Conv. Lincei, Vol. 17, Accademia Nazionale dei Lincei, Rome, 1973, pp. 481–511.

Jan Kratochvíl
Charles University, Prague
KA MFF UK, Sokolovská 83,
186 00 Praha 8, Czechoslovakia

Jaroslav Nešetřil, Ondřej Zýka
Charles University, Prague
KAM MFF UK, Malostranské nám. 25,
118 00 Praha 1, Czechoslovakia

Fourth Czechoslovakian Symposium on
Combinatorics, Graphs and Complexity
J. Nešetřil and M. Fiedler (Editors)
© 1992 Elsevier Science Publishers B.V. All rights reserved.

The Harmonious Chromatic Number of a Graph

ANTON KUNDRÍK

Let us start with the following definitions. A *k-coloring of the graph* G is a mapping of $V(G)$ onto the set $\{1, 2, \ldots, k\}$. The *color* of the edge $e = uv$ is $f(e) = \{f(u), f(v)\}$. A *harmonious k-coloring* is defined as a k-coloring with adjacent vertices receiving different colors and all edges receiving different color pairs. The *harmonious chromatic number of a graph* G (denoted by $h(G)$) is the minimum k for which G has a harmonious k-coloring. The number $h(G)$ has been determined for paths and cycles (see [2]). Lee and Mitchem have found an upper bound for $h(G)$ where G is an arbitrary graph (see [1]). In [2] Mitchem has found an upper bound for the harmonious chromatic number of the complete binary tree on n levels.

It is easy to see the following assertion holds.

Proposition 1. *Let G be a graph. Then $h(G) = \min\{k\colon$ there exists a partition $V_1, V_2, \ldots, V_k$ of $V(G)$ such that $E(\langle V_l \rangle) = \emptyset$ and $|E(\langle V_i \cup V_j \rangle)| \leqslant 1$ for each i, j, $l \in \{1, 2, \ldots, k\}, i \neq j\}$, where the symbol $\langle W \rangle$ means the subgraph of G induced by a subset W of $V(G)$.*

Let $h(G) = p$. A partition $V_1, V_2, \ldots, V_p$ of $V(G)$ fulfilling the conditions introduced in Proposition 1 is said to be a harmonious partition of $V(G)$.

Proposition 2. *If G is a graph then $h(G) \geqslant \psi(G)$ where $\psi(G)$ denotes the achromatic number of G.*

Proof. Assume $h(G) = p$. Then $|E(G)| \leqslant \binom{p}{2}$ which implies the desired result. □

Theorem 1. *Let G be a graph of order n and let $diam(G)$ denote the diameter of G. Then $h(G) = n$ if and only if $diam(G) \leqslant 2$.*

Proof. Define $N_G(x) = \{w\colon w \in V(G), xw \in E(G)\}$ for each $x \in V(G)$. The fact $h(G) = n$ implies that $N_G(u) \cap N_G(v) = \emptyset$ or $uv \in E(G)$ for any two vertices of G. Hence $diam(G) \leqslant 2$. The sufficiency can be proved similarly. □

In 1956 Nordhaus and Gaddum proved their well known theorem for the chromatic number of a graph and of its complement [3]. A similar result is presented in the following theorem.

Theorem 2. *If G is a graph of order n then*

$$n + 1 \leqslant h(G) + h(\bar{G}) \leqslant 2n, \tag{1}$$
$$n \leqslant h(G) \cdot h(\bar{G}) \leqslant n^2. \tag{2}$$

Proof. The upper bounds immediately follow from $h(G) \leqslant h(K_n) = n$. Assume $h(G) = p$, $h(\bar{G}) = q$. Let $V_1, V_2, \ldots, V_p$ be a harmonious partition of $V(G)$ and let $W_1, W_2, \ldots, W_q$ be a harmonious partition of $V(\bar{G})$. Let us first prove the lower bound of (1). Denote the number $|\{i : i \in \{1, 2, \ldots, p\}, |V_i| = 1\}|$ by s. Assume $s > 0$, $|V_i| = 1$ for $i \in \{1, 2, \ldots, s\}$. If $s = p$ then $p = n$. Because $q \geqslant 1$ we have $p + q \geqslant n + 1$. So assume $s < p$. Then it is not difficult to verify that the following assertion holds

$$(\forall j)(j \in \{s+1, s+2, \ldots, p\})(\forall x)(x \in V_j)(\exists k)(k \in \{1, 2, \ldots, q\})(W_k = \{x\}). \tag{C}$$

Hence $q \geqslant n - s$. The assumption $s + 1 \leqslant p$ and the preceding inequality imply $p + q \geqslant n + 1$. In the case $s = 0$ a condition like (C) holds and we can proceed similarly. Using a similar technique as in [3] we can easily prove the lower bound of (2). □

Theorem 3. *Let G be a graph of order n. Then the following assertions hold*
(a) $h(G) \cdot h(\bar{G}) = n$ if and only if G or $\bar{G}$ is the complete graph on n vertices,
(b) The bounds of Theorem 2 are attained for any $n \geqslant 4$.

Proof. Assertion (a). The sufficiency is simple. Suppose $h(G) \cdot h(\bar{G}) = n$. Distinguish the possibilities:
    1. $1 < h(G) < n$,
    2. $h(G) = 1$,
    3. $h(G) = n$.

    1. Let $h(G) = p$. Then $h(\bar{G}) \geqslant n + 1 - p$. Hence $h(G) \cdot h(\bar{G}) \geqslant p(n + 1 - p)$. The assumption implies the inequality $p(n + 1 - p) > n$.

    2. The assumption $h(G) = 1$ implies $E(G) = \emptyset$. Hence $\bar{G}$ is complete.

    3. If $E(\bar{G}) \neq \emptyset$ then $h(\bar{G}) \geqslant 2$ which implies $h(G) \cdot h(\bar{G}) \geqslant 2n$, impossible. So G is complete.

Assertion (b). If G is complete then it is simple that $h(G) + h(\bar{G}) = n + 1$, $h(G) \cdot h(\bar{G}) = n$. Now we are going to show a graph fulfilling $h(G) = h(\bar{G}) = n$ for an arbitrary integer n, $n \geqslant 4$. Let G_4 be a cycle C_4 with the vertex-set $V(G_4) = \{x_1, x_2, x_3, x_4\}$ and let G_5 be a cycle C_5 with the vertex-set $V(G_5) = \{y_1, y_2, y_3, y_4, y_5\}$. Suppose G_k is defined and $V(G_k) = \{v_1, v_2, \ldots, v_k\}$. Define the graph G_{k+2} by

$$V(G_{k+2}) = V(G_k) \cup \{v_{k+1}, v_{k+2}\},$$
$$E(G_{k+2}) = E(G_k) \cup \{v_{k+1} v_{k+2}\} \cup E_1 \cup E_2,$$

where $E_1 = \{v_{k+1} v_i : i \in \{1, 2, \ldots, k\}, i \equiv k \pmod 3\}$ and $E_2 = \{v_{k+2} v_i : i \in \{1, 2, \ldots, k\}, i \equiv 1 \pmod 3\}$. Every graph G_k has the property $\text{diam}(G) \leqslant 2$ and $\text{diam}(\bar{G}) \leqslant 2$. Then by Theorem 1, $h(G) = h(\bar{G}) = n$. □

It is easy to find the best upper bound for $h(G) + h(\bar{G})$ and for $h(G) \cdot h(\bar{G})$ for graphs of order n, $n \leqslant 3$.

Theorem 4. $(\forall\, n \geqslant 7)(\forall\, i \in \{1, 2, \ldots, n\})\ (\exists\, G)\ (|V(G)| = n)\bigl(h(G) + h(\bar{G}) = n + i\bigr)$.

Proof. Define the desired graph G in the following way:
$$V(G) = V_1 \cup V_2, \quad \text{where } V_1 = \{x_1, x_2, \ldots, x_i\},\ V_2 = \{y_1, y_2, \ldots, y_{n-i}\},$$
$$E(G) = \{x_k y_l : k \in \{1, 2, \ldots, i\}, l \in \{1, 2, \ldots, n-i\}\} \cup$$
$$\cup\, \{y_k y_l : k, l \in \{1, 2, \ldots, n-i\}, k \neq l\}.$$

It is easy to see that $h(G) = n$, $h(\bar{G}) = i$ which concludes the proof. $\square$

The harmonious chromatic number of a graph having a special property is investigated in the following theorem.

Theorem 5. *If G is a planar graph of order n, $n \geqslant 9$, then $h(\bar{G}) \geqslant n - 2$.*

Proof. The fact $\Delta(\bar{G}) \geqslant n - 6$ implies $h(\bar{G}) \geqslant n - 5$. Distinguishing all possibilities it can be proved that $h(\bar{G}) \geqslant n - 2$. In the following one case is discussed only. The other cases are similar. So to get a contradiction suppose $h(\bar{G}) = n - 5$. Let $V_1, V_2, \ldots, V_{n-5}$ be a harmonious partition of $V(\bar{G})$. Assume $|V_i| = n_i$ for $i \in \{1, 2, \ldots, n-5\}$, $n_i \geqslant n_{i+1}$ for $i \in \{1, 2, \ldots, n-6\}$. Suppose $n_1 = n_2 = 3$, $n_3 = 2$, $n_{3+i} = 1$ for $i \in \{1, 2, \ldots, n-8\}$. Then the definition of the harmonious chromatic number implies that the number of edges of G is at least $25 + 5(n - 8)$ which is greater than $3n - 6$ for $n \geqslant 5$. This is a contradiction. $\square$

In Fig. 1 planar graphs G_0, G_1, G_2 fulfilling $h(\bar{G}) = n - i$ for $i \in \{0, 1, 2\}$ are indicated.

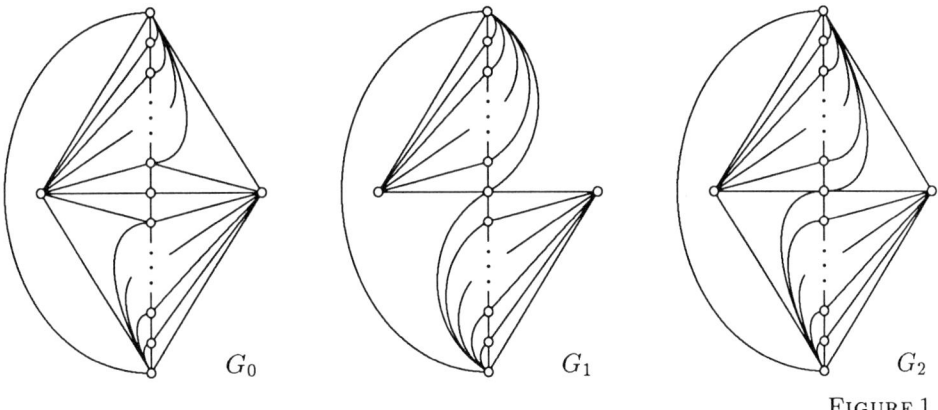

FIGURE 1

References

[1] S. M. Lee, J. Mitchem, *An upper bound for the harmonious chromatic number of a graph*, J. Graph Theory **11** (1987), 565–567.

[2] J. Mitchem, *On the harmonious chromatic number of a graph*, Disc. Math. **74** (1989), 151–157.

[3] E. A. Nordhaus, J. W. Gaddum, *On complementary graphs*, Amer. Math. Monthly **63** (1956), 175–177.

Anton Kundrík
Department of Geometry and Algebra,
Faculty of Science,
P. J. Šafárik University, Jesenná 5, 041 54 Košice,
Czechoslovakia

Fourth Czechoslovakian Symposium on
Combinatorics, Graphs and Complexity
J. Nešetřil and M. Fiedler (Editors)
© 1992 Elsevier Science Publishers B.V. All rights reserved.

Arboricity and Star Arboricity of Graphs

ANDRZEJ KUREK

The arboricity $A(G)$ of the graph G is the minimum number of forests whose union covers all edges of G. A star forest is a forest in which every connected component is a star. The star arboricity $\text{st}(G)$ of a graph G is the minimum number of star forests whose union covers all edges of G. Since every tree can be decomposed into two star forests, $\text{st}(G) \leqslant 2A(G)$ for every graph G. In [1] the question about the maximum star arboricity of a graph with given arboricity was posed. We show that for any k there exists a graph with $A(G) = k$ and $\text{st}(G) = 2k$. The same result was obtained independently by Alon, McDiarmid and Reed (see [2]).

1. Introduction

All graphs considered here are finite, undirected and simple. A star forest is a forest whose connected components are stars. The arboricity [star arboricity] of a graph G is the minimum number of forests [star forests] in G whose union covers the set of edges of G. For a graph (G), $A(G)$ and $\text{st}(G)$ stand for arboricity and star arboricity of G, respectively. It follows from the definition that $A(G) \leqslant \text{st}(G)$. On the other hand, it can be easily shown (see [1]) that every tree can be decomposed into two star forests and hence $\text{st}(G) \leqslant 2A(G)$. The authors of [1] asked what is the maximum star arboricity among graphs with a given arboricity. We prove that, for every natural k, there exists a graph with $A(G) = k$ and $\text{st}(G) = 2k$. The construction of such graphs was also found by Noga Alon, Colin McDiarmid and Bruce Reed [2].

2. Construction

We shall describe the construction of graphs with arboricity k and star arboricity $2k$ (for any natural k). We assume that $k \geqslant 2$ since the construction is clear for $k = 1$. We begin with two lemmas.

Lemma 1. *Let $A(G) \leqslant k$ and let G' be a graph obtained from G by adding a new vertex v of degree at most k. Then $A(G') \leqslant k$.*

Proof. By definition, G can be decomposed into k forests. Assigning the new edges to different forests of this decomposition we obtain a decomposition of G' into k forests. □

Corollary. *If $A(G) \leqslant k$ then any graph G' obtained from G by adding arbitrarily many vertices of degree k satisfies $A(G') \leqslant k$ as well.*

Lemma 2. *For the complete bipartite graph $K_{k,l}$ with $l \geqslant (k-1)\lfloor k/2 \rfloor + 1$, $\text{st}(K_{k,l}) = k$.*

Proof. It is clear that $K_{k,l}$ can be decomposed into k stars. Let us suppose that a decomposition into $k-1$ star forests exists. The graph $K_{k,l}$ contains two independent sets A and B, $|A| = k$, $|B| = l$. Each vertex of B has degree k and so has degree at least two in one of the star forests. But for a particular star forest F, at most $\lfloor k/2 \rfloor$ vertices of B may have degree at least two in F. Hence, the set B contains at most $(k-1)\lfloor k/2 \rfloor$ vertices—a contradiction. □

Now the construction.

Assume k, n, l to be given natural numbers. Let us start with the complete bipartite graph $K_{k,n}$ with the independent sets A and B of cardinality k and n, respectively. For each k-element subset S of B we add a set C_S of cardinality l and connect each vertex of S with each vertex of C_S by an edge. Thus we have a complete bipartite graph $K_{k,l}$ attached to each set S. We denote the obtained graph by $G(k, n, l)$.

Theorem. *For any natural k, choose $n \geqslant (2k-1)\lfloor k/2 \rfloor + \dfrac{(2k-1)!}{(k-1)!}(k-1)+1$ and $l \geqslant (k-1)\lfloor k/2 \rfloor + 1$. Let $G_k = G(k, n, l)$. Then $A(G_k) = k$ and $\text{st}(G_k) = 2k$.*

Proof. Because $\text{st}(G) \leqslant 2A(G)$, it is enough to prove that 1. $A(G_k) \leqslant k$ and 2. $\text{st}(G_k) \geqslant 2k$.

1. The graph G_k is composed of $K_{k,n}$ and a number of added vertices of degree k. It is clear that $A(K_{k,n}) \leqslant k$. According to the Corollary we conclude that $A(G_k) \leqslant k$.

2. Let us suppose that $\text{st}(G_k) \leqslant 2k-1$, i.e. that there exists a decomposition of G_k into $2k-1$ star forests. Let $\Pi = \{F_1, F_2, \ldots, F_{2k-1}\}$ be such a decomposition. Consider the restriction Π_0 of Π to $K_{k,n}$. Similarly to the proof of Lemma 2 one can show that there are at most $(2k-1)\lfloor k/2 \rfloor$ vertices of B with degree at least two in at least one star forest of Π_0. Hence, each of the remaining at least $\dfrac{(2k-1)!}{(k-1)!}(k-1)+1$ vertices of B has the property that every edge (of $K_{k,l}$) incident with it belongs to a different star forest. Let $A = \{a_1, a_2, \ldots, a_k\}$. We characterize each remaining vertex $b \in B$ by a sequence $(x_1, x_2, \ldots, x_k)$, where x_i is the index of the star forest containing the edge (b, a_i). There are at most $\dfrac{(2k-1)!}{(k-1)!}$ such sequences. By the pigeon-hole principle certain k vertices are assigned the same sequence, i.e. edges joining them to the same vertex of set A belong to the same star forest. Consider now the bipartite graph $K_{k,l}$ attached to these k vertices. This graph cannot contain edges belonging to certain k star forests of decomposition Π. Thus, this graph must be composed of $k-1$ star forests, which is in conflict with Lemma 2.

The proof is completed. □

References

[1] I. Algor, N. Alon, *The star arboricity of graph*, to appear.
[2] N. Alon, C. McDiarmid, B. Reed, *Star arboricity*, to appear.

Andrzej Kurek
Department of Discrete Mathematics,
Adam Mickiewicz University,
Poznań, Poland

Fourth Czechoslovakian Symposium on
Combinatorics, Graphs and Complexity
J. Nešetřil and M. Fiedler (Editors)
© 1992 Elsevier Science Publishers B.V. All rights reserved.

Extended 4-Profiles of Hadamard Matrices

CANTIAN LIN, W. D. WALLIS, ZHU LIE

Profiles of an Hadamard matrix have been discussed in [1], [4], [5] and [6] in an investigation of equivalence classes of Hadamard matrices. If H is an Hadamard matrix of order n $(n \geqslant 8)$, we define P_{ijkl} as follows:

$$P_{ijkl} = \left| \sum_{x=1}^{n} h_{ix} \cdot h_{jx} \cdot h_{kx} \cdot h_{lx} \right|$$

where i, j, k, l are mutually different. If we write $\pi(m)$ for the number of sets $\{i, j, k, l\}$ of four distinct rows such that $P_{ijkl} = m$, then the function π is called the 4-profile of H. The k-profile is defined similarly (see [6]).

The following basic results were proven in [1].

Theorem 1. $P_{ijkl} \equiv n \pmod{8}$.

Theorem 2. $\pi(m) = 0$ unless $m \geqslant 0$ and $m \equiv n \pmod{8}$.

Theorem 3. *Equivalent Hadamard matrices have the same 4-profiles.*

From Theorem 3, 4-profiles can be used to give lower bounds on the number of inequivalent Hadamard matrices. (This observation also applies to k-profiles in general as was observed in [4]).

As an extension of this work on profiles of Hadamard matrices, we now define the extended 4-profile of an Hadamard matrix and discuss its use as another tool to investigate equivalence classes of Hadamard matrices.

If H is an Hadamard matrix of order n $(n \geqslant 4)$, we define the *extended 4-profile* $\chi(m)$ (or $\chi_H(m)$) as follows.

The numbers P_{ijkl} are defined as before. We write

$$B_m = \{\{i, j, k, l\} : P_{ijkl} = m\}.$$

The sets B_m form a partition of the set of all 4-subsets of set $X = \{1, 2, \ldots, n\}$. Finally, we consider the vector $\psi(m)$ of length n whose i-th entry equals the number of 4-sets S on X such that $i \in S$ and $S \in B_m$. $\chi(m)$ is defined as the set of elements of $\psi(m)$. (The distinction is that $\chi(m)$ is not an ordered structure.) It is easy to prove

Theorem 4. *Equivalent Hadamard matrices have the same extended 4-profiles.*

Thus extended 4-profiles can be used to give a lower bound on the number of inequivalent Hadamard matrices.

To illustrate these measures, we use the case of order 24. Hadamard matrices of order 24 have been completely classified up to equivalence; for details, see [2] and [3]. We follow the notation in [3] for the sixty equivalence classes $H1, H2, \ldots, H60$. The 4-profiles and extended 4-profiles are tabulated.

There are 33 different 4-profiles of Hadamard matrices of order 24. If a "profile class" is defined to be a set of equivalence classes with the same 4-profile, then the profile classes are

$$\{H1, H9, H17, H34, H35, H49, H50, H58\},$$
$$\{H25, H33, H38, H39, H40, H54\}, \quad \{H30, H32, H43, H44, H45\},$$
$$\{H2, H7\}, \quad \{H4, H6\}, \quad \{H11, H12\}, \quad \{H13, H14\},$$
$$\{H15, H16\}, \quad \{H18, H19\} \quad \{H21, H24\}, \quad \{H26, H29\},$$
$$\{H36, H37\}, \quad \{H41, H42\} \quad \{H51, H52\},$$

nineteen singleton sets.

The same calculation was carried out for 8-profiles [4]. There are 45 classes:

$$\{H43, H44, H45\}, \quad \{H2, H7\}, \quad \{H4, H6\}, \quad \{H11, H12\},$$
$$\{H15, H16\}, \quad \{H18, H19\}, \quad \{H21, H24\}, \quad \{H25, H33\}, \quad \{H30, H32\},$$
$$\{H34, H35\}, \quad \{H38, H39\}, \quad \{H41, H42\}, \quad \{H49, H50\}, \quad \{H51, H52\},$$

thirty-one singleton sets,

Under extended 4-profile, the classes are

$$\{H1, H9, H17, H34, H35, H49, H50, H58\},$$
$$\{H2, H7\}, \quad \{H4, H6\}, \quad \{H11, H12\}, \quad \{H13, H14\}, \quad \{H15, H16\},$$
$$\{H18, H19\} \quad \{H21, H24\}, \quad \{H33, H54\}, \quad \{H41, H42\} \quad \{H43, H45\},$$

thirty-two singleton sets,

so there are 43 classes.

Thus extended 4-profiles give about the same improvement as 8-profiles. However, there is no significant difference in the time required to compute extended 4-profiles (for order 24, about 20 seconds on a PC with 8088 processor and 8087 coprocessor running at 7.16 MHz), while 8-profiles take 40 times as long for order 24. Moreover, extended 4-profiles can distinguish inequivalent classes of Hadamard matrices which are not distinguishable using ordinary profiles. For example, $H43$ and $H44$ (where $H44 = H43^T$) are separated by extended 4-profiles but not by 4-profiles or 6-profiles or 8-profiles.

4-Profiles of Hadamard Matrices

	$\pi(0)$	$\pi(8)$	$\pi(16)$	$\pi(24)$		$\pi(0)$	$\pi(8)$	$\pi(16)$	$\pi(24)$
$H\,1:$	6600	3960	0	66	$H\,31:$	6480	4020	120	6
$H\,2:$	7140	3150	324	12	$H\,32:$	6492	4014	108	12
$H\,3:$	7080	3240	288	18	$H\,33:$	6504	4008	96	18
$H\,4:$	7200	3060	360	6	$H\,34:$	6600	3960	0	66
$H\,5:$	6960	3420	216	30	$H\,35:$	6600	3960	0	66
$H\,6:$	7200	3060	360	6	$H\,36:$	6528	3996	72	30
$H\,7:$	7140	3150	324	12	$H\,37:$	6528	3996	72	30
$H\,8:$	7260	2970	396	0	$H\,38:$	6504	4008	96	18
$H\,9:$	6600	3960	0	66	$H\,39:$	6504	4008	96	18
$H\,10:$	6672	3804	120	30	$H\,40:$	6456	4072	80	18
$H\,11:$	6696	3752	160	18	$H\,41:$	6432	4104	72	18
$H\,12:$	6696	3752	160	18	$H\,42:$	6432	4104	72	18
$H\,13:$	6708	3726	180	12	$H\,43:$	6492	4014	108	12
$H\,14:$	6708	3726	180	12	$H\,44:$	6492	4014	108	12
$H\,15:$	6720	3700	200	6	$H\,45:$	6492	4014	108	12
$H\,16:$	6720	3700	200	6	$H\,46:$	6444	4078	92	12
$H\,17:$	6600	3960	0	66	$H\,47:$	6408	4116	96	6
$H\,18:$	6576	3932	88	30	$H\,48:$	6432	4094	88	12
$H\,19:$	6576	3932	88	30	$H\,49:$	6600	3960	0	66
$H\,20:$	6552	3944	112	18	$H\,50:$	6600	3960	0	66
$H\,21:$	6564	3918	132	12	$H\,51:$	6480	4060	56	30
$H\,22:$	6600	3880	128	18	$H\,52:$	6480	4060	56	30
$H\,23:$	6528	3956	136	6	$H\,53:$	6408	4136	64	18
$H\,24:$	6564	3918	132	12	$H\,54:$	6504	4008	96	18
$H\,25:$	6504	4008	96	18	$H\,55:$	6384	4168	56	18
$H\,26:$	6624	3828	168	6	$H\,56:$	6360	4200	48	18
$H\,27:$	6528	3976	104	18	$H\,57:$	6348	4206	60	12
$H\,28:$	6516	3982	116	12	$H\,58:$	6600	3960	0	66
$H\,29:$	6504	3988	128	6	$H\,59:$	6336	4212	72	6
$H\,30:$	6492	4014	108	12	$H\,60:$	6072	4554	0	0

Extended 4-Profiles of Hadamard Matrices

	$\chi(0)$	$\chi(8)$	$\chi(16)$	$\chi(24)$		$\chi(0)$	$\chi(8)$	$\chi(16)$	$\chi(24)$
$H\,1:$	1100:24	660:24	0:24	11:24	$H\,22:$	1108: 8	652:16	24: 8	3:24
$H\,2:$	1190:24	525:24	54:24	2:24		1096:16	636: 8	20:16	
$H\,3:$	1180:24	540:24	48:24	3:24	$H\,23:$	1104: 8	670:16	28: 8	1:24
$H\,4:$	1200:24	510:24	60:24	1:24		1080:16	638: 8	20:16	
$H\,5:$	1160:24	570:24	36:24	5:24	$H\,24:$	1106: 8	661:16	26: 8	2:24
$H\,6:$	1200:24	510:24	60:24	1:24		1088:16	637: 8	20:16	
$H\,7:$	1190:24	525:24	54:24	2:24	$H\,25:$	1084:24	668:24	16:24	3:24
$H\,8:$	1210:24	495:24	66:24	0:24	$H\,26:$	1104:24	638:24	28:24	1:24
$H\,9:$	1100:24	660:24	0:24	11:24	$H\,27:$	1096: 8	676: 8	20: 8	3:24
$H\,10:$	1124: 8	666: 4	24: 8	5:24		1090: 8	660: 8	18: 8	
	1112:12	634:12	20:12			1078: 8	652: 8	14: 8	
	1088: 4	618: 8	12: 4		$H\,28:$	1094: 8	677: 8	22: 8	2:24
$H\,11:$	1132: 4	668: 4	32: 4	3:24		1088: 8	661: 8	20: 8	
	1120:16	620:16	28:16			1076: 8	653: 8	16: 8	
	1084: 4	604: 4	16: 4		$H\,29:$	1092: 8	678: 8	24: 8	1:24
$H\,12:$	1132: 4	668: 4	32:24	3:24		1086: 8	662: 8	22: 8	
	1120:16	620:16	28:16			1074: 8	654: 8	18: 8	
	1084: 4	604: 4	16: 4		$H\,30:$	1088: 6	677: 6	20: 6	2:24
$H\,13:$	1136: 2	669: 2	36: 2	2:24		1082:12	669:12	18:12	
	1124:18	613:18	32:18			1076: 6	661: 6	16: 6	
	1082: 4	597: 4	18: 4		$H\,31:$	1080:24	670:24	20:24	1:24
$H\,14:$	1136: 2	669: 2	36: 2	2:24	$H\,32:$	1088:12	677:12	20:12	2:24
	1124:18	613:18	32:18			1076:12	661:12	16:12	
	1082: 4	597: 4	18: 4		$H\,33:$	1084:24	668:24	16:24	3:24
$H\,15:$	1128:20	670: 4	36:20	1:24	$H\,34:$	1100:24	660:24	0:24	11:24
	1080: 4	606:20	20: 4		$H\,35:$	1100:24	660:24	0:24	11:24
$H\,16:$	1128:20	670: 4	36:20	1:24	$H\,36:$	1106: 4	690: 2	18: 4	5:24
	1080: 4	606:20	20: 4			1088:12	674: 6	12:12	
$H\,17:$	1100:24	660:24	0:24	11:24		1082: 6	666:12	10: 6	
$H\,18:$	1112: 8	666:16	20: 8	5:24		1070: 2	642: 4	6:2	
	1088:16	634: 8	12:16		$H\,37:$	1106: 6	690: 6	18: 6	5:24
$H\,19:$	1112: 8	666:16	20: 8	5:24		1088:12	666:12	12:12	
	1088:16	634: 8	12:16			1070: 6	642: 6	6: 6	
$H\,20:$	1108: 8	668:16	24: 8	3:24	$H\,38:$	1090:16	684: 8	18:16	3:24
	1084:16	636: 8	16:16			1072: 8	666:16	12: 8	
$H\,21:$	1106: 8	661:16	26: 8	2:24	$H\,39:$	1108: 2	700: 6	24: 2	3:24
	1088:16	637: 8	20:16			1090:16	660:16	18:16	
						1060: 6	636: 2	8: 6	

	$\chi(0)$	$\chi(8)$	$\chi(16)$	$\chi(24)$		$\chi(0)$	$\chi(8)$	$\chi(16)$	$\chi(24)$
$H\,40$:	1090: 4	692: 4	18: 4	3:24	$H\,48$:	1076: 8	685:16	16: 8	2:24
	1084: 4	684:12	16: 4			1070:16	677: 8	14:16	
	1072:12	668: 4	12:12		$H\,49$:	1100:24	660:24	0:24	11:24
	1066: 4	660: 4	10: 4		$H\,50$:	1100:24	660:24	0:24	11:24
$H\,41$:	1078:12	692:12	14:12	3:24	$H\,51$:	1088: 8	682:16	12: 8	5:24
	1066:12	676:12	10:12			1076:16	666: 8	8:16	
$H\,42$:	1078:12	692:12	14:12	3:24	$H\,52$:	1082:20	690: 4	10:20	5:24
	1066:12	676:12	10:12			1070: 4	674:20	6: 4	
$H\,43$:	1091:18	705: 6	21:18	2:24	$H\,53$:	1072:16	700: 8	12:16	3:24
	1055: 6	657:18	9: 6			1060: 8	684:16	8: 8	
$H\,44$:	1091: 6	673:18	21: 6	2:24	$H\,54$:	1084:24	668:24	16:24	3:24
	1079:18	657: 6	17:18		$H\,55$:	1066:16	700: 8	10:16	3:24
$H\,45$:	1091:18	705: 6	21:18	2:24		1060: 8	692:16	8: 8	
	1055: 6	657:18	9: 6		$H\,56$:	1060:24	700:24	8:24	3:24
$H\,46$:	1085: 2	689: 6	19: 2	2:24	$H\,57$:	1058:24	701:24	10:24	2:24
	1079: 6	681:10	17: 6		$H\,58$:	1100:24	660:24	0:24	11:24
	1073:10	673: 6	15:10		$H\,59$:	1012:24	759:24	0:24	0:24
	1067: 6	665: 2	13: 6		$H\,60$:	1056:24	702:24	12:24	1:24
$H\,47$:	1068:24	686:24	16:24	1:24					

References

[1] J. Cooper, J. Milas and W. D. Wallis, *Hadamard equivalence*, Combinatorial Mathematics, Proceedings of the Second International Conference, Lecture Notes in Mathematics No. 686, Springer-Verlag, Berlin, 1978, pp. 126–135.

[2] N. Ito, J. S. Leon and J. Q. Longyear, *Classification of 3-(24,12,5) Designs and 24-Dimensional Hadamard Matrices*, J. Combin. Theory **A31** (1981), 66–93.

[3] H. Kimura, *New Hadamard matrix of order 24*, Graphs and Combinatorics **5** (1989), 235–242.

[4] C. Lin and W. D. Wallis, *Profiles of Hadamard matrices of order 24*, Congressus Num. **66** (1988), 93–102.

[5] W. D. Wallis, *Hadamard equivalence*, Congressus Num. **28** (1980), 15–25.

[6] ———, *Combinatorial Designs*, Marcel Dekker, New York-Basel, 1988.

Cantian Lin, W. D. Wallis
Department of Mathematics,
Southern Illinois University,
Carbondale, IL 62901-4408, USA

Zhu Lie
Department of Mathematics,
Suzhou University,
People's Republic of China

Good Family Packing

M. LOEBL, S. POLJAK

The aim of this paper is to extend the results presented in [1]. We define *good families* as collections of graphs satisfying certain axioms and we show an efficient algorithm for good family packing.

1. Introduction

We consider a natural generalization of matching theory, called F-factor problem or F-packing problem.

F-factor problem. Given a graph and a family F of its subgraphs, decide whether all the vertices may be covered by vertex-disjoint members of F.

We present in this paper an efficient algorithm to solve the F-factor problem for good families.

The general F-factor problem was studied extensively, for a survey of the results see e.g. [2]. Actually, this paper extends the positive result of our paper [1] and we decided to write it separately in the interest of the compactness of [1], where we gave a complete complexity classification of the case that the family F consists of all edges and all copies of a fixed connected graph H (we misled readers by using the incorrect name K_2, H-*factor problem*). We proved the K_2, H-*factor problem* is polynomially solvable if H is either a hypomatchable graph or a graph with a perfect matching or a propeller, and is NP-complete otherwise. A *propeller* is a graph obtained from a hypomatchable graph by adding one vertex-disjoint edge and by linking one of its end-vertices to a nonempty subset of vertices of the hypomatchable graph (hence a propeller has an odd number of vertices, and if it is not P_2 then it has exactly one vertex of degree 1).

We start by generalizing by the notion of propeller.

1.1 Definition

We call a connected graph a k-*propeller*, $k \geq 1$, if it has a vertex (called *center of the k-propeller*) such that when this vertex is deleted then $k+1$ connectivity components $D_0, D_1, \ldots, D_k$ remain. Moreover $|D_0| = 1$ and each D_i, $i = 0, \ldots, k$, is hypomatchable. The only one vertex of D_0 is called *root*, the components $D_0, \ldots, D_k$ are called *blades*. Finally the edge (center, root) is called *stick*.

Further "propeller" will mean a k-propeller for some k. The basic and simplest example of a k-propeller is a star with $k+1$ end-vertices. The following notion generalizes sequential sets of stars.

1.2 Definition (Closed Family)

Let G be a graph and let F be a family of propellers of G. We call the family F *closed* if it satisfies the following 2 axioms.

1. Stick Exchange. Let H be a propeller of F with center c and root r. Let $\bar{r}$ be a vertex of $G-H$ joined by an edge (in G) to c. Then the propeller obtained by replacing (r, c) by $(\bar{r}, c)$ also belongs to F.

2. Blade Exchange. For each vertex x of G consider the set $B_x = \{B; B$ is a blade of some propeller of F with its center at $x\}$, and a set-system $\tilde{B}_x$ on B_x defined by the rule

$x \in \tilde{B}_x$ iff either X consists of exactly one neighbour of x, or X is the set of all blades of a propeller of F with its center at x.

Then $(B_x, \tilde{B}_x)$ forms a matroid.

1.3 Definition (Good Family)

A family of subgraphs of a graph G is called *good* if it consists of all edges of G, some hypomatchable subgraphs and a closed family of propellers of G.

2. Efficient Good Family Packing

In this section we describe an algorithm which finds a good family-packing which covers the maximum number of vertices. The complexity of the algorithm is polynomial in the size of G and the size of the good family.

One of the basic steps of the algorithm is the recognition of F-*critical graphs*. A graph G is called F-critical if it does not have an F-factor, but $G - x$ already has one for any vertex x. The following characterization works satisfactorily.

2.1 Theorem [1]

Let F be a good family of subgraphs of a graph G. Then G is F-critical iff G is hypomatchable and does not have an F-factor which uses exactly one propeller or hypomatchable graph.

Our plan is to proceed further as follows. First we informally describe the algorithm, then we introduce an optimality criterium called *capacity* (*of G and F*), and we write the algorithm more precisely and prove its properties.

2.2 Informal Description of an Algorithm Which Finds a Maximum Good Family Packing

Given a graph G and a good family F of its subgraphs, we start by the Edmonds-Gallai decomposition of G, i.e. we partition $V = V(G)$ into three classes $V = C \,\dot\cup\, A \,\dot\cup\, D$, where D is the set of vertices omitted by at least one maximum matching, A is the set of all neighbours of vertices from D and C contains the remaining vertices.

It follows from Edmonds-Gallai theorem that the part C has a perfect matching. Hence we leave such a matching in C and further we will try to cover maximum number of vertices of $A \cup D$.

It follows once more from Edmonds-Gallai theorem that the connectivity components of D are hypomatchable. Hence these components are either F-critical, or have an F-factor. Let Y be the set of all F-critical components of D. Again we leave an F-factor in $D - Y$, and further we will try to cover maximum number of vertices of $A \cup Y$.

Using the Blade Exchange Axiom of good families and a matroid partition algorithm we cover perfectly (i.e. all vertices of) the set A and maximum number of components of Y by edges between A and Y, edges in components of Y, and propellers of F satisfying:
(i) The center belongs to A,
(ii) The blades belong to distinct components of Y, and if the blades are deleted, the corresponding components have a perfect matching.

Finally we complete the packing by almost perfect matching in components of Y, which haven't been covered yet.

This completes the description of the algorithm.

2.3 Definition (Capacity)

Let G be a graph and let F be a good family of its subgraphs. Let $\mathcal{A}$ be a collection of mutually disjoint subsets of $V = V(G)$, and let x be a vertex out of all subsets of $\mathcal{A}$. We define *the capacity of x with respect to $\mathcal{A}$* (denoted by $\mathrm{cap}(x, \mathcal{A})$), as the maximum k for which there is a $(k-1)$-propeller satisfying:
(i) The center is x,
(ii) The blades lie in k distinct sets of $\mathcal{A}$, and if the blades are deleted, the corresponding sets of $\mathcal{A}$ have a perfect matching.

In this definition we formally suppose that each edge is a 0-propeller consisting of a center and 1 one-vertex blade.

We say that a packing has *defect d*, if it covers all but d vertices. Further we explain the unclear details of algorithm 2.2. This algorithm either finds a packing with defect at most d, or it finds an obstacle in the form of subset $A \in V$ and a system $\mathcal{A}$ of F-critical components of $G - A$ such that

$$|\mathcal{A}| > \sum_{x \in A} \mathrm{cap}(x, \mathcal{A}) + d.$$

2.4 The Explicit Form of 2.2

The beginning of 2.2 does not need any explanation. We construct an Edmonds-Gallai decomposition $V = C \mathbin{\dot\cup} A \mathbin{\dot\cup} D$, we fix a perfect matching of C and an F-factor of $D - Y$.

Further we will try to cover a maximum number of vertices of $A \cup Y$. We proceed in 2 steps.

1. For each $x \in A$ we introduce a system M_x on the underlying set Y given by its rank function
$$r_x(\tilde{Y}) = \text{cap}(x, \tilde{Y}) \text{ for each } \tilde{Y} \subseteq Y.$$
It follows from 1.2.2 that each M_x is a matroid.

Further we solve the matroid partition problem for the collection of matroids $(M_x; x \in A)$. The output of the algorithm is a base B of $\bigcup M_x$ and a set $\tilde{Y} \subseteq Y$ such that
$$|B| = \sum_{x \in A} r_x(\tilde{Y}) + |Y - \tilde{Y}|.$$

As $r_x(\tilde{Y}) = \text{cap}(x, \tilde{Y})$, we get that either $|B| \geq |Y| - d$ or $|\tilde{Y}| > \sum_{x \in A} \text{cap}(x, \tilde{Y}) + d$. Hence suppose that $|B| \geq |Y| - d$.

We have that B is a disjoint union of some B_x, $x \in A$, where each B_x is independent in M_x. Hence $|B_x| = r_x(B_x) = \text{cap}(x, B_x)$ and thus each nonempty B_x is saturated by a propeller (see the definition 2.3 of capacity). Let such a propeller be denoted by H_x.

2. The result of step 1 is a perfect F-packing of at least $|Y| - d$ components of Y, and some vertices of A. Further we indicate how to modify this packing so that it covers all vertices of A. We take a matching $(x, \varphi(x))$, $x \in A$ between A and the vertices of components of D satisfying:
   (i) The matching covers all vertices of A,
   (ii) The vertices $\varphi(x)$, $x \in A$, lie in different components of D,
   (iii) The matching covers a vertex of each component of Y which contains a root of a propeller H_x (see step 1).

This matching satisfying (i)–(iii) may be found as follows: Consider the matroid M on the underlying set of all vertices of the components of D where a subset Z is independent iff the vertices of Z may be matched to A. Hence the set of all vertices containing a root of a propeller H_x is independent in M, thus it may be extended to a base. Further it follows from Edmonds-Gallai theorem that any base-matching covers all vertices of A.

Further we will modify the collection of propellers constructed in step 1 as follows:
   (i) We replace the stick of each propeller H_x by the edge $(x, \varphi(x))$,
   (ii) Let a blade of propeller H_X belong to a component, which also contains $\varphi(x')$, $x' \neq x$. Then this blade is deleted from H_x.

At last, our packing with defect at most d consists of:
- all modified propellers H_x,
- all edges $(x, \varphi(x))$ which are not used as the sticks of propellers H_x,
- a perfect matching of the remaining parts of components containing either a blade of H_x or $\varphi(x)$ for some $x \in A$,
- an almost perfect matching of at most d F-critical components which are disjoint with all H_x and $\varphi(x)$, $x \in A$,
- a perfect F-packing of the components of $D - Y$, which are disjoint with all H_x and $\varphi(x)$, $x \in A$.

We conclude the paper by a MIN-MAX theorem.

2.5 Theorem

Let G be a graph and let F be a family of its subgraphs. Then G has an F-packing with defect at most d if and only if

$$|\mathcal{A}| \leqslant \sum_{x \in A} \operatorname{cup}(x, \mathcal{A}) + d$$

for every subset $A \subset V$ and every family $\mathcal{A}$ of F-critical components of $G - A$.

Proof. The "if part" follows directly from 2.4. Hence let A be a subset of V and let $\mathcal{A}$ be a collection of F-critical components of $G - A$. Moreover let Q be an F-packing which omits at most d vertices of G.
We construct an F-packing $\bar{Q}$ from Q as follows:
(i) If there is a subgraph B which is a blade of a propeller of Q, or a hypomatchable graph of Q, such that B intersects A, then choose arbitrarily a vertex x of such an intersection and replace B by a perfect matching of $B \setminus \{x\}$ in Q. The resulting packing will be denoted by Q'.
(ii) Further we will once more assume that each edge is a 0-propeller with 1 one-vertex blade. Let $X \in \mathcal{A}$ have all vertices covered by Q'. Put

$$F' = \{H \in F;\ H \subset X\} \cup \{B;\ B \subset X, B \text{ is a blade of a propeller of } Q'\}.$$

Observe that F' is a good family and that X has an F'-factor. Hence by 2.1 the component X has an F'-factor which uses exactly one propeller of F' or a hypomatchable graph of F'. Moreover this graph may not belong to F as X is chosen to be F-critical, hence it is a blade, say B, of a propeller of Q'. Hence we will modify Q' so that the vertices of X are covered by B and a perfect matching of $X - B$. The resulting packing will be denoted by $\bar{Q}$.

As every component of $\mathcal{A}$ saturated before the modification remains saturated, $\bar{Q}$ still does not cover at most d vertices of components of $\mathcal{A}$. Each saturated component of $\mathcal{A}$ either contains a blade of a propeller with its center x in A, or a vertex y with $(x, y) \in \bar{Q}$ and $x \in A$. We say that such a component is saturated from $x \in A$. By the construction of Q at most $\operatorname{cap}(x, \mathcal{A})$ components of $\mathcal{A}$ are saturated from any vertex of A. Hence the inequality

$$|\mathcal{A}| \leqslant \sum_{x \in A} \operatorname{cap}(x, \mathcal{A}) + d$$

follows. This completes the proof of 2.5. □

References

[1] M. Loebl, S. Poljak, *Efficient Subgraph Packing*, KAM Series, Prague, preprint of KAM MFF UK Charles University.

[2] M. Loebl, S. Poljak, *Subgraph Packing—A survey*, Topics in Combinatorics and Graph Theory (R. Bodendiek, R. Henn, eds.), Physica-Verlag, Heidelberg, 1990.

M. Loebl, S. Poljak
KAM MFF UK,
Charles University,
Malostranské nám. 25,
118 00 Praha 1
Czechoslovakia

Solution of an Extremal Problem Concerning Edge-Partitions of Graphs

ZBIGNIEW LONC

A collection of k-element sets $\mathcal{C} = \{C_1, \ldots, C_m\}$ is said to be a Δ-*system* if there is a set K such that $C_i \cap C_j = K$, for every $i \neq j$. A *matching* is a Δ-system with $K = \emptyset$. For $k = 2$, collections of k-element sets are edge sets of graphs. Clearly, there are two types of graphs induced by Δ-systems: stars and matchings.

Lonc and Truszczyński [6] proved that if $m \equiv 0 \pmod{c}$ and m is sufficiently large given k and c then every collection of m k-element sets can be partitioned into subcollections consisting of c sets, each of them being either a Δ-system or a Δ-system with a defect (see [6] for the exact definition of a Δ-system with a defect). Surprisingly enough, it turned out that the number of types of subcollections into which a collection of k-element sets can be partitioned, does not depend on c. In particular, in the case of graphs, if $m \equiv 0 \pmod{c}$ and m is sufficiently large given c then the edge set of every graph with m edges can be partitioned into c-element subsets, each of them inducing either a star, a matching or a graph being a disjoint union of a $(c-1)$-edge star and a single edge.

In view of the above result, it seems interesting to characterize those collections of k-element sets that can be partitioned into Δ-systems alone. However, this task seems to be hopeless because the characterization of such collections in the simplest nontrivial case of $k = 2$ and $c = 3$ is very complicated (cf. Favaron et al. [4]). Nevertheless some sufficient conditions for such a partition to exist, satisfied by a large family of collections of k-element sets, have been found.

Let $\nu(\mathcal{C})$ be the *matching number*, i.e. the maximum size of a matching contained in $\mathcal{C}$. Lonc [5] has shown that if $\nu(\mathcal{C}) \geq k!\,c(c-1)^{k-1}$ then $\mathcal{C}$ can be partitioned into Δ-systems of size c if and only if $|\mathcal{C}| \equiv 0 \pmod{c}$. This bound for $\nu(\mathcal{C})$ is not the best possible in general. In this paper we consider the case of graphs ($k = 2$) and we find, for every positive integer c, the least number $\nu_0(c)$ such that if $\nu \geq \nu_0(c)$ then the edge set E of any graph with $\nu(E) = \nu$ can be partitioned into Δ-systems of size c if and only if $|E| \equiv 0 \pmod{c}$. Our main result is the following. (We assume that $c \geq 3$ to avoid trivial cases.)

Theorem 1. $\nu_0(c) = g(c) = \begin{cases} c^2 - 3c + 4 & \text{for } c \geq 5 \\ c^2 - 3c + 5 & \text{for } c = 3, 4. \end{cases}$

In the paper, we denote by K_k, $K_{1,k}$ and kK_2 the complete k-vertex graph, the k-edge star and the k-edge matching, respectively. Moreover, by $\Delta(G)$, $\chi'(G)$ and $\deg_G v$ we mean the maximum degree of a vertex in G, the chromatic index of G and the degree of a vertex v in G, respectively.

We shall need three lemmas.

Lemma 2. (Caro [1].) *Let G be a graph with the edge set E. If $|E| \equiv 0 \pmod{c}$ and $|E| \geqslant c\chi'(G)$ then E can be partitioned into subsets inducing matchings cK_2.*

Lemma 3. (Chetwynd and Hilton [2] and [3].) *If a connected graph G has $r \leqslant 3$ vertices of maximum degree then it is a Class 2 graph (i.e. $\chi'(G) = \Delta(G)+1$) if and only if $r = 3$ and, for some k, G is obtained from K_{2k+1} by removing a matching $(k-1)K_2$.*

Lemma 4. *If there are at least s vertices of degree $\Delta = \Delta(G)$ in G then*

$$|E| - \nu(E) \geqslant \frac{s}{2}(\Delta - 1) + \frac{s}{2}\max(\Delta - s, 0), \qquad (1)$$

where E is the edge set of G.

Proof. Let D be the set of vertices of degree Δ in G and $|D| = t \geqslant s$. Denote by E_1 and E_2 the set of edges with at least one endvertex in D and the set of edges of a maximum-sized matching in G, respectively. Notice that

$$|E| \geqslant |E_1 \cup E_2| = |E_1| + |E_2| - |E_1 \cap E_2| \geqslant t\Delta - \binom{t}{2} + \nu(E) - t \qquad (2)$$

so

$$|E| - \nu(E) \geqslant \frac{t}{2}(\Delta - 1) + \frac{t}{2}(\Delta - t). \qquad (3)$$

Let G' be the graph induced by the set of edges $E - E_2$. Then

$$|E| - \nu(E) = |E - E_2| \geqslant \frac{1}{2}\sum_{v \in D} \deg_{G'} v \geqslant \frac{1}{2}t(\Delta - 1). \qquad (4)$$

The lemma follows by (3) and (4) because the right hand side of (1) is a nondecreasing function of the integer variable s. □

Proof of Theorem 1. The graphs depicted in Figure 1 show that $\nu_0(c) \geqslant g(c)$ for $c \geqslant 3$. We shall give a general proof of the inequality $\nu_0(c) \leqslant g(c)$ for $c \geqslant 7$ beneath. The case $c = 3$ follows immediately by Theorem 3.1 in [4]. We leave a routine case elimination proof in the remaining cases to the reader.

Let $c \geqslant 7$ and assume that G is a graph with the edge set E such that $|E| \equiv 0 \pmod{c}$ and $\nu(E) \geqslant g(c) = c^2 - 3c + 4$. We shall show that E can be partitioned into Δ-systems of size c.

Remove from E the set E_2 of edges of a maximum-sized matching in G. Delete from $E - E_2$ the sets of edges inducing graphs isomorphic to either $K_{1,c}$ or cK_2 until it is impossible to continue. Add the edges of E_2 to the resulting graph and denote by G' the graph obtained this way. It suffices to show that the edge set E' of G' can be partitioned into sets inducing $K_{1,c}$ or cK_2. Clearly, $\Delta(G') \leqslant c$. If $|E'| > c^2$ then $|E'| \geqslant c^2 + c \geqslant c(\Delta(G') + 1) \geqslant c\chi'(G')$, so we are done by Lemma 2. On the other hand $|E'| \geqslant \nu(E) > c^2 - 3c$.

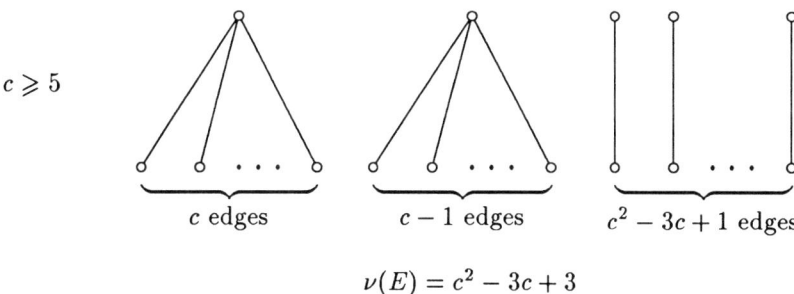

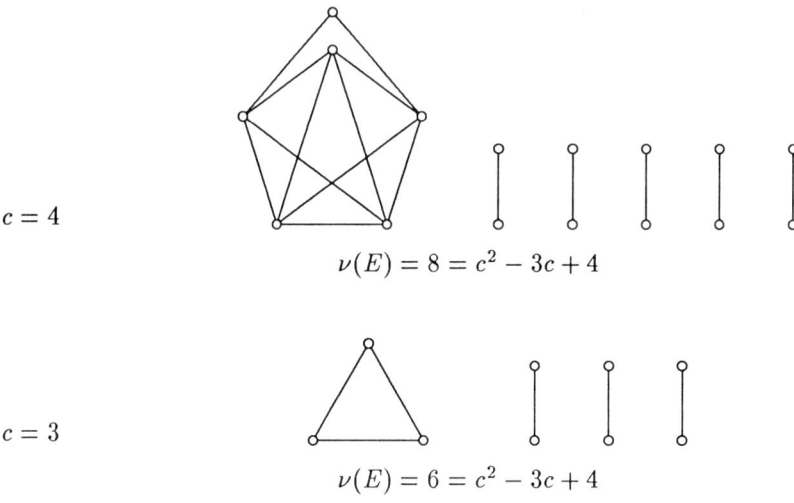

FIGURE 1

It remains to consider the cases $|E'| = c^2 - 2c$, $c^2 - c$ and c^2.

Case 1. $|E'| = c^2 - 2c$.

Notice that $|E' - E_2| \leqslant c - 4$ so $\chi'(G') \leqslant c - 3 \leqslant |E'|/c$ and we are done by Lemma 2 again.

Case 2. $|E'| = c^2 - c$.

If $\Delta(G') = c$ then delete a copy of $K_{1,c}$ from G' and denote the edge set of the resulting graph G'' by E''. Since $|E'' - E_2| \leqslant c - 3$, $\chi'(G'') \leqslant |E''|/c$ and the theorem follows by Lemma 2.

If $\Delta(G') < c - 1$ then $\chi'(G') \leqslant c - 1 = |E'|/c$ and we are done by Lemma 2 as before. Thus, assume that $\Delta(G') = c - 1$. Notice that G' cannot have 3 or more vertices of degree $c - 1$. Indeed, if it had then, by Lemma 4, $2c - 4 \geqslant$

$|E'| - \nu(E') \geq 3c - 9$, a contradiction for $c \geq 6$. Thus, according to Lemma 3, $c\chi'(G') = c\Delta(G') = |E'|$ and the assertion follows by Lemma 2.

Case 3. $|E'| = c^2$.

We can assume that $\Delta(G') = c$ and G' is a Class 2 graph for otherwise $|E'| \geq c\chi'(G')$.

First suppose that G' has at least 4 vertices of degree c. Then, by Lemma 4,

$$3c - 4 \geq |E'| - \nu(E') \geq 2(c-1) + 2(c-4) = 4c - 10,$$

a contradiction for $c \geq 7$. Thus, assume that G' has at most 3 vertices of degree c. According to Lemma 3, c must be even and G' contains a component H isomorphic to a graph obtained from K_{c+1} by removing a matching $(\frac{c}{2} - 1)K_2$. Let E_H be the edge set of H. Then, for $c \geq 5$ we get

$$|E'| \geq |E_2| + |E_H - E_2| \geq c^2 - 3c + 4 + \frac{c^2}{2} + 1 - \frac{c}{2} > c^2 = |E'|,$$

a contradiction. This completes the proof for $c \geq 7$. □

References

[1] Y. Caro, *Chromatic decomposition of a graph*, Ars Combinatoria **21** (1986), 29–32.

[2] A. G. Chetwynd and A. J. W. Hilton, *Regular graphs of high degree are 1-factorizable*, Proc. London Math. Soc. **50** (1985), 193–206.

[3] ———, *The chromatic index of graphs with at most four vertices of maximum degree*, Congr. Numer. **43** (1984), 221–248.

[4] O. Favaron, Z. Lonc and M. Truszczyński, *Decompositions of graphs into graphs with three edges*, Ars Combinatoria **20** (1985), 125–146.

[5] Z. Lonc, *On decomposition of hypergraphs into Δ-systems*, J. Combin. Theory Ser. A **52** (1989), 158–162.

[6] Z. Lonc and M. Truszczyński, *Decompositions of large uniform hypergraphs*, Order **1** (1985), 345–350.

Zbigniew Lonc
Institute of Mathematics,
Warsaw University of Technology,
00-661 Warsaw, Poland

Balanced Extensions of Spare Graphs

Tomasz Łuczak, Andrzej Ruciński

> Let $d(H)$ be the ratio of the number of edges to the number of vertices of a graph H and let $d^*(H)$ be obtained from $d(H)$ by decreasing the denominator by 1. Furthermore, let $m(G) = \max_{H \subseteq G} d(H)$ and $m^*(G) = \max_{H \subseteq G} d^*(H)$. We prove that there are numbers n_0 and C such that for every graph G with $n \geq n_0$ vertices and $1 < m(G) < \frac{10}{9}$ [$m(G) > 4.25$] there is a supergraph F with at most $Cn/(m(G)-1)$ [$203n$] vertices such that $m(F) = d(F) = m(G)$ and we also prove a similar result for $m^*(G)$. The former partially solves a problem raised by Erdős. The latter can be reformulated as follows: for every graph G with $n \geq n_0$ vertices and $m^*(G) = p/q$, p and q integers, $1 < p/q < \frac{10}{9}$ [$p/q > 4.25$], the smallest graph which contains G and whose edge set can be covered by p spanning trees so that each edge belongs to precisely q of them has at most $Cn/(m^*(G)-1)$ [$203n$] vertices. Our method involves random graphs and therefore is nonconstructive.

§1. Introduction

This paper is concerned with an extremal problem on graphs, born a few years ago within the theory of random graphs. While the problem itself is purely deterministic, our solution to it makes use of properties of random graphs and therefore is probabilistic in nature. To formulate the problem only the knowledge of the notions of a graph and a subgraph is required.

For a graph G, let $|G|$ and e_G stand for the number of vertices and edges of G, respectively. We define *the density of* G as $d(G) = e_G/|G|$ and *the strong density of* G as $d^*(G) = e_G/(|G|-1)$. (Throughout the paper we consider only graphs and subgraphs with at least one edge.) Furthermore, we define $m(G) = \max_{H \subseteq G} d(H)$ and $m^*(G) = \max_{H \subseteq G} d^*(H)$ and say that G is *balanced* [*strongly balanced*] if $m(G) = d(G)$ [$m^*(G) = d^*(G)$]. A graph F is said to be [*a strongly*] *balanced extension of* G if $G \subseteq F$ and $m(F) = d(F) = m(G)$ [$m^*(F) = d^*(F) = m^*(G)$]. Let

$$\text{ext}(G) = \min\{|F|: F \text{ is a balanced extension of } G\}$$

and let $\text{ext}^*(G)$ be defined similarly. Our goal is to estimate these two parameters with respect to the [strong] density of G.

The notions of balanced graph and balanced extension appeared in relation to a problem on subgraphs of random graphs (see Erdős and Rényi (1960), Győri,

Rothschild, and Ruciński (1985), Ruciński and Vince (1985, 1986)). In particular, in response to a conjecture of Karoński and Ruciński (1982), it was proved by Győri, Rothschild and Ruciński (1985), and independently by Payan (1986), that every graph does have a balanced extension. The first proof was later simplified by Ruciński and Vince (1986) by introducing strongly balanced graphs and proving the existence of such graphs with v vertices and e edges for all $1 \leqslant v - 1 \leqslant e \leqslant \binom{v}{2}$. Payan also proved that every graph has a strongly balanced extension. Applying the well known Nash-Williams arboricity theorem he noticed that, setting $m^*(G) = p/q$, one can always cover the edge set of the graph G by p forests so that each edge belongs to precisely q forests, and that the same can be done with forests replaced by spanning trees *if and only if* G is strongly balanced. Moreover, due to a matroid theorem of Edmonds, the statement remains true after replacing $m^*(G)$ by $m(G)$, strong balance by balance, forests by graphs with $m(G) \leqslant 1$ (i.e. each component has at most one cycle) and trees by graphs with $d(G) = m(G) = 1$ (i.e. each component has exactly one cycle). (Observe that, for forests, $m^*(G) \leqslant 1$ and, for trees, $m^*(G) = d^*(G) = 1$. For further developments concerning packing and covering of matroids and for their applications to various graph problems the reader may consult Catlin et al. (1988). For both, the Nash-Williams and Edmonds' theorems, see Berge (1973).) From the above viewpoint, $\text{ext}^*(G) - |G|$ may be interpreted as the smallest number of vertices one has to add to the graph G (with a bunch of new edges structured suitably) to be able to cover the edge set of the new graph by p spanning trees so that each edge belongs to precisely q of them. A similar interpretation of $\text{ext}(G)$ is obvious, though awkward to express.

Let us now briefly resume what has been known about the parameters $\text{ext}(G)$ and $\text{ext}^*(G)$ by the time of this writing. First of all, setting $m(G) = p/q$, where p and q are relatively prime,

$$\text{ext}(G) = \begin{cases} |G|, & \text{if } q = 1 \text{ or } q = 2 \text{ and } |G| \text{ is even} \\ |G| + 1, & \text{if } q = 2 \text{ and } |G| \text{ is odd.} \end{cases}$$

But already for $q = 3$, $\text{ext}(G)$ can be as large as $\frac{3}{2}(|G| - 1)$ (take a forest whose all but one tree are just edges and the largest one is a path of length 2). Note that only for forests $m(G) < 1$ and that $m(G) = 1$ if and only if the densest component is a unicyclic one. In both cases $\text{ext}(G)$ remains linear in $|G|$. This changes dramatically when $m(G)$ exceeds 1. No balanced graph with density larger than 1 has a vertex of degree 1 and therefore $\text{ext}(G) \geqslant \frac{1}{2}[\Delta(G) - 2]/[m(G) - 1]$, where $\Delta(G)$ is the maximum degree in G. Thus there are graphs, rather sparse, with $\text{ext}(G) > \frac{1}{8}|G|^2$. But things cannot go much worse, since the construction employed for proving the existence of balanced extensions gives, for every $\varepsilon > 0$, the inequality $\text{ext}(G) < \frac{1}{4}(1 + \varepsilon)|G|^2$ (see Ruciński and Vince (1988, 1989) for details). The other parameter, $\text{ext}^*(G)$, behaves similarly, except that already for $q = 2$ it can go arbitrarily large (but linearly in $|G|$). The lower bound when $m^*(G) > 1$ is now of the form $\text{ext}^*(G) \geqslant \frac{1}{2}\Delta(G)/[m^*(G) - 1]$ and a cycle C with $|C|$ pendant vertices hanging at the same cyclic vertex gives $\text{ext}^*(G) > \frac{1}{8}|G|^2$. No upper bound of order $|G|^2$ is explicit in the construction of Payan (1986).

In 1985, during a discussion in Gainesville, Florida, Pál Erdős conjectured that for sufficiently dense graphs ext(G) should be linear in $|G|$. Indeed, Ruciński and Vince (1988) proved that if $e_G > c|G|^2$, $0 < c < \frac{1}{2}$, then ext(G) $< c'|G|$, where c' depends on c only. In this paper we show that both ext(G) and ext*(G) remain linear in $|G|$ even for much sparser graphs, thereby partially solving Problem 2 from Ruciński and Vince (1989).

Theorem 1. *There exist numbers n_0 and C such that for all graphs G with $|G| = n > n_0$*

$$\text{ext}(G) < \begin{cases} \dfrac{Cn}{(m(G)-1)}, & \text{if } 1 < m(G) < 1\frac{1}{9} \\ 203n, & \text{if } m(G) > 4\frac{1}{4}. \end{cases}$$

Theorem 2. *There exist numbers n_0 and C such that for all graphs G with $|G| = n > n_0$*

$$\text{ext}^*(G) < \begin{cases} \dfrac{Cn}{(m^*(G)-1)}, & \text{if } 1 < m^*(G) < 1\frac{1}{9} \\ 203n, & \text{if } m^*(G) > 4\frac{1}{4}. \end{cases}$$

Comments.

(i) The irritating gap in the covered range of $m(G)$ and $m^*(G)$ is due, we believe, to some technicalities which can be improved in the "near future" — see Section 5 for the prospects on that.

(ii) Observe that $m(G) > 1$ really means $m(G) - 1 \geqslant 1/n$. Moreover, when $m(G) = 1 + r/q$, $r = r(n)$, $q = q(n)$, $r = o(q)$ then the upper bound is of order $qn/r \gg n$. On the other hand, for the graph G obtained from any balanced graph with q vertices and $q+r$ edges by attaching to one of its vertices $n - q$ pendant vertices, ext(G) $\geqslant \frac{1}{2}q(n-q+r)/r$, which is of the same order of magnitude as the upper bound as long as $n - q \neq o(n)$. In the case $q = n - s$, $s = s(n) = o(n)$, the construction from Ruciński and Vince (1986) improves the upper bound to $C \min(qn/r, sr)$. But this is of the same order as the lower bound only if $r = O(1)$. Otherwise the bounds diverge apart. However, there is always a sequence $\bar{q} = \bar{q}(n)$, $|\bar{q} - q| < r$ such that r divides $\bar{q}$ (note that $r/\bar{q} \sim r/q$) and so the lower bound for $m(G) = r/\bar{q}$ jumps to $\bar{q}n/r$ ($\bar{q}/r$ assumes the role of q). We believe that the same comment applies to ext*(G), though the upper bound sn does not appear explicitly in the literature.

(iii) In contrast to the results of Ruciński and Vince (1988), ours are *nonconstructive*, since we apply random graphs in the proof.

(iv) We show the existence of [strongly] balanced extensions F of the required order with the additional property that G is an induced subgraph of F.

The paper is organized as follows. In Section 2 we establish some properties of random graphs needed later for the choice of graphs on which our constructions will be based. Section 3 and 4 contain the proof of Theorem 1 in the dense and sparse case, respectively. The final section consists of two comments indicating how to derive the proof of Theorem 2 from that of Theorem 1 and how to close the gap in the theorems.

§2. Probabilistic lemmas

In this section we establish some expanding properties shared by almost all graphs with given number of edges and by almost all regular graphs. Actually, the properties we have in mind say that not too large subgraphs have relatively not too many edges but this just means that many edges must go outside the subgraph. In addition, for the first class of graphs, we shall show that the minimum degree, δ, is close to the doubled density. In the forthcoming sections we shall only need to know that there is at least one graph in each family having the properties, so the results here might seem unnecessarily strong. However, using the probabilistic method one always ends up with proving the existence of many objects of the required type.

Let $\mathcal{G}(n, M)$ be the equiprobable space of all graphs on vertex set $[n] = \{1, \ldots, n\}$ with M edges. Below we shall identify a subset $S \subset [n]$ with the subgraph induced by S.

Lemma 2.1. *Let $a \in (0,1)$. If $M = M(n) \geq \frac{1}{2}(\log n)^2 n$ and $n \to \infty$ then almost all $G \in \mathcal{G}(n, M)$ are such that*

(i) $\delta(G) \geq (.99)2M/n$,
(ii) *for all $S \subset [n]$, $|S| \leq an$, $d(S) \leq (a + .01)M/n$.*

Proof. It is more convenient to work with the space $\mathcal{G}(n, p)$ consisting of all graphs on $[n]$, where the probability assigned to a graph with l edges is $p^l(1-p)^{\binom{n}{2}-l}$, and rely on Bollobás (1985), Theorem II.2(iii), which says that if $p = M/\binom{n}{2}$ and Q is a graph property then

$$\text{Prob}[\mathcal{G}(n, M) \cap Q] \leq 3\sqrt{M}\,\text{Prob}[\mathcal{G}(n, p) \cap Q].$$

We will write, for short, $\text{Prob}(Q)$ instead of $\text{Prob}[\mathcal{G}(n, p) \cap Q]$. We have, by Bollobás (1985), Theorem I.7(i),

$$\text{Prob}[\delta(G) < (.99)(n-1)p] \leq n\,\text{Prob}[\deg_G(1) < (.99)(n-1)p] \leq n\,e^{-cnp}$$

for some constant $c > 0$, where $\deg_G(v)$ stands for the degree of vertex v in G. Hence part (i) is proved.

For a fixed $S \subset [n]$, $an/E^2 \leq s = |S| \leq an$, the same bound on the deviation of a binomial distribution gives

$$\text{Prob}\left[d(S) > \frac{1}{2}(a + .01)pn\right] \leq \text{Prob}\left[|es - \binom{s}{2}p| > .01\binom{s}{2}p\right] \leq e^{-c_1 n^2 p}.$$

Actually, the above mentioned bound applies only for $p \leq \frac{1}{2}$. However, for larger p one can make use of

$$\text{Prob}[X > (1+\varepsilon)np] \leq \text{Prob}[n - X < (1-\varepsilon)n(1-p)]$$

and
$$\text{Prob}[X < (1-\varepsilon)np] \leqslant \text{Prob}[n - X > (1+\varepsilon)n(1-p)],$$

valid for all $\text{Bi}(n,p)$-distributed random variables X, $p \geqslant \frac{1}{2}$. For $s < an/e^2$ we apply Bollobás (1985), Theorem I.7(ii), to get

$$\text{Prob}\left[d(S) > \frac{1}{2}anp\right] \leqslant \text{Prob}\left[e_S > a\frac{n}{s}\binom{s}{2}p\right] < e^{-c_2 snp}.$$

Altogether,

$$\text{Prob}\left[\exists S, |S| \leqslant an : d(S) > \frac{1}{2}(a + .01)np\right]$$

$$\leqslant 2^n e^{-c_1 n^2 p} + \sum_{s=\log n}^{bn/e^2} n^2 e^{-c_2 snp} = o(M^{-\frac{1}{2}}).$$

$\square$

Let us turn now to the regular graphs. Although the lemma bellow follows from Bollobás (1988), we present its proof here for completeness. We say that an r-regular graph is α-sparse if each subgraph of it, with at most half of the vertices, has density at most αr, $0 < \alpha < .5$.

Lemma 2.2. *For r and α specified below, almost all r-regular graphs on n vertices are α-sparse:*

(i) $20 < r = r(n) < 2(\log n)^2$, $\alpha = .45$;
(ii) $r > 7$, $\alpha = .5 - (.75)r^{-1}$;
(iii) $r = 3$, $\alpha = .47$.

Proof. In Frieze (1988) it is proved that for $3 \leqslant r \leqslant n^{1/6}$ almost all n-vertex r-regular graphs are r-connected, what was earlier known for fixed r (see Wormald (1981)). Thus the first step in right direction has been done: no subgraph on $s < n$ vertices has all $\frac{1}{2}rs$ edges in it. More seriously, we will use this fact to avoid some inconvenience in the forthcoming estimates.

Let $V_1, \ldots, V_n$ be disjoint r-element sets and let $\mathcal{M}(n,r)$ be the family of all $(nr)!/(\frac{1}{2}nr)!\,2^{\frac{1}{2}nr}$ matchings of the set $V_1 \cup \cdots \cup V_n$. By contracting the sets V_i to single points, each element of $\mathcal{M}(n,r)$ corresponds to an n-vertex r-regular pseudograph (loops and multiple edges allowed) and every such pseudograph is the image of exactly $(r!)^n$ matchings. It was also proved in Frieze (1988) that, for $3 \leqslant r \leqslant n^{1/6}$,

$$|\mathcal{R}(n,r)| \sim e^{-r^2/4} \frac{|\mathcal{M}(n \cdot r)|}{(r!)^n},$$

extending an earlier result from Bender and Canfield (1978) valid for fixed r. Hence, it is enough to prove that a random matching has the corresponding properties with probability $1 - o(e^{-r^2/4})$. (The last restriction applies only in case (i); for the other two properties we require that the probability tends to 1 as $n \to \infty$.) This approach, originated in Bollobás (1980), has become standard by now.

The corresponding properties of a matching are defined in a natural way. We say that a matching M is α-sparse if for each subset $S \subset [n] = \{1,\ldots,n\}$, $|S| \leqslant \frac{1}{2}n$, no more than $\alpha r|S|$ pairs of M are contained in $U(S) = \bigcup_{i \in S} V_i$. M is connected if there is no $S \subset [n]$, $|S| < n$, with $U(S)$ containing $\frac{1}{2}|S|r$ pairs of M.

Let $\mathbf{M}$ be a random element of $\mathcal{M}(n,r)$. Denote by $A_{s,t}$ the event that there is an $S \subseteq [n], |S| = s$, with $|\mathbf{M} \cap [U(S)]^2| = t$ and let $E(s,t)$ be the expected number of such sets, $s = 1, \ldots, n$, $t = 0, \ldots, \frac{1}{2}sr$. The probability that $\mathbf{M}$ is not α-sparse but connected is not greater than

$$\sum_{s=2\alpha r+1}^{.5n} \sum_{t=\alpha rs}^{\frac{1}{2}(sr-1)} \text{Prob}(A_{s,t}). \tag{2.1}$$

But $\text{Prob}(A_{s,t}) \leqslant E(s,t)$ and

$$E(s,t) = \binom{n}{s}\left[\binom{rs}{2t}\binom{rn-rs}{rs-2t}\mu(2t)(rs-2t)!\frac{\mu(rn-2rs+2t)}{\mu(rn)}\right],$$

where $\mu(2x) = (2x)!/(x!2^x)$. The expression in square brackets equals the probability that, for a fixed s-element set $S \subset [n]$, there are exactly t pairs of $\mathbf{M}$ in $U(S)$. Using the inequalities

$$c_1 \left(\frac{N}{e}\right)^N \sqrt{N} < N! < c_2 \left(\frac{N}{e}\right)^N \sqrt{N},$$

valid for all natural N with suitably chosen constants c_1 and c_2, we further obtain

$$E(s,t) = \frac{n!(rs)!(rn-rs)!(\frac{1}{2}rn)!\, 2^{rs-2t}}{s!(n-s)!(rs-2t)!t!(\frac{1}{2}rn-rs+t)!(rn)!}$$

$$\leqslant C\sqrt{rn}\left(\frac{b^{b(r-1)}(1-b)^{(1-b)(r-1)}r^{\frac{1}{2}r}}{(br-2a)^{br-2a}(2a)^a(r-2br+2a)^{\frac{1}{2}r-br+a}}\right)^n,$$

where $s = bn$, $t = an$ and C is a constant. Substituting further $a = cr$, we get

$$E(s,t) \leqslant C\sqrt{rn}[f(b,c)]^n,$$

where

$$f(b,c) = \frac{\left(b^b(1-b)^{1-b}\right)^{r-1}}{\left((b-2c)^{b-2c}(2c)^c(1-2b+2c)^{5-b+c}\right)^r}.$$

Regarding f as a function of c while b is fixed, it decreases for $c > \frac{1}{2}b^2$, which is our case. Hence,

$$E(s,t) \leqslant C\sqrt{rn}\bigl(f(b,\alpha b)\bigr)^n.$$

Set $g(b) = \log f(b, \alpha b)$ and observe that

$$g''(b) = \frac{r\alpha - 1}{b} + \frac{r - 1}{1 - b} - \frac{2r(1 - \alpha)^2}{1 - 2b + 2\alpha b} \geq 0$$

for $\alpha \geq r/(3r - 2)$, which is our case. Moreover,

$$\lim_{b \to 0+} g'(b) = -\infty$$

and

$$\lim_{b \to .5-} g'(b) = r(1 - 2\alpha) \log \frac{2\alpha}{1 - 2\alpha}.$$

Thus, the whole sum (2.1) is not greater than

$$Cr^{3/2}n^{5/2}\big(\exp[\max(g(2\alpha r/n), g(.5))]\big)^n.$$

But in all cases (i)–(iii) $\exp[g(.5)] <$ constant < 1 and $g(2\alpha r/n) = -(\alpha r - 1) \times 2\alpha r \log n/n + O(r \log r/n)$, so it is further bounded, in cases (i) and (ii), by

$$Cr^{3/2}n^{5/2} \exp[-(\alpha r - 1)2\alpha r \log n + O(r \log r)] = o(e^{-r^2/4}).$$

In case (iii) we split (2.1) into two sums, the second one ranging over $\log n \leq s \leq .5n$, and bounded by

$$C\sqrt{n}(\log n)^2 \, e^{-1.15 \log n + O(1)} + C n^{3/2} \, e^{-0.41(\log n)^2 + O(\log n \log \log n)} = o(1).$$

$\square$

§3. Proof of Theorem 1—the dense case

In this section we shall prove Theorem 1 for graphs with $m = m(G) > 4.25$. Although we consider 3 subcases separately, the underlying idea is common to all of them, as well as to the sparse case to be considered in Section 4. The idea is as follows. For a given graph G we construct a balanced extension F by picking a graph R disjoint from G (and much bigger than G) and joining it to G in a special way. In order to have $d(F) = m$, the size of R and of the junction with G must satisfy the equation

$$e_G + e_R + e(G, R) = m(|R| + |G|), \tag{3.1}$$

where $e(G, R)$ is the number of edges with one endpoint in G and the other in R. To ensure that F is balanced, one imposes some structural restrictions on R corresponding to the properties in Lemmas 2.1 and 2.2.

To verify that F is balanced we introduce the deficit function

$$f(H) = m|H| - e_H$$

defined on graphs. Thus our goal is to prove that f is nonnegative for all $H \subset F$. We shall distinguish 2 subcases in respect to the size of $|H|$ for which we shall argue differently.

I. For "not too large" H we set $H_1 = H \cap G$, $H_2 = H \cap R$ and observe that, by the modularity of f,

$$f(H) = f(H_1) + f(H_2) - e(H_1, H_2) \geqslant f(H_2) - e(G, H_2).$$

Hence, we are to show that

$$|H_2|(m - d(H_2)) \geqslant e(G, H_2). \tag{3.2}$$

II. For "too large" H we use the fact that

$$0 = f(F) = f(H) + f(\bar{H}) - e(H, \bar{H}),$$

where $\bar{H}$ is the subgraph of F induced by all vertices outside H. Thus $f(H) \geqslant 0$ is equivalent to

$$e(H, \bar{H}) + e_{\bar{H}} \geqslant m|\bar{H}|. \tag{3.3}$$

But the left-hand side of (3.3) is at least

$$e(\bar{H}_1, R) + e(\bar{H}_2, H_2) + e_{\bar{H}_2},$$

and

$$e(\bar{H}_2, H_2) = \sum_{v \in \bar{H}_2} \deg_R(v) - 2e_{\bar{H}_2},$$

where $\bar{H}_1 = \bar{H} \cap G$, $\bar{H}_2 = \bar{H} \cap R$ and $\deg_R(v)$ is the degree of vertex v in R. Hence (3.3) is implied by the conjunction of

$$e(\bar{H}_1, R) \geqslant m|\bar{H}_1| \tag{3.4}$$

and

$$\sum_{v \in \bar{H}_2} \deg_R(v) - e_{\bar{H}_2} \geqslant |\bar{H}_2|. \tag{3.5}$$

Because of the big disproportion between G and R, the phrases "not too large" and "too large" refer as well to $\bar{H}_2$ as a subgraph of R. Throughout Section 3, by a "not too large" subgraph H of G we mean that for which $|H_2| \leqslant .5|R|$. Thus, in all cases we will be justifying (3.2) and (3.5) for all subgraphs of R with at most $.5|R|$ vertices, and (3.4) for all subgraphs of G.

Case $m \geqslant (\log n)^2$.

Let $N = |R|$, $n = |G|$ and $e(G, R) = \lfloor 2mn \rfloor$. Express $m = p/q$, where p and q are relatively prime integers, and pick $N = 100n + \gamma$, $0 \leqslant \gamma < q$, so that $n + N$ is divisible by q. Let us join G and R in such a way that each vertex of G has either

$\lfloor e/n \rfloor$ or $\lceil e/n \rceil$ neighbors in R and each vertex of R has either $\lfloor e/N \rfloor$ or $\lceil e/N \rceil$ neighbors in G. Thus, from (3.1),

$$e_R = m(N+n) - e_G - e(G,R) \geqslant m(N-2n)$$

and we may assume that R satisfies the properties (i) and (ii) of Lemma 2.1 with $a = .5$. Now let K be a subgraph of R with $|K| \leqslant .5|R|$. We have

$$e(G,K) \leqslant |K| \lceil 2mn/N \rceil \leqslant (.02m+1)|K|,$$
$$f(K) = (m - d(K))|K| \geqslant m|K|(1 - .51),$$

and

$$\sum_{v \in K} \deg_R(v) - e_K \geqslant |K|(\delta(R) - d(K)) \geqslant |K|m((.99)2(.98) - .51).$$

Also, for every $K \subset G$, $e(K,R) \geqslant \lfloor 2m \rfloor |K|$ and so, (3.2), (3.4), and (3.5), all hold with a big margin.

Case $30 \leqslant m \leqslant (\log n)^2$.

Let R be a $2r$-regular, .45-sparse graph, $r = \lfloor .98m \rfloor$, with $N = 100n + \gamma$ vertices. The choice of γ and the way R is adjoint to G are the same as in the previous case, but this time (3.1) acquires the form

$$e = e(G,R) = m(N+n) - e_G - Nr.$$

Thus

$$\lfloor e/n \rfloor \geqslant \lfloor N(m-r)/n \rfloor \geqslant \lfloor 2m \rfloor > m$$

and (3.4) holds. On the other hand, $\lceil e/N \rceil \leqslant .03m + 2$, and thus (3.2) reduces to $d(K) \leqslant .47m - 2$, which follows from $d(K) \leqslant \frac{47}{98}r - 2$. Similarly, (3.5) is implied by $d(K) < 0.96r - 1$. Both lower bounds on $d(K)$ follow from the fact that R is 0.45-sparse.

Case $4.25 \leqslant m \leqslant 30$.

This time we need n to be even. If n happens to be odd we add a pendant vertex to G, thereby not changing m. Then any balanced extension of the new graph serves for G as well. So, without loss of generality we may assume that n is even and set $N = 200n + 2\gamma$, $0 \leqslant \gamma \leqslant q-1$, so that N is even and $N+n$ is divisible by q. Let R be a $2r$-regular, $(.5 - .75/r)$-sparse graph on N vertices, where, setting $m' = m - \lfloor m \rfloor$,

$$r = \begin{cases} \lfloor m \rfloor - \frac{1}{2}, & \text{if } m' < \frac{1}{4} \\ \lfloor m \rfloor, & \text{if } \frac{1}{4} \leqslant m' < \frac{3}{4} \\ \lfloor m \rfloor + \frac{1}{2}, & \text{if } m' \geqslant \frac{3}{4}. \end{cases}$$

Then $\frac{1}{4} \leqslant m - r \leqslant \frac{3}{4}$ and so $e/n \geqslant 50$ implying (3.4). Also $e/N \leqslant \frac{3}{4} + mn/N < 1$ and both (3.2) and (3.5) follow by the fact that $d(K) \leqslant r - \frac{3}{4}$.

§4. Proof of Theorem 1—the sparse case

In this section we prove Theorem 1 for graphs with $m = m(G) < \frac{10}{9}$. Set $|G| = n$, $m = 1 + \varepsilon$, $\varepsilon = p/q$, p and q relatively prime. Let C be an enormous constant to be specified later and let N be an even integer bigger than Cn and such that q divides $N' = \frac{1}{2}N + n + e_G$. Let R_0 be an N-vertex 3-regular .47-sparse graph. To obtain R, we equidistribute $X = N'/\varepsilon - n - N$ new vertices of degree 2 on the edges of R_0. Hence, R has vertices of two types: *old vertices* of degree 3 and *new vertices* of degree 2. Clearly, $\frac{3}{2}Nk'$ edges get $\lceil k \rceil$ new vertices (we call these edges *large*) and the remaining ones (we call them *small*) get $\lfloor k \rfloor$ vertices each, where $k = \frac{2}{3}X/N$ and $k' = k - \lfloor k \rfloor$. The parameter k is about $1/(3\varepsilon)$. More precisely,

$$\frac{1}{3\varepsilon} - \frac{2}{3} < k < \frac{1}{3\varepsilon} - \frac{2}{3} + \frac{4}{3\varepsilon C}.$$

To complete the construction of F, we join each vertex of G with 2 old vertices of R in such a way that all $2n$ neighbors are within distance at least $C_0 = \log_3 C - 1$ in R_0 from each other. This is possible, since $|R_0| = N > Cn$. Hence we have $d(F) = 1 + \varepsilon$ and we claim that F is balanced. The proof is analogous to that in Section 3 but a little bit more involved. The distribution of large and small edges over R_0 will play a crucial role at the end and we come back to it later. Right now we assume it is arbitrary. First of all notice that (3.4) is trivially satisfied since $m(G) < 2$, so we focus on (3.2) and (3.5). There is one thing we should say now and which can be verified by simple calculations (using the above bounds on k). In order to prove that F is balanced, it suffices to check whether $f(H) \geqslant 0$ only for induced subgraphs of F with minimum degree greater than 1 and with the property that together with any pair of old vertices which form an edge of R_0 they contain all small vertices put on that edge. Let us denote by $\mathcal{H}$ the set of such subgraphs. Observe that for the members of $\mathcal{H}$ also the following holds: together with a small vertex they contain both large vertices nearest to it. Let $\mathcal{H}_2 = \{H \cap R : H \in \mathcal{H}\}$. Then for each $K \in \mathcal{H}_2$ there is a unique subgraph $T = T(K)$ of R_0 induced in R_0 by the old vertices of K and this correspondence is one-to-one. Following the general pattern described at the beginning of Section 3, we say that a subgraph H of F is "not too large" if $|T(H \cap R)| \leqslant .5N$. By the modularity of f it suffices to prove (3.2) only for connected K. But then $e(G, K) \leqslant 2|T(K)|/C_0$ by the construction of F. (See Ruciński and Vince (1988) for a simple graph-theoretic lemma.) Turning to the case of "too large" H, set $\bar{\mathcal{H}}_2 = \{\bar{K} : K \in \mathcal{H}_2\}$ and observe that each $K \in \bar{\mathcal{H}}_2$ has the following property: if there is an edge in R_0 such that one of its endpoints is in K and the other is not, then K contains all small vertices put on that edge. Thus the mapping $T(.)$ is one-to-one on $\bar{\mathcal{H}}_2$ as well.

We are approaching the very important moment of the proof when we shall shift our attention from F and R to the properties of R_0. For a subgraph T of R_0, let us denote by x_T and y_T the number of small and large edges of T, respectively, and by x_T^+ and y_T^+ the number of small and large edges with at least one endpoint in T, respectively. For $K \in \mathcal{H}_2$, setting $T = T(K)$, $|K| - |T| = e_K - e_T = x_T \lfloor k \rfloor + y_T \lceil k \rceil$ and the same is true for $K \in \bar{\mathcal{H}}_2$ with x_T and y_T replaced by x_T^+ and y_T^+. Moreover, in the last case $e(K,\bar{K}) = 3|T| - 2e_T$. Hence (3.2) and (3.5) will follow if, for all $T \subset R_0$, $4|T| \leqslant .5N$,

$$(1+\varepsilon)|T| + \varepsilon[e_T(k-k') + y_T] - e_T \geqslant 2|T|/C_0 \qquad (4.1)$$

and

$$e_T + \varepsilon[(3|T| - e_T)(k-k') + y_T^+] \leqslant (2-\varepsilon)|T|. \qquad (4.2)$$

Using just the crude bounds $0 \leqslant k' \leqslant 1$, $y_T \geqslant 0$ and $y_T^+ \leqslant 3|T| - e_T$ one can now immediately complete the proof for $\varepsilon < \frac{1}{26}$. However, by a special distribution of small and large edges we shall gain the proof for about 3times larger ε. It is known that, for large n, a positive fraction of n-vertex 3-regular graphs are hamiltonian (cf. Robinson and Wormald (1984)). Therefore we may assume that R_0 has, in addition to being .47-sparse, a 1-factorization (M_1, M_2, M_3), say. Now we can partition the edges of R_0 into small and large in such a way that the following requirements are met. (Below S and L stand for the sets of small and large edges, respectively.)

(i) If $k' \leqslant \frac{1}{3}$ then $M_1 \cup M_2 \subset S$;
(ii) if $\frac{1}{3} \leqslant k' \leqslant \frac{2}{3}$ then $M_1 \subset S$, $M_3 \subset L$;
(iii) if $k' \geqslant \frac{2}{3}$ then $S \subset M_1$.

In case (i), every vertex is incident to at least 2 small edges. Thus, setting $t = |T|$, for all $T \subset R_0$, $x_T^+ \geqslant t$, and so $y_T^+ \leqslant 2t - e_T$. In case (ii), y_T is at least as large as half of the number of vertices of degree 3 in T, i.e. $y_T \geqslant e_T - t$. Also, $x_T^+ \geqslant \frac{1}{2}t$, so $y_T^+ \leqslant 2.5t - e_T$. In case (iii), $x_T \leqslant \frac{1}{2}t$, so $y_T \geqslant e - \frac{1}{2}t$. Using these refined bounds plus the fact that $e_T \leqslant 1.41t$ we can verify (4.1) and (4.2) for $\varepsilon \leqslant \frac{1}{9}$ provided $C_0 > 260$ and so $C > 3^{261}$. Indeed, the left-hand side of (4.1) is, in cases (i)–(iii) resp., at least $(.06 - .41\varepsilon)t$, $(.06 - .47\varepsilon)t$, and $(.06-.44\varepsilon)t$, and (4.2) is implied in corresponding cases by $.53(\varepsilon + 4/C) \leqslant .06$, $.5\varepsilon + (.53)4/C \leqslant .06$, and $.47\varepsilon + (.53)4/C \leqslant .06$. We omit the details. $\square$

§5. Comments

I. How to prove Theorem 2. The proof of Theorem 2 mimics that of Theorem 1. The construction of F is identical except that the equation $d(F) = m^*(G)$ now takes the form $e_G + e_R + e(G,R) = m^*(G)(N+n-1)$. However, this difference is not essential, since we always apply the inequalities $0 < m^*(G) \times (n-1) - e_G < m^*(G)n$, as we applied $0 < m(G)n - e_G < m(G)n$ before. We define another deficit function $f^*(H) = m^*(G)(|H|-1) - e_H$ and want to prove that it is nonnegative for all subgraphs of F. In the "not too large" case we have

$$f^*(H) = f^*(H_1) + f^*(H_2) + m^*(G) - e(H_1, H_2)$$
$$\geqslant m^*(G)|H_2| - e_{H_2} - e(H_1, H_2)$$

which is analogous to (3.2). For "too large" H we also argue as before:

$$0 = f^*(H) + f^*(\bar{H}) + m^*(G) - e(H, \bar{H})$$

and $f^*(H) \geqslant 0$ is equivalent to $e(H, \bar{H}) + e_{\bar{H}} \geqslant m^*(G)|\bar{H}|$, an analog of (3.3). The rest of the proof is just a boring repetition.

II. How to close the gap? To attack the gap from the left side, one should use the construction of Section 4 but with small and large edges distributed even more equally around R_0. We saw how much progress has been achieved by distributing them along the perfect matchings. However we think that one has to apply a random distribution so that all sufficiently large but "not too large" induced subgraphs have about $k'e_T$ large edges. Our optimism is planted on the fact that in the ideal case when $k' = 0$ as well as in two other special cases $k' = \frac{1}{3}$ and $k' = \frac{2}{3}$ or if k' is very close to any of these three values, the argument of Section 4 goes through for all $\varepsilon < \frac{1}{2}$, i.e. for whole range for which that method has been designed.

To close the remainder of the gap, i.e. to cover the interval (1.5, 4.25) one has to use, instead of regular graphs, graphs with density much closer to $m(G)$. Thus, graphs whose vertex degrees take only two consecutive values, r and $r+1$, $r = 3$, ..., 7, are the first natural candidates. (Note that our construction in Section 4 was of this sort, with $r = 2$.) But our attempts to prove an analog of Lemma 2.2 for random graphs of such type failed due to some technical difficulties. We hope we shall be able to overcome them in the "near future".

References

[1] E. A. Bender, E. R. Canfield, *The asymptotic number of labelled graphs with given degree sequences*, J. Combinatorial Theory, Series A **24** (1978), 296–307.

[2] C. Berge, *Graphs and Hypergraphs*, North-Holland, 1973.

[3] B. Bollobás, *A probabilistic proof of an asymptotic formula for the number of labelled regular graphs*, European J. Combinatorics **1** (1980), 311–316.

[4] _____, *Random Graphs*, Academic Press, London, 1985.

[5] _____, *The isoperimetric number of random regular graphs*, European J. Combinatorics **9** (1988), 241–244.

[6] P. A. Catlin, J. W. Grossman, A. M. Hobbs, H.-J. Lai, *Fractional arboricity, strength, and principal partitions in graphs and matroids*, Research Report CORR, Faculty of Mathematics, University of Waterloo, 1989, pp. 89–13.

[7] P. Erdős, A. Rényi, *On the evolution of random graphs*, Publ. Math. Inst. Hung. Acad. Sci. **5** (1960), 17–61.

[8] A. Frieze, *On random regular graphs with non-constant degree*, Research Report No. 88-2, Dept. Math. Carnegie Mellon Univ..

[9] E. Győri, B. Rothschild, A. Ruciński, *Every graph is contained in a sparsest possible balanced graph*, Math. Proc. Cambridge Phil. Soc. **98** (1985), 397–401.

[10] M. Karoński, A. Ruciński, *Problem 4*, Proceedings of the Third Czechoslovak Symposium on Graph Theory, held in Prague in 1982, 1982, p. 350.

[11] C. Payan, *Graphes équilibrés et arboricité rationnelle*, European J. Combinatorics **7** (1986), 263–270. (French)

[12] R. W. Robinson, N. C. Wormald, *Existence of long cycles in random cubic graphs*, Progress in Enumeration and Design, Proceedings of Waterloo Conference on Combinatorics, 1984, pp. 251–270.

[13] A. Ruciński, A. Vince, *Balanced graphs and the problem of subgraphs of random graphs*, Congres. Numerantium **49** (1985), 181–190.

[14] _____, *Strongly balanced graphs and random graphs*, J. Graph Theory **10** (1986), 251–264.

[15] _____, *Balanced extensions of graphs*, Proceedings of the New York Graph Theory Conference, held in 1985, Annals of the New York Academy of Sciences, 1989, pp. 347–351.

[16] _____, *Balanced extensions of graphs and hypergraphs*, Combinatorica **8** (1988), 279–291.

[17] N. C. Wormald, *The asymptotic connectivity of labelled regular graphs*, J. Combinatorial Theory, Series B **31** (1981), 156–167.

Tomasz Łuczak, Andrzej Ruciński
Department of Discrete Mathematics,
Adam Mickiewicz University,
Poznań, Poland

Two Results on Antisocial Families of Balls

A. MALNIČ AND B. MOHAR

A *self-centered* covering of a metric space X is a collection of open balls $\{B(x,r_x) : x \in X\}$ where x is the center and r_x is the radius of $B(x,r_x)$. Two balls are *antisocial* if neither contains the center of the other. A family of balls is antisocial if its balls are pairwise antisocial. Two problems posed by Krantz and Parsons concerning antisocial subcoverings and antisocial intersection graphs are solved.

1. In [1] Krantz and Parsons asked whether every self-centered covering of a compact metric space contains an antisocial subcovering. The answer is yes for compact subspaces of $\mathbb{R}$, or if the mapping $x \mapsto r_x$ is continuous (or at least well-behaved enough). We show by an example that this does not hold in general.

Consider the unit circle S^1 in the complex plane with induced metric, i.e., distance measured in angles on S^1. Let $z_1 := 1$, and $z_{n+1} := z_n \cdot \exp(i\varphi_n)$, where $\varphi_n = \pi - \pi/2^{n+1}$. Now, construct a self-centered covering $\mathcal{C}$ of S^1 by taking:
(a) the open balls $B(z_n, \varphi_n)$ for all $n \geq 1$,
(b) the open balls $B(i, \pi/4)$ and $B(-i, \pi/4)$, and
(c) the open balls $B(x, r_x)$ with their centers x at all other points of S^1 and each r_x so small that $B(x, r_x)$ contains neither a point z_n, $n \geq 1$, nor $\pm i$.

It is obvious that no antisocial subcovering of $\mathcal{C}$ contains balls from (b) and (c) only, since balls from (b) and (c) do not cover all points, e.g., z_1. To obtain an antisocial subcovering of $\mathcal{C}$ we have to take at least one ball $B(z_n, \varphi_n)$. It can be shown that $B(z_n, \varphi_n)$ contains $\pm i$ and all the points z_k except z_{n+1}. Hence, the only possibility to cover z_{n+1} is to take $B(z_{n+1}, \varphi_{n+1})$ which is not antisocial with $B(z_n, \varphi_n)$. This argument proves the nonexistence of an antisocial subcover of $\mathcal{C}$.

The above example can be extended to a self-centered covering of the closed unit disc—which is simply connected—by adding to the described covering of S^1 for each interior point x a ball $B(x, r_x)$ containing no boundary points. Obviously, the obtained covering has no antisocial subcover.

2. Krantz and Parsons posed in [1] another question. We give a partial result on this.

An *antisocial* graph is a graph which can be realized as the intersection graph of some antisocial family of balls in $\mathbb{R}^2$. It is mentioned in [1] that the chromatic number $\chi(G)$ of any such graph satisfies $\chi(G) \leq 20$, and there are examples with $\chi(G) \geq 5$. It is asked what is the least upper bound for chromatic numbers of antisocial graphs. By showing that the complete graph K_8 is antisocial we increase the lower bound for this least upper bound to 8. Moreover, we conjecture that the least upper bound is actually equal to 8.

Take a regular pentagon $P = A_0 A_1 A_2 A_3 A_4$ and at each point A_i take the closed ball $B(A_i, r)$ where $r = \frac{1}{2} |\overline{A_0 A_2}|$. Clearly, these balls are antisocial. Let B_i be the intersection point of the boundaries of $B(A_{i-2}, r)$ and $B(A_{i+2}, r)$ which lies in the interior of P (indices are taken modulo 5). See Figure 1.

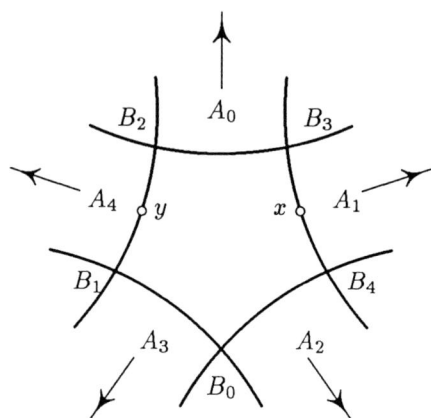

FIGURE 1

Now, let x be the middle point of the arc between B_4 and B_3 on $B(A_1, r)$. Similarly, let y be on the arc between B_2 and B_1, see Figure 1. Clearly, the closed ball centered at x tangent to $B(A_4, r)$ and $B(A_3, r)$ does not contain y and B_0. The analogous property holds for the symmetric ball centered at y. Denote by q the common radius of these two balls. Let x' be a point "very close" to x which is in the direction towards B_1. Define similarly y'. It is clear that the closed balls $B(A_i, r)$, $i = 0, 1, \ldots, 4$, $B(x', q)$, and $B(y', q)$ are antisocial and realizing the graph K_7. Moreover, near B_0 there is some space which is not covered by these balls. Put a point, say z, into this space, and take the radius r_z which is "very close" but less than the minimum of $\text{dist}(z, x')$ and $\text{dist}(z, y')$. It is easily seen that the ball $B(z, r_z)$ intersects all formerly chosen balls except $B(A_0, r)$. But $B(A_0, r)$ can be blown up to be very close to x' and y' (but not containing them), and preserving the antisocialness. Now, if r_0 is the obtained radius, it is clear that $B(A_0, r_0)$, $B(A_i, r)$, $i = 1, \ldots, 4$, $B(x', q)$, $B(y', q)$ and $B(z, r_z)$ are antisocial and have K_8 as their intersection graph.

References

[1] S. G. Krantz and T. D. Parsons, *Antisocial subcovers of self-centered coverings*, Amer. Math. Monthly **93** (1986), 45–48.

A. Malnič and B. Mohar
University of Ljubljana,
Department of Mathematics,
Jadranska 19, 61111 Ljubljana,
Yugoslavia

Fourth Czechoslovakian Symposium on
Combinatorics, Graphs and Complexity
J. Nešetřil and M. Fiedler (Editors)
© 1992 Elsevier Science Publishers B.V. All rights reserved.

Hamiltonicity of Vertex-transitive pq-Graphs

DRAGAN MARUŠIČ

It is shown that every connected vertex transitive graph with pq vertices and an imprimitive subgroup of automorphisms, other than the Petersen graph, contains a Hamilton cycle.

1. Introduction

In this paper we always let p and q denote prime numbers. L. Lovász [4, p. 497] conjectured that every connected vertex-transitive graph has a Hamilton path. This conjecture has been verified for graphs of order p, $2p$, $3p$, p^2, p^3 and $2p^2$ (in which case the graph has a Hamilton cycle, unless it is the Petersen graph O_3) and $4p$, $5p$ (see [6, 7]). The aim of this paper is to show that every connected vertex-transitive graph with pq vertices, other than the Petersen graph, and an imprimitive subgroup of automorphisms is hamiltonian.

We shall assume that the reader is familiar with the basic graph-theory terminology. By an n-*graph* we mean a graph with n vertices. We shall say that a vertex-transitive graph is *imprimitive* if its automorphism group contains an imprimitive subgroup. A special case of imprimitive graphs are the so called metacirculants. An (m, n)-*metacirculant* is a graph which has an automorphism f with m orbits of size $n > 1$ and another automorphism g which cyclically permutes the orbits of f. Of course, the orbits of f form a complete block system for the group $\langle f, g \rangle$.

For a graph Y and an integer $m > 1$, let $\sigma \colon A(Y) \to S_m$ be a labeling of the arcs of Y by permutations in S_m such that $\sigma(u,v)\sigma(v,u) = \mathrm{id}$ for all pairs of adjacent vertices u, v in Y. The graph $X = (Y, \sigma)$ with vertex set $V(Y) \times \mathbb{Z}_m$ and edges of the form $(u, r)(v, s)$, where $u, v \in E(Y)$ and $s = \sigma(u,v)(r)$ is said to be an *m-fold covering* of Y and Y is called the *base graph* of X.

Given positive integers k and m greater than 1, let $\varrho \in S_m$ be the m-cycle $(0\,1\,\ldots\,m-1)$, $r \in S_m$ be the involution which fixes 0 and interchanges r with $-r$ ($r \in \mathbb{Z}_m$) and let $\tau_r = \varrho^r \tau \varrho^{-r}$ ($r \in \mathbb{Z}_m$). With a fixed generator w of $GF(2^k)^*$ we let $X(k, m)$ denote the graph (Y, σ), where Y is the complete graph with vertex-set

---

Supported in part by the Research Council of Slovenija

$GF(2^k) \cup \{\infty\}$ and the labeling σ is defined as follows

$$\sigma(x,y) = \begin{cases} \text{id}, & \infty \in \{x,y\} \\ \tau_r, & x+y = w^r \end{cases} \tag{1}$$

Our result on hamiltonicity of imprimitive graphs is based on the following proposition which may be deduced from [9, Theorem].

Proposition 1.1. *Let X be an imprimitive pq-graph, where $p > q$ are primes. Then either X is a metacirculant or $p = 2^{2^s} + 1$ is a Fermat prime, q divides $2^{2^s} - 1$ and X has a subgraph isomorphic to $X(2^s, q)$.* □

As for the hamiltonian properties of metacirculants, the following result was proved independently by Alspach and Parsons [2, Theorem 1] and by the author [5, Theorem 7.4].

Proposition 1.2. *Let p and q be primes. Then every connected metacirculant with pq vertices, other than the Petersen graph, is hamiltonian.* □

In fact, it seems reasonable to conjecture that the Petersen graph is the only connected metacirculant without a Hamilton cycle. In a recent paper Alspach [1] proved this to be the case under the additional assumption that the number of vertices in a block is a prime power.

In view of Propositions 1.1 and 1.2 the hamiltonicity of vertex-transitive pq-graphs with imprimitive subgroups of automorphisms will be established if we show that the graphs $X(2^N, q)$ above are hamiltonian. This is taken care of in the next section.

2. Main result

We first prove the following general lemma about the existence of Hamilton cycles in the graphs $X(k,m)$.

Lemma 2.1. *Let k and m be positive integers greater than 1 and let m divide $2^k - 1$. If $m < 2^k - 1$, then $X(k,m)$ is hamiltonian.*

Proof. Let $X = X(k,m)$ and let Y be the base graph of X with the corresponding labeling function σ. Moreover, let ϱ, τ and τ_r ($r \in \mathbb{Z}_m$) have the meaning described in the previous section. Fix a generator w in $GF(2^k)^*$ and let $C = 01ww^2\ldots w^n 0$, with $n = 2^k - 2$, be a cycle in Y which leaves out the vertex ∞. Then the respective labels of the corresponding arcs $(0,1), (1,w), (w,w^2), \ldots, (w^{n-1}, w^n), (w^n, 0)$ of C are $\tau_0, \tau_a, \tau_{a+1}, \ldots, \tau_{a+n-1}, \tau_n$, where a satisfies the equation $1 + w = w^a$. Clearly, $m < n$ and so

$$\text{each } \tau_r \ (r \in \mathbb{Z}_m) \text{ is a label of at least one arc in } C. \tag{2}$$

Moreover, since $\tau_b \tau_c = \varrho^{2(b-c)}$ for all $b, c \in \mathbb{Z}_m$, it follows that the value of the labeling σ computed on the cycle C equals

$$\sigma(C) = \tau_0 \overbrace{\varrho^{-2} \ldots \varrho^{-2}}^{n/2} \tau_n = \tau_0 \varrho^{-n} \tau_n = \varrho.$$

But $\varrho^m = \text{id}$ and so, since m divides $n+1$, it follows that $\sigma(C) = \varrho^{-(n+1)}\varrho = \varrho$. This implies that the cycle C lifts to a Hamilton cycle, call it H, in the graph $X - \{(\infty, r) : r \in \mathbb{Z}_m\}$.

The hamiltonicity of X will now follow if we show that, for each $r \in \mathbb{Z}_m$, the vertex (∞, r) is a common neighbour of two successive vertices in H. Fix $r \in \mathbb{Z}_m$. Then, by (2), there exist $x, y \in GF(2^k)$ such that $\sigma(x, y) = \tau_r$. Of course, $\tau_r(r) = r$ and so (y, r) is a successor of (x, r) in H. But, by (1), both (x, r) and (y, r) are neighbours of (∞, r) and so the cycle H can be extended by inserting (∞, r) between (x, r) and (y, r). Doing this for each $r \in \mathbb{Z}_m$ we obtain a Hamilton cycle in X. □

Using the above lemma we may now establish the hamiltonicity of imprimitive pq-graphs.

Theorem 2.2. *Let $X \neq O_3$ be a connected imprimitive pq-graph, where p and q are primes. Then X is hamiltonian.*

Proof. If X is a metacirculant, then it is hamiltonian by Proposition 1.2. In view of Proposition 1.1 we may therefore assume that $p = 2^{2^s} + 1$ is a Fermat prime and q divides $2^{2^s} - 1$ and that X contains a subgraph isomorphic to $X(2^s, q)$.

Since 3 divides $2^{2^s} - 1$, it follows that the latter is a prime if and only if $p = 5$ and $q = 3$. The graph $X(2, 3)$ turns out to be the line graph $L(O_3)$ of the Petersen graph and contains a Hamilton cycle (see [8, Theorem 5.6]). We may therefore assume that $q < 2^{2^s} - 1$ and so X is hamiltonian by Lemma 2.1. This completes the proof of Theorem 2.2. □

As a final comment, let us say that the Petersen graph is very likely the only nonhamiltonian graph among connected vertex-transitive pq-graphs. In order to see whether this really is the case, a sufficient knowledge about the structure of these graphs is needed. The list of primitive groups of degree mp, where $m < p$, by Liebeck and Saxl [3] may prove to be useful in that respect.

References

[1] B. Alspach, *Hamilton cycles in metacirculant graphs with prime cardinal blocks*, Graph theory in memory of G. A. Dirac, Sandbjerg, 1985; Ann. Discrete Math. **41**, 7–16.

[2] B. Alspach, T. D. Parsons, *A construction for vertex-transitive graphs*, Canad. J. Math. **34** (1982), 307–318.

[3] M. W. Liebeck, J. Saxl, *Primitive permutation groups cantaining an element of large prime order*, J. London Math. Soc. (2) **31** (1985), 237–249.

[4] L. Lovász, *Combinatorial Structures and Their Applications* (R. Guy, H. Hanani, N. Sauer and J. Schonheim, eds.), Gordon and Breach, New York, 1970.

[5] D. Marušič, *On vertex-symmetric digraphs*, PhD thesis, University of Reading, England, 1981.

[6] ———, *Some problems in vertex-symmetric graphs*, Colloq. Math. Soc. Janos Bolyai 37 (1983), 1–11, Finite and Infinite Sets, Eger, Hungary, 1981.

[7] _____, *Hamiltonian cycles in vertex-symmetric graphs of order* $2p^2$, Discrete Math. **66** (1987), 169–174.

[8] _____, *On vertex-transitive graphs of order* qp, J. Combin. Math. Combin. Comp. **4** (1988), 97–114.

[9] D. Marušič, R. Scapellato, *Characterizing vertex-transitive pq-graphs*, submitted to J. London Math. Soc.

Dragan Marušič
Inštitut za matematiko, fiziko in mehaniko,
Univerza v Ljublani,
Jadranska 19, 61111 Ljubljana,
Slovenija, Yugoslavia

Fourth Czechoslovakian Symposium on
Combinatorics, Graphs and Complexity
J. Nešetřil and M. Fiedler (Editors)
© 1992 Elsevier Science Publishers B.V. All rights reserved.

On Nodes of Given Out-Degree in Random Trees

A. MEIR AND J. W. MOON

Let $\mathcal{F}$ denote a simply generated family of rooted trees whose generating function $y(x)$ satisfies the relation $y = x\theta(y)$; suppose that $\tau\theta'(\tau) = \theta(\tau)$ and that certain technical conditions are met. Let $H(j,n)$ denote the expected height of the jth endnode of a tree selected at random from the trees in $\mathcal{F}$ with n nodes (where the endnodes are labelled from left to right). The author shows that if $j, n \to \infty$ in such a way that $\nu = j/n$ is fixed and $\nu\theta(\tau) < 1$, then $H(j,n) \sim \lambda n^{1/2}$ where $\lambda = 8\theta'(\tau) \cdot \{\nu(1 - \nu\theta(\tau))/2\pi\theta''(\tau)\}^{1/2}$.

1. Introduction

The *out-degree* of a node v in a rooted tree T_n is the number $d(v)$ of edges incident with v that lead away from the root of T_n (for definitions not given here see, e.g., [3]). Let $N_m = N_m(T_n)$ denote the number of nodes of out-degree m in the tree T_n. Our object is to investigate the distribution of N_m over all trees T_n in certain families $\mathcal{F}$ of rooted trees. In §2 we define the families we shall be considering here, the simply generated families, and in §3 we establish some preliminary results of a technical nature. Our main results are in §4 where we determine the limiting behaviour of $\Pr\{N_m = j\}$ for three ranges of values of m. In the first two cases, where m is fixed or is increasing fairly slowly as a function of n, the distribution of N_m is asymptotically normal; in the third case, where m is increasing somewhat more rapidly, the limiting distribution is Poisson.

We remark that when $\mathcal{F}$ is the family of labelled trees some of these results follow from earlier results obtained in connection with the classical occupancy problem. In particular, the asymptotic normality of N_m for this family follows from a result of Weiss [58] when $m = 0$ and from a result of Sevast'yanov and Chistyakov [13] when m is an arbitrary fixed integer; see also [14] and [4; p.144]. (The arguments in these papers make use of the method of moments or the saddlepoint method; our approach is based on an application of the dominated convergence theorem.) The cases $m = 0$ and $m = 1$ for labelled trees were also considered in [11] and [5], respectively; and the case $m = 1$ was considered for general families of simple generated trees in [7].

2. Simply Generated Families

We recall that *plane* trees — or *ordered* trees — are rooted trees with an ordering specified for the branches incident with each node (cf. [3; p. 306]). Suppose there exists a sequence of non-negative constants $c_0(=1), c_1, c_2, \cdots$ such that each plane tree T_n is assigned the *weight* $\omega(T_n)$ where

$$\omega(T_n) = \prod_0^\infty c_k^{N_k(T_n)}. \qquad (2.1)$$

The collection of plane trees with such an assignment of weights will be called a *simply generated family* of trees. Let y_n denote the number of trees T_n in such a family $\mathcal{F}$ where (here and elsewhere) the weights are taken into account; that is

$$y_n = \sum \omega(T_n) \qquad (2.2)$$

where the sum is over all plane trees T_n with n nodes. It is not difficult to see (cf. [6; p.999] or [15; p. 24]) that the generating function $y = \sum y_n x^n$ of a simply generated family $\mathcal{F}$ satisfies the relation

$$y = x\varphi(y) \qquad (2.3)$$

where $\varphi(t) = 1 + \sum_1^\infty c_k t^k$. Two familiar examples of simply generated families are the ordinary plane trees for which $\varphi(t) = (1-t)^{-1}$ and the labelled trees for which $\varphi(t) = e^t$.

We shall assume henceforth that $\mathcal{F}$ is some given simply generated family such that the function $\varphi(t)$ appearing in (2.3) is regular when $|t| < R \leq \infty$. We further assume that

$$c_k \geq 0 \quad \text{for} \quad k \geq 1; \qquad (2.4)$$
$$\gcd\{k : c_k > 0\} = 1; \quad \text{and} \qquad (2.5)$$
$$\tau\varphi'(\tau) = \varphi(\tau) \quad \text{for some } \tau, \text{ where } 0 < \tau < R. \qquad (2.6)$$

It follows from these assumptions (cf. [10; p. 216], [6; p. 999], or [15, p.32]) that

$$y_n = c(\varphi(\tau)/\tau)^n \cdot (n^{-3/2} + \mathcal{O}(n^{-5/2})) \qquad (2.7)$$

as $n \to \infty$, where $c = (\varphi(\tau)/2\pi\varphi''(\tau))^{1/2}$.

3. Preliminary Results

For any non-negative integer m let $\varphi_m(t) = \varphi(t) - c_m t^m$, $p = p_m = c_m \tau^m/\varphi(\tau)$, and $q = q_m = 1 - p_m$. (It will be seen later that the expected number of nodes of out-degree m in a tree T_n is asymptotically equal to np_m.) We shall tacitly assume henceforth that $c_m > 0$ for any value of m being considered; for, if $c_m = 0$ then

$N_m(T_n) = 0$ for all trees T_n in $\mathcal{F}$. We now derive a formula for $\Pr\{N_m = j\}$ where the probability is taken over all the trees T_n with n nodes in the family $\mathcal{F}$.

Lemma 1. *If $0 \leqslant mj \leqslant n-1$ then*

$$\Pr\{N_m = j\} = \tau(ny_n)^{-1} \cdot (\varphi(\tau)/\tau)^n \cdot \binom{n}{j} p^j q^{n-j} \cdot I \qquad (3.1)$$

where

$$I = (2\pi)^{-1} \cdot \int_{-\pi}^{\pi} \left\{ \frac{\varphi_m(\tau e^{i\theta})}{\varphi_m(\tau)} \right\}^{n-j} \cdot e^{(1+mj-n)i\theta} d\theta.$$

Proof. Let $y_{nj} = \Pr\{N_m = j\} y_n$ denote the number of trees T_n in $\mathcal{F}$ such that $N_m(T_n) = j$ for a given value of m. It is not difficult to see that the generating function $Y_m = Y_m(x, z) = \sum y_{nj} z^j x^n$ satisfies the relation $Y_m = x\Phi_m(Y_m)$ where $\Phi_m(t) = \varphi_m(t) + c_m z t^m$. Thus it follows from Lagrange's inversion formula [1; p. 148] that

$$\sum_j n y_{nj} z^j = \mathcal{C}_{n-1}\{(\varphi_m(t) + c_m z t^m)^n\}$$

where $\mathcal{C}_\nu\{f(t)\}$ denotes the coefficient of t^ν in the power series expansion of $f(t)$. When we compare coefficients of powers of z and appeal to Cauchy's theorem we find that

$$n y_{nj} = \binom{n}{j} c_m^j \mathcal{C}_{n-1-mj}\{(\varphi_m(t))^{n-j}\}$$

$$= \binom{n}{j} c_m^j (2\pi i)^{-1} \int_C \{\varphi_m(t)\}^{n-j} \cdot t^{mj-n} dt$$

where C may be taken to be the circle $|t| = \tau$. Consequently,

$$n y_{nj} = \binom{n}{j} c_m^j \tau^{1+mj-n} (\varphi_m(\tau))^{n-j} (2\pi)^{-1} \int_{-\pi}^{\pi} \left\{ \frac{\varphi_m(\tau e^{i\theta})}{\varphi_m(\tau)} \right\}^{n-j} \cdot e^{(1+mj-n)i\theta} d\theta.$$

This implies formula (3.1) upon appealing to the relation $\varphi_m(\tau) = q\varphi(\tau)$ and the definitions of p and y_{nj}. $\square$

We observe, for later use, that if n and j (and perhaps m) tend to infinity in such a way that $pqn \to \infty$ and $x_j = (j - pn) \cdot (pqn)^{-1/2}$ remains bounded by some absolute constant, then it follows from Stirling's formula (see, e.g., [2; p. 170] or [12; p. 151]) that

$$\binom{n}{j} p^j q^{n-j} = (2\pi pqn)^{-1/2} \cdot e^{-1/2\, x_j^2} \cdot \{1 + \mathcal{O}((pqn)^{-1/2})\}$$

where the constant implicit in the $\mathcal{O}$-term is independent of m, n, and j. Moreover, it follows from (2.7) that

$$\tau(ny_n)^{-1} \cdot (\varphi(\tau)/\tau)^n = (2\pi An)^{1/2} \cdot (1 + \mathcal{O}(n^{-1})) \qquad (3.2)$$

where $A = \tau^2 \varphi''(\tau)/\varphi(\tau)$. Thus, under these conditions, Lemma 1 implies that

$$\Pr\{N_m = j\} = (A/pq)^{1/2} e^{-1/2\, x_j^2} \cdot I \cdot \{1 + \mathcal{O}((pqn)^{-1/2})\}. \tag{3.3}$$

Assumption (2.5) implies that there exists a *finite* set $\mathcal{H}$ of positive integers such that $c_h > 0$ if $h \in \mathcal{H}$ and $\gcd\{h: h \in \mathcal{H}\} = 1$. Let $\ell = \min\{k: k \geqslant 1 \text{ and } c_k > 0\}$. For technical convenience we shall henceforth further assume that

$$\gcd\{k - \ell: k > \ell \text{ and } c_k > 0\} = 1 \tag{3.4a}$$

and that

$$\gcd\{k: k \neq m \text{ and } c_k > 0\} = 1 \tag{3.4b}$$

for $m = 1, 2, \cdots$. Notice that the existence of the set $\mathcal{H}$ ensures that (3.4b) necessarily holds for all sufficiently large m without any additional assumptions.

Lemma 2. *Suppose that $\varphi(t)$ satisfies conditions (2.4), (2.5), and (3.4) and suppose that $0 < r < R$. Then there exists a positive constant $Q = Q_\varphi(r)$ such that the inequality*

$$|\varphi_m(re^{i\theta})| \leqslant \varphi_m(r) e^{-Q\theta^2} \tag{3.5}$$

holds for all integers m and all $\theta \in [-\pi, \pi]$.

Proof. It was shown in [8] that if $\varphi(t)$ satisfies (2.4) and (2.5) then there exists a positive constant S, depending only on the set $\mathcal{H}$ defined above and the coefficients c_h where $h \in \mathcal{H}$, such that $|\varphi(re^{i\theta})| \leqslant \varphi(r) e^{-S\theta^2/\varphi(r)}$ for all $\theta \in [-\pi, \pi]$. Since (3.4) holds we may apply this result to each function $\varphi_m(t)$ separately and conclude that for each m there exists a constant $Q_m = S_m/\varphi_m(r)$ such that $|\varphi_m(re^{i\theta})| \leqslant \varphi_m(r) e^{-Q_m \theta^2}$ for all $\theta \in [-\pi, \pi]$. (Strictly speaking, when $m = 0$ we apply the earlier result to the function $\varphi_0(t)/c_\ell t^\ell$, but essentially the same conclusion holds.) It is not difficult to see that we may take S_m to be the same constant S^* for all $m > m_0 = \max\{h: h \in \mathcal{H}\}$; thus we may replace Q_m by $Q^* = S^*/\varphi(r)$ when $m > m_0$. But then conclusion (3.5) holds for all m with $Q = \min\{Q_0, Q_1, \ldots, Q_{m_0}, Q^*\}$.

In order to estimate the integral I in Lemma 1 we shall need an expansion for the function $\varphi_m(\tau e^{i\theta})$ in the neighbourhood of $\theta = 0$. For any given value of m let $\sigma^2 = \sigma_m^2 = pq - (m-1)^2 p^2 / A$ where $A = \tau^2 \varphi''(\tau)/\varphi(\tau)$. We observe that it follows from our definitions and assumption (2.5) that

$$\tau \varphi_m'(\tau) = \tau \varphi'(\tau) - mc_m \tau^m = \varphi(\tau)(1 - pm) \tag{3.6}$$

and

$$\tau^2 \varphi_m''(\tau) = \tau^2 \varphi''(\tau) - m(m-1) c_m \tau^m = \varphi(\tau)(A - pm(m-1)). \tag{3.7}$$

$\square$

Lemma 3. *There exists a positive constant δ such that if $|\theta| < \delta$ then*

$$\log \varphi_m(\tau e^{i\theta}) = \log \varphi_m(\tau) + a_m i\theta - \tfrac{1}{2} A_m \theta^2 + \mathcal{O}(|\theta|^3) \tag{3.8}$$

for $m \geqslant 0$, where $a_m = (1-pm)/q$, $A_m = A\sigma^2/pq^2$, and the $\mathcal{O}$-term is independent of m.

Proof. For any given value of m let $f(t) = \log \varphi_m(t)$; and for any given value of $\theta \in (-\pi, \pi)$ let Γ denote the (shorter) arc of the circle $|t| = \tau$ from $t = \tau$ to $t = \tau e^{i\theta}$. Then

$$f(\tau e^{i\theta}) = f(\tau) + \tau f'(\tau)(e^{i\theta} - 1) + \tfrac{1}{2}\tau^2 f''(\tau)(e^{i\theta} - 1)^2 + R_3 \tag{3.9}$$

where

$$R_3 = \frac{1}{2}\int_\Gamma f'''(t) \cdot (\tau e^{i\theta} - t)^2 dt,$$

by Taylor's formula. Let $M_m = \max\{|f^{(k)}(t)| : t \in \Gamma,\ 1 \leqslant k \leqslant 3\}$. Then it follows from (3.9), (3.6), and (3.7) that

$$f(\tau e^{i\theta}) = f(\tau) + ia_m\theta - \tfrac{1}{2}A_m\theta^2 + R_3^*$$

where

$$\begin{aligned}
a_m &= \tau f'(\tau) = \tau \varphi_m'(\tau)/\varphi_m(\tau) = (1 - pm)/q, \\
A_m &= \tau^2 f''(\tau) + \tau f'(\tau) = \tau^2 \varphi_m''(\tau)/\varphi_m(\tau) - a_m^2 + a_m \\
&= (A - pm(m-1))/q + p(m-1)(1-pm)/q^2 \\
&= (Aq - p(m-1)^2)/q = A\sigma^2/pq^2,
\end{aligned}$$

and

$$|R_3^*| \leqslant (\tau + \tau^2/2 + \tau^3/6)M_m \cdot \mathcal{O}(|\theta|^3),$$

with the $\mathcal{O}$-factor being independent of m. To complete the proof it suffices to show that there exist positive constants δ and M, independent of m, such that if $|\alpha| < \delta$ then

$$|f^{(k)}(\tau e^{i\alpha})| \leqslant M \tag{3.10}$$

for $1 \leqslant k \leqslant 3$ and $m = 0, 1, 2, \ldots$.

To prove (3.10) we first observe that $|\varphi_m^{(h)}(\tau e^{i\alpha})| \leqslant \varphi_m^{(h)}(\tau) \leqslant \varphi^{(h)}(\tau)$ for all m, h, and α since the coefficients of $\varphi_m^{(h)}(t)$ are non-negative and $\varphi_m(t) = \varphi(t) - c_m t^m$. Now, if $1 \leqslant k \leqslant 3$ then $f^{(k)}(\tau e^{i\alpha})$ can be expressed as a fraction whose numerator can be bounded by a cubic polynomial in $\max\{\varphi^{(h)}(\tau): 0 \leqslant h \leqslant k\}$ and, hence, by a positive constant N_k. The denominator of this fraction is $(\varphi_m(\tau e^{i\alpha}))^{2k}$. Since $\varrho = \min\{\varphi_m(\tau): m \geqslant 0\} > 0$, there exists a constant $\delta > 0$ such that if $|\alpha| < \delta$ then $|\varphi(\tau e^{i\alpha}) - \varphi(\tau)| < \tfrac{1}{2}\varrho$, whence

$$\begin{aligned}
|\varphi_m(\tau e^{i\alpha})| &= |\varphi(\tau) - c_m\tau^m e^{im\alpha} + \varphi(\tau e^{i\alpha}) - \varphi(\tau)| \\
&\geqslant \varphi(\tau) - c_m\tau^m - |\varphi(\tau e^{i\alpha}) - \varphi(\tau)| \\
&> \varphi_m(\tau) - \tfrac{1}{2}\varrho \geqslant \tfrac{1}{2}\varrho
\end{aligned}$$

for all m. Thus (3.10) holds with $M = \max\{(2/\varrho)^{2k} \cdot N_k : 1 \leqslant k \leqslant 3\}$, as required. □

4. Main Result

We now determine the limiting behaviour of the distribution of N_m for three ranges of values of m. In what follows u denotes an arbitrary constant and, as before, $p = 1 - q = c_m \tau^m / \varphi(\tau)$ and $\sigma^2 = pq - (m-1)^2 p^2 / A$ where $A = \tau^2 \varphi''(\tau)/\varphi(\tau)$.

Theorem 1. *(i) Suppose the integers m and j are such that m remains fixed and $j = pn + u\sigma n^{1/2} + o(n^{1/2})$ as $n \to \infty$. Then*

$$\Pr\{N_m = j\} \sim (2\pi\sigma^2 n)^{-1/2} \cdot e^{-1/2\, u^2} \quad \text{as} \quad n \to \infty. \tag{4.1}$$

(ii) Suppose the integers n, m and j tend to infinity in such a way that $pn \to \infty$ and $j = pn + u(pn)^{1/2} + o((pn)^{1/2})$ as $m, n \to \infty$. Then

$$\Pr\{N_m = j\} \sim (2\pi pn)^{-1/2} \cdot e^{-1/2\, u^2} \quad \text{as} \quad m, n \to \infty. \tag{4.2}$$

(iii) Suppose the integers m and n tend to infinity in such a way that $pn \to \lambda$ for some positive constant λ. Then for every fixed integer j

$$\Pr\{N_m = j\} \to \frac{1}{j!} \lambda^j e^{-j} \quad \text{as} \quad m, n \to \infty. \tag{4.3}$$

Proof. We begin by considering the integral I that appears in Lemma 1 and (3.2). If we let

$$G_{n,m,j}(\theta) = \left\{ \frac{\varphi_m(\tau e^{i\theta})}{\varphi_m(\tau)} \right\}^{n-j} \cdot e^{(1+mj-n)i\theta},$$

then it follows readily from Lemma 3 that

$$G_{n,m,j}(\theta) = e^{i\beta\theta - B\theta^2 + \mathcal{O}(1)\cdot(n-j)|\theta|^3} \tag{4.4}$$

when $|\theta| < \delta$ where

$$\beta = 1 + (m-1)(j-pn)/q, \quad B = \tfrac{1}{2} A_m(n-j),$$

and the $\mathcal{O}$-factor is independent of m, n, and j. We now set $\theta = s(n-j)^{1/2}$ in this relation and introduce the function

$$F_{n,m,j}(s) = \begin{cases} G_{n,m,j}\big(s(n-j)^{-1/2}\big), & \text{if } |s| \leqslant \pi(n-j)^{1/2}; \\ 0, & \text{if } |s| > \pi(n-j)^{1/2}. \end{cases}$$

Then the earlier expression for I can be rewritten as

$$I = \frac{1}{2\pi(n-j)^{1/2}} \int_{-\infty}^{\infty} F_{n,m,j}(s)\, ds. \tag{4.5}$$

We now consider case (i) where m is fixed (and, hence, so are p, q, σ, and A_m) and $j - pn = u\sigma n^{1/2} + o(n^{1/2})$. Then $n - j = qn + \mathcal{O}(n^{1/2})$ and

$$\frac{\beta}{(n-j)^{1/2}} = \frac{q + (m-1)(j-pn)}{q(n-j)^{1/2}} = (m-1)u\sigma q^{-3/2} + o(1).$$

So it follows from (4.4) that

$$\lim_{n,j \to \infty} F_{n,m,j}(s) = e^{-i\gamma s - 1/2\, A_m s^2}$$

for every *fixed* value of s, where $\gamma = (m-1)u\sigma q^{-3/2}$ and $A_m = A\sigma^2/q^2 p$. Since

$$|F_{n,m,j}(s)| \leq e^{-Qs^2} \qquad (4.6)$$

for all n, m, j, and s, by Lemma 2, we may appeal to Lebesgue's dominated convergence theorem [16; p. 291] and conclude that

$$\lim_{n,j \to \infty} \int_{-\infty}^{\infty} F_{n,m,j}(s)\, ds = \int_{-\infty}^{\infty} e^{-i\gamma s - 1/2\, A_m s^2}\, ds = (2\pi/A_m)^{1/2} \cdot e^{-\gamma^2/2A_m}.$$

Consequently,

$$I \sim (2\pi q A_m n)^{-1/2} \cdot e^{-\gamma^2/2A_m} = (2\pi A\sigma^2 n/pq)^{-1/2} \cdot e^{-(m-1)^2 u^2 p/2qA} \qquad (4.7)$$

as $n, j \to \infty$. Now $x_j = (j - pn) \cdot (pqn)^{-1/2} \to u\sigma/pq$. Thus it follows from (3.2) and (4.7) that in case (i)

$$\Pr\{N_m = j\} \sim (2\pi\sigma^2 n)^{-1/2} \cdot e^{-1/2\, u^2\{\sigma^2/pq + (m-1)^2 p/qA\}}$$
$$= (2\pi\sigma^2 n)^{-1/2} \cdot e^{-1/2\, u^2} \qquad (4.8)$$

as required.

We next consider case (ii) where n, m, and j tend to infinity in such a way that $pn \to \infty$ and $j - pn = u(pn)^{1/2} + o((pn)^{1/2})$. Notice that $p = c_m \tau^m/\varphi(\tau)$ is the mth term of the power series expansion of the function $\varphi(t)/\varphi(\tau)$ evaluated at a point inside its circle of convergence; hence p, mp, and $m^2 p$ all tend to zero as $m \to \infty$ and, consequently, $mp^{1/2} \to 0$, $q \to 1$, and $A_m \to A$. Now

$$\frac{\beta}{(n-j)^{1/2}} = \frac{q + (m-1)(j-pn)}{q(n-j)^{1/2}} = \mathcal{O}\big((m-1)up^{1/2}\big)\big) \to 0,$$

so it follows from (4.4) that $F_{n,m,j}(s) \to e^{-1/2\, As^2}$ for every fixed value of s as $n, m, j \to \infty$. The bound (4.6) still holds and, after appealing to the dominated convergence theorem again, we find that

$$I \sim \frac{1}{2\pi n^{1/2}} \int_{-\infty}^{\infty} e^{-1/2\, As^2}\, ds = (2\pi An)^{-1/2} \qquad (4.9)$$

as $n, m, j \to \infty$. This time $x_j = (j - pn) \cdot (pqn)^{-1/2} \to u$. Hence it follows from (3.2) and (4.9) that in case (ii)

$$\Pr\{N_m = j\} \sim (2\pi pn)^{-1/2} \cdot e^{-1/2\, u^2} \tag{4.10}$$

as required.

Finally, we consider case (iii) where j is fixed and $pn \to \lambda$ as $m, n \to \infty$. We assert that the condition $pn \to \lambda$ or, equivalently, $nc_m \tau^m \to \lambda \varphi(\tau) > 0$, implies that $m = \mathcal{O}(\log n)$. For, if $R < \infty$ then $\overline{\lim}\, c_m^{1/m} = R^{-1}$ so $c_m < (2/(R+\tau))^m$ and $nc_m \tau^m > \tfrac{1}{2}\lambda \varphi(\tau)$ when m is sufficiently large. But then

$$\tfrac{1}{2}\lambda \varphi(\tau) < nc_m \tau^m < n\bigl(2\tau/(R+\tau)\bigr)^m$$

and the conclusion $m = \mathcal{O}(\log n)$ follows upon taking logarithms. The same argument applies when $R = \infty$ if R is replaced by 2τ in the inequalities. Hence, in this last case

$$\frac{b}{(n-j)^{1/2}} = \frac{q + (m-1)(j - pn)}{q(n-j)^{1/2}} = \mathcal{O}(m/n^{1/2}) \to 0$$

and we find, just as in case (ii), that $F_{n,m,j}(s) = e^{-1/2\, As^2}$ and that

$$I \sim (2\pi An)^{-1/2} \tag{4.11}$$

as $n, m \to \infty$. If j is fixed and $pn \to \lambda$ then

$$\binom{n}{j} p^j q^{n-j} \to \frac{1}{j!}\lambda^j e^{-\lambda} \tag{4.12}$$

as $m, n \to \infty$. Thus it follows from Lemma 1 and relations (3.2), (4.11), and (4.12) that

$$\Pr\{N_m = j\} \to \frac{1}{j!}\lambda^j e^{-\lambda} \tag{4.13}$$

in case (iii), as required. This completes the proof of the theorem. □

The following result describes the asymptotic normality of the variable N_m when m is not too large. We let (a, b) denote any fixed subinterval of the real line.

Theorem 2. (i) *Suppose that m remains fixed as $n \to \infty$. Then*

$$\lim_{n \to \infty} \Pr\left\{a < \frac{N_m - pn}{\sigma n^{1/2}} < b\right\} = (2\pi)^{-1/2} \int_a^b e^{-1/2\, x^2} dx. \tag{4.14}$$

(ii) *Suppose that m and n tend to infinity in such a way that $pn \to \infty$. Then*

$$\lim_{m,n \to \infty} \Pr\left\{a < \frac{N_m - pn}{(pn)^{1/2}} < b\right\} = (2\pi)^{-1/2} \int_a^b e^{-1/2\, x^2} dx. \tag{4.15}$$

Proof. Let j be any integer such that

$$pn + a\sigma n^{1/2} < j < pn + b\sigma n^{1/2} \tag{4.16}$$

or

$$pn + a(pn)^{1/2} < j < pn + b(pn)^{1/2}, \tag{4.17}$$

according as case (i) or (ii) holds. We first observe that it follows from (4.5) and (4.6) that

$$I \leqslant \frac{1}{2\pi(n-j)^{1/2}} \int_{-\infty}^{\infty} e^{-Qs^2} ds = \left(4\pi Q(n-j)\right)^{-1/2}$$

where Q is independent of m, n, and j. Now let $\max_m \{c_m \tau^m / \varphi(\tau)\} = 1 - \kappa$; then $\kappa > 0$ and $j \leqslant (1 - 1/2\,\kappa)n$ if n is sufficiently large. Hence there exists a positive constant C_1 such that

$$I \leqslant C_1 n^{-1/2} \tag{4.18}$$

for all m, n, and j satisfying the hypothesis. Thus it follows from (3.3) and (4.18) that

$$\Pr\{N_m = j\} \leqslant C_2 \left((pn)^{-1/2}\right) \tag{4.19}$$

for some positive constant C_2 whose value depends on a, b, and φ but which is independent of m, n, and j.

We now consider case (i) where m is fixed. For any real x let

$$f_n(x) = \Pr\left\{N_m = [pn + x\sigma n^{1/2}]\right\} \cdot \sigma n^{1/2}.$$

Then $f_n(x) \to (2\pi)^{-1/2} \cdot e^{-1/2\,x^2}$ as $n \to \infty$ for every fixed x by (4.1). Moreover, $f_n(x) = \mathcal{O}(1)$ uniformly in x throughout any fixed finite interval (a, b) by (4.19). Hence

$$\int_a^b f_n(x) dx \to (2\pi)^{-1/2} \int_a^b e^{1/2\,x^2} dx$$

as $n \to \infty$. This implies conclusion (4.14) since the first integral is asymptotic to $\Sigma' \Pr\{N_m = j\}$ where the sum is over all j satisfying (4.16). Conclusion (4.15) follows by essentially the same argument with σ replaced by $p^{1/2}$. □

We remark in closing that by considering the distribution of the number of nodes of out-degree *greater* than m, where $pn \to \lambda$ as m, $n \to \infty$, it is possible to extend some of the results in [9] on the distribution of the maximum out-degree occuring in a tree T_n. We hope to pursue this in a future paper.

5. Acknowledgement

The preparation of this paper was assisted by grants from the Natural Sciences and Engineering Research Council of Canada.

References

[1] L. Comtet, *Advanced Combinatorics*, Reidel, Dordrecht, 1974.

[2] W. Feller, *An Introduction to Probability Theory and its Applications*, Vol. I, Wiley, New York, 1960.

[3] D. E. Knuth, *The Art of Computer Programming*, Vol. I, Addison-Wesley, Reading, 1973.

[4] V. F. Kolchin, *Random Mappings*, Optimization Software, New York, 1986.

[5] A. Meir and J. W. Moon, *On nodes of degree two in random trees*, Mathematika **15** (1968), 188–192.

[6] _____, *On the altitude of nodes in random trees*, Can. J. Math. **30** (1978), 997–1015.

[7] _____, *On nodes of out-degree one in random trees*, Colloq. Math. Soc. J. Bolyai **52** (1987), 405–416.

[8] _____, *The asymptotic behaviour of coefficients of powers of certain generating functions*, Europ. J. Comb. **11** (1990), 581–587.

[9] _____, *On the maximum out-degree in random trees*, Australas. J. Combin. **2** (1990), 147–156.

[10] R. Otter, *The multiplicative process*, Ann. Math. Statist. **20** (1949), 206–224.

[11] A. Rényi, *Some remarks on the theory of trees*, Publi. Math. Inst. Hung. Acad. Sci. **4** (1959), 73–85.

[12] _____, *Probability Theory*, North-Holland, Amsterdam, 1970.

[13] B. A. Sevast'yanov and V. P. Chistyakov, *Asymptotic normality in the classical ball problem*, Th. Prob. Appl. **9** (1964), 198–211.

[14] V. E. Stepanov, *Random Graphs*, Proceedings of the Seminar on Combinatorial Mathematics, Izd-vo Sovetskoe Radio, Moscow, 1973, pp. 164–185.

[15] J. M. Steyaert and P. Flajolet, *Patterns and pattern-matchings in trees: an analysis*, Inf. and Control **58** (1983), 19–58.

[16] K. R. Stromberg, *Introduction to Classical Real Analysis*, Wadsworth, New York, 1981.

[17] I. Weiss, *Limiting distributions in some occupancy problems*, Ann. Math. Statist. **29** (1958), 878–884.

A. Meir and J. W. Moon
Department of Mathematics,
University of Alberta,
Edmonton, Alberta T6G 2GI,
Canada

Fourth Czechoslovakian Symposium on
Combinatorics, Graphs and Complexity
J. Nešetřil and M. Fiedler (Editors)
© 1992 Elsevier Science Publishers B.V. All rights reserved.

All Leaves and Excesses Are Realizable for $k = 3$ and All λ

ERIC MENDELSOHN, NABIL SHALABY, SHEN HAO

We prove that for every leave (excess), numerically possible, for $k = 3$, all v, λ, there is a maximum packing (minimum covering) achieving this leave.

Introduction

We wish to construct a table for all possible leaves (excesses) for $k = 3$, all v, λ, and show that these graphs are the only realizable ones.

Formally let us define:

(a) The pair (V, P) is a v, k, λ *packing* design (a $PD(v; k, \lambda)$) iff $|V| = v$, P is a collection of k-subsets of V, so that every 2-subset is a subset of at most λ elements of P. $|P|$ as large as possible.
(b) The pair (V, C) is a v, k, λ *covering* design (a $CD(v; k, \lambda)$) iff $|V| = v$, C is a collection of k-subsets of V, so that every 2-subset is a subset of at least λ elements of C. $|C|$ as small as possible.

The leave of a packing (V, P) is a graph (V, E) where $(xy) \in E$ with multiplicity m if $\{x, y\}$ is a subset of a $\lambda - m$ blocks of P. The excess of a covering (V, C) is a graph (V, E) where $(xy) \in E$ with multiplicity m if $\{x, y\}$ is a subset of $m - \lambda$ blocks of C. The following necessary conditions hold for a leave of (V, P)

(a) $\frac{1}{2}v(v-1)\lambda - |E| \equiv 0 \mod(12k(k-1))$
(b) for all $x \in V$, $\deg(x) \equiv \lambda(v-1) \mod(k-1)$
(c) $|E|$ is minimal w.r.t. (a) and (b)
(d) for all $x \in V$, $\deg(x) \leqslant \lambda(v-1)$.

Similarly for the excess we have

(a) $12v(v-1)\lambda + |E| \equiv 0 \mod(12k(k-1))$
(b) for all $x \in V$, $\deg(x) \equiv \lambda(1-v) \mod(k-1)$
(c) $|E|$ is minimal w.r.t. (a) and (b)

We note that excesses and leaves are not necessarily unique as there may be many non-isomorphic graphs on $|E|$ edges with the degree of each vertex in the same congruence class mod $(k-1)$. For example the following are graphs with 4 edges and each degree even. (a) AB, BC, CD, DA, (b) AB, AB, BC, BC, (c) AB, CD, AB, CD and (d) AB, AB, AB, AB.

The following conjecture which is a natural generalization of the Wilson's [W] theorem would be of great help.

Conjecture. *There is a n_0 such that for all $n > n_0$, if (V, E) is a graph with $|V| = n$, and (V, E) satisfies the necessary conditions above to be a leave (excess), then there is a packing (covering) with a leave (excess) of (V, E).*

It is believed that this conjecture is not harder than Wilson's Theorem. Theorem 1 provides a proof for $k = 3$, $n_0 = 3$.

Theorem 1. *For $k = 3$ and any v, $v \geqslant 3$, λ the only graphs which can be leaves or excesses are as in Table 1 with the following abbreviations and all are realizable.*

Graphs of odd degrees

1F			A 1-factor on $6t$ vertices
1FY			A 1-factor on $6t - 4$ vertices and a tree on 4 vertices with one vertex of degree 3
06	(a)	1FH	a 1-factor on $6t - 6$ vertices and a graph AB, BC, BD, DF, DG
	(b)	1F5	a 1-factor on $6t - 6$ vertices and a tree on 6 vertices with one vertex of degree 5
	(c)	1FYY	a 1-factor on $6t - 8$ vertices and two trees each on 4 vertices with one vertex of degree 3
	(e)	1F3	a 1-factor on $6t - 2$ vertices and a triple edge AB, AB, AB
	(f)	1F-0-	a 1-factor on $6t - 4$ vertices and a graph AB, BC, BC, CD

Graphs of even degrees

2			A double edge AB, AB
E4	(a)	Q	a quadrilateral AB, BC, CD, DA
	(b)	4	a quadruple edge AB, AB, AB, AB
	(c)	2^2	2 double edges AB, AB, CD, CD
	(d)	∞	AB, AB, BC, BC

		\multicolumn{6}{c}{$v \bmod 6$}
		0	1	2	3	4	5
	0	0;0	0;0	0;0	0;0	0;0	0;0
	1	1F;1F	0;0	1F;1FY	0;0	1FY;1FY	E4;2
$\lambda \bmod 6$	2	0;0	0;0	2;E4	0;0	0;0	2;E4
	3	1F;1F	0;0	06;06	0;0	1FY;1FY	0;0
	4	0;0	0;0	E4;2	0;0	0;0	E4;2
	5	1F;1F	0;0	1FY;1FY	0;0	1FY;1F	2;E4

TABLE 1

Proof. The papers [FH], [H1] provide all the entries in the table below—when a leave (excess) is unique or unique for $\lambda = 1$ it *must* be achieved. Some of the entries can be obtained by simply adding the blocks of two packings (coverings) on the same set together. We can of course take the packings or coverings independently in order to achieve the desired leave (excess). We shall use the notation $(\lambda \equiv 1) + (\lambda \equiv 4)$ for example, to mean take the blocks of a design with $\lambda \equiv 1 \pmod{6}$ and add them to the blocks of a design with $\lambda \equiv 4 \pmod{6}$.

		\multicolumn{6}{c}{$v \bmod 6$}
		0	1	2	3	4	5
$\lambda \bmod 6$	0	0;0	0;0	0;0	0;0	0;0	0;0
	1	1F;1F	0;0	1FY;1FY	0;0	1F;1F	Q;2
	2	0;0	0;0	2;	0;0	0;0	2;
	3	1F;1F	0;0		0;0	1FY;1FY	0;0
	4	0;0	0;0	;2	0;0	0;0	;2
	5	1F;1F	0;0		0;0	1FY;1FY	2;

TABLE 2

It is easy to see that the necessary conditions for the fifth column are

$$(0\,;\,0 \quad E4\,;\,2 \quad 2\,;\,E4 \quad 0\,;\,0 \quad E4\,;\,2 \quad 2\,;\,E4)^T,$$

and the necessary conditions on 2nd column are

$$(0\,;\,0 \quad 1F\,;\,1FY \quad 2\,;\,E4 \quad 06\,;\,06 \quad E4\,;\,2 \quad 1FY\,;\,1F)^T.$$

The only non-trivial calculation is $v \equiv 2 \pmod{6}$, $\lambda \equiv 3 \pmod{6}$; the necessary conditions force the leave (excess) to be a graph with $3t + 3$ edges and all degrees odd. It is easily seen that no degree can exceed 6 and the 5 graphs of 06 are the only possible ones.

Sufficiency of the 5th column
Packing

$\lambda \equiv 4 \pmod{6}$. For Q take $(\lambda \equiv 1) + (\lambda \equiv 3)$; for 4, 2^2, ∞ take two copies of $(\lambda \equiv 2)$.

$\lambda \equiv 1 \pmod{6}$, $\lambda > 1$. The rest of E4 is obtained by $(\lambda \equiv 3) + (\lambda \equiv 4)$.

Covering

$\lambda \equiv 2 \pmod{6}$. Q and ∞ can be obtained by adding 2 blocks to a packing. For 4, 2^2 we take $2 \times (\lambda \equiv 4) + (\lambda \equiv 3)$ for $\lambda > 2$. For $\lambda = 2$ we take two copies of a cover for $\lambda = 1$.

$\lambda \equiv 5 \pmod{6}$. We take $(\lambda \equiv 3) + (\lambda \equiv 2)$.

Sufficiency for the second column
Packing

$\lambda \equiv 3 \pmod{6}$. The graphs 1F-0-, and 1F3 can be obtained by $(\lambda \equiv 1) + (\lambda \equiv 2)$. To obtain the remaining three graphs we note that 3 1-factors on $v - 8$

points can be united to form a configuration which consists of disjoint copies G where G is a hexagon $abcdef$ and edges ae, bd, fc. Removing two triangles from G leaves the 1-factor fc, ab, ed. Let us take 3 copies of a packing with $\lambda = 1$ and ensure that $v-8$ have a leave in the form of this configuration. We add also to the packing the triangles afe, bdc for each copy. Let two of the leaves have union which consists of the 4-cycles $ABCD$, and $A'B'C'D'$ on the remaining points.

To obtain 1FYY, let the third factor be AD, BC, $A'B'$, $D'C'$ and add $B'C'D'$, BCD to the packing.

To obtain 1FH, let the third factor be AA', DD', CB, $C'B'$ and add $A'B'C'$, ABC to the packing.

To obtain 1F5, we start with 1FH (obtained from previous step) leave but ensure that $D'CB$ is a block. We then remove $D'CB$ and add DCB. (See Diagram 1).

For $\lambda \equiv 4 \pmod{6}$ (a) ∞, 2^2, 4 can all be obtained by doubling ($\lambda \equiv 2$).
(b) Q can be obtained by $(\lambda \equiv 1) + (\lambda \equiv 3)$.

For $\lambda \equiv 5 \pmod{6}$, take $(\lambda \equiv 2)$ and 1F-0-, to get a graph $AB\ BC\ BC\ CD\ BD\ BD$ + a one 1F on $6t-2$ points, add BCD to the packing to get a 1FY.

Covering

$\lambda \equiv 2 \pmod{6}$. Q, ∞ can be obtained by adding two blocks to a packing. For $\lambda > 2$; 4, 2^2 can be obtained by doubling a $(\lambda \equiv 4)$. This leaves the case $\lambda = 2$, and excesses 2^2, 4.

Excess 2^2. In this case we use a $(\lambda \equiv 1)$ nuclear design [MSS] and a $(\lambda \equiv 1)$ packing. It will be seen that a $(\lambda \equiv 1)$ nuclear design has a leave which consists for $v = 6t+2$ of t copies of the graph on $abcdef$ whose edges are ab, bc, ac, ad, be.

Let us look at $t-1$ of these. Take the 1-factor from the packing to have edges ad, be, cf. Thus adding the triangles abd, bec, acf gives an empty excess on $v-8$ points. For the remaining copy of the hexagon add the triangle abc to the cover. We make sure that on the 8 remaining points the union of the leaves is two 4-cycles $ABCD$, $EFGH$; we add ABC, ADC, EFG, FHG to get an excess of 2^2.

Excess 4. We handle this case as follows, first we will exhibit a packing for $v = 8$. We then note that by Stern and Lenz [SL] K_{6t}, $t > 1$ can be decomposed into $t-1$ orbits of triangles and 5 1-factors. This is also an easy exercise using Rosa's Skolem sequence technique. Thus $2K_{6t}$ can be decomposed into $2t-3$ orbits of triples and 16 1-factors. We can thus find a covering on $6t+8$ for $t > 1$ whose excess is 4. $v = 14$ we do separately.

$v = 8$: $V = \{0,1,2,3,4,5\} \cup \{A,B\}$
  $B = $ (a) iAB, $i = 0, \ldots, 5$
  (b) $A03$, $B03$, $A14$, $B14$, $A25$, $B25$
  (c) $0+i$, $1+i$, $2+i$, $i \in \mathbb{Z}_5$
  (d) 024, 135

$v = 14$: $V = \{0,1,2,3,4,5\} \cup \{0',1',2',3',4',5'\} \cup \{A,B\}$.

All Leaves and Excesses Are Realizable For k = 3 and All λ   227

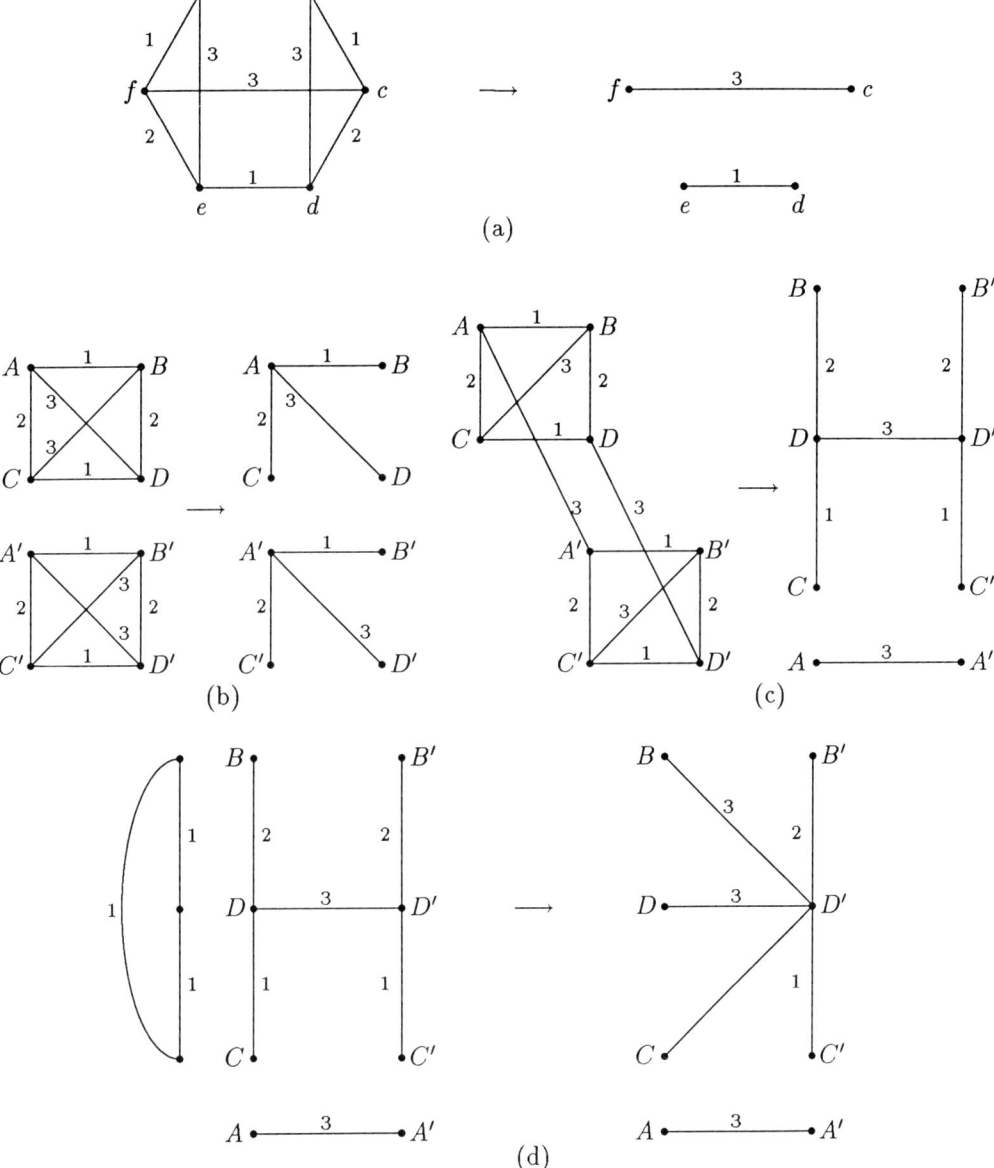

The number on the edge indicates which λ = 1 packing it comes from

DIAGRAM 1

a) Let F_1 through F_{10} be a 1-factorization of $2K_6$ on $\{0', 1', 2', 3', 4', 5'\}$. We form triples by attaching these 1-factors to the points 0, 1, 2, 3, 4, 5, A, B, A, B, respectively.

Add the triples

b) ABi, $i = 0, \ldots, 5$.
c) $[i, i+1, i+2]$, $i = 0, \ldots, 5$, and $[0,2,4]$, $[1,3,5]$.
d) $[A, i, i+3]$, $[B, i, i+3]$, $i = 0, 1, 2$.

$\lambda \equiv 3 \pmod{6}$. 1FY from $\lambda \equiv 1$, and Q from $\lambda \equiv 2$ with the removal of two triangles gives a 1FY, 1FH, 1FF. [We must be careful in coverings to make sure that if we wish to remove a triangle from an excess that the cover actually contains that triangle. It is easy to ensure that here]. A 1FY and ∞ will give the 1F; removing two triangles will give the 1F3 and 1F-0-.

$\lambda \equiv 5 \pmod{6}$. Take ($\lambda \equiv 3$) cover with excess 1FH and ($\lambda \equiv 2$) cover with excess ∞, and remove two triangles, to obtain a 1FY.

This completes the proof.
For $v = 0, 1, 2$ and all λ the results are obvious. □

References

[C] C. J. Colbourn, *Leaves, excesses, and neighborhoods*, Acta Univ. Car. **28** no. 2 (1987), 41–47.

[FH] M. K. Fort, Jr. and G. A. Hedlund, *Minimal coverings of pairs by triples*, Pacific J. Math. **8** (1958), 709–719.

[H1] H. Hanani, *Balanced incomplete block designs and related designs*, Discrete Math. **11** (1975), 225–369.

[LR] C. C. Lindner, A. Rosa, *Steiner triple systems having a prescribed number of triples in common*, Canad. J. Math. **27** (1975), 1166–1175.

[MR] E. Mendelsohn, A. Rosa, *Embedding maximal packings of triples*, Cong. Numerantium **40** (1983), 235–247.

[MSS] E. Mendelsohn, N. Shalaby, Shen Hao, *Nuclear designs*, to appear in Ars. Combinatoria.

[SL] G. Stern, H. Lenz, *Steiner triple systems with given subspaces; another proof of the Doyen-Wilson theorem*, Bull. Un. Mat. Ital. A(5) **17** (1980), 109–114.

[W] R. M. Wilson, *An existence theorem for pairwise balanced designs I, II, III*, J. Combinat. Theory (A) **13** (1972), 220–273; (A) **18** (1975), 71–79.

Eric Mendelsohn, Nabil Shalaby
Department of Mathematics,
University of Toronto, Toronto,
Ontario, Canada, M5A 1A7

Shen Hao
Department of Applied Mathematics,
Shanghai Jiao Tong University,
Shanghai 200030, P. R. China

The Binding Number of k-Trees

Danuta Michalak

The notion of k-trees for $k \geq 2$ has been introduced by Harary and Palmer [2]. This notion is a natural generalization of an ordinary tree (1-tree). The binding number was introduced by Woodall [6]. Kane, Mohanty and Straus [3] have proved general results about realizing sets for the binding number in the case if the binding number is less than or equal 1. Faragó [1] has proved that the problem of calculating the binding number in that case is polynomially solvable. We shall establish some special properties of realizing sets for the binding number of k-trees in the case if $\text{bind}(G) < 1$.

By a graph we mean a finite, undirected graph without multiple edges and denote it by $G = (V(G), E(G))$. For a vertex $v \in V(G)$ we denote by $\Gamma_G(v)$ the set of all vertices of G adjacent to v in G. For a subset $X \subseteq V(G)$ we define $\Gamma_G(X) = \bigcup_{v \in X} \Gamma_G(v)$ (shortly $\Gamma(X)$). Put $\mathcal{F}_G = \{X \subseteq V(G) \colon X \neq \emptyset \text{ and } \Gamma(X) \neq V(G)\}$.

The *binding number* of G denoted by $\text{bind}(G)$ is defined as follows

$$\text{bind}(G) = \min\{|\Gamma(X)|/|X| \colon X \in \mathcal{F}_G\}.$$

A set $X \in \mathcal{F}_G$ is called *a realizing set* for $\text{bind}(G)$ if $\text{bind}(G) = |\Gamma(X)|/|X|$.

Proposition 1 [3]. *Let G be a graph with n vertices.*
(a) *If $\text{bind}(G) < 1$, then every realizing set for $\text{bind}(G)$ is independent and $|\Gamma(X)| + |X| \leq n - 2$ or $= n$.*
(b) *If $\text{bind}(G) = 1$ and G is connected, then G has a realizing set for $\text{bind}(G)$ which is an independent set in G.*

A k-tree, $k \geq 1$, we define in the following way:
(i) The complete graph K_k is a k-tree.
(ii) Given a k-tree H of order n $(n \geq k)$, a k-tree of order $n+1$ is constructed by adding a new vertex adjacent to k mutually adjacent vertices of H.

Let us state some well-known properties of k-trees.
Let G be a k-tree on p vertices, $p \geq k+1$. Then
(1) G is triangulated and K_{k+2}-free.
(2) $\delta(G) = k$, and if $p \geq k+2$ then all vertices of degree k form an independent set.

A vertex of degree k is called *endvertex*.

A graph G is called *Hallian* if $|\Gamma(X)| \geqslant |X|$ for any set $X \subseteq V(G)$. This implies $\text{bind}(G) \geqslant 1$. We denote the class of non-Hallian k-trees by $\mathcal{H}$. Obviously, if $k \geqslant 2$ and $G \in \mathcal{H}$ then G has at least $k+2$ vertices.

By Propositon 1 for $G \in \mathcal{H}$ the realizing set for $\text{bind}(G)$ is an independent set of G.

Theorem 1. *Let G be a k-tree on $p \geqslant k+2$ vertices. If X is an independent set of vertices of G such that $\deg(v) \geqslant k+1$ for every $v \in X$, then $|\Gamma(X)| \geqslant |X| + k$.*

Proof. It is well-known (Rose [5]) that a graph is triangulated if and only if it has a *perfect elimination scheme* or PES, i.e., there exists an order Π of eliminating the vertices of G such that each vertex is simplicial at the time of its elimination.

Let $X = \{v_1, \ldots, v_t\}$ be an independent set of vertices and $\deg(v_i) \geqslant k+1$, $1 \leqslant i \leqslant t$. Let $\varphi(v_i)$ be the vertex of G such that eliminating $\varphi(v_i)$ in PES we obtain v_i as endvertex. Since X is independent, $\varphi(v_i) \neq \varphi(v_j)$ for $i \neq j$ and there is no vertex v_i, $1 \leqslant i \leqslant t$ such that $v_i = \varphi(v_j)$, $1 \leqslant j \leqslant t$.

Let $Y = \{\varphi(v_i) : 1 \leqslant i \leqslant t\}$. It is clear that $Y \subseteq \Gamma(X)$, $Y \cap X = \emptyset$ and this implies $|\Gamma(X)| \geqslant |X|$. Moreover, if we stopped the PES at the moment when the last vertex of X is obtained as endvertex, we notice that this vertex has k adjacent vertices in the obtained graph. Finally $|\Gamma(X)| \geqslant |X| + k$. □

Corollary 1.1. *If $G \in \mathcal{H}$, then each realizing set for $\text{bind}(G)$ contains endvertices.*

Now we present some general properties of realizing sets for $\text{bind}(G)$, when G is a non-Hallian graph.

Lemma 2. *Let G be a non-Hallian graph. If X is a realizing set for $\text{bind}(G)$ then for every partition of X, $X = (X_1, X_2)$, such that $Z = \Gamma(X_1) \cap \Gamma(X_2) \neq \emptyset$, the following inequality holds*

$$\frac{|\Gamma(X_2 \setminus Z)|}{|X_2|} \leqslant \frac{|\Gamma(X_1)|}{|X_1|}.$$

Proof. Let $\text{bind}(G) = |\Gamma(X)|/|X|$. Suppose that there exists a partition $X = (X_1, X_2)$ of X, such that $|\Gamma(X_2) \setminus Z|/|X_2| > |\Gamma(X_1)|/|X_1|$. This implies

$$\frac{|\Gamma(X)|}{|X|} = \frac{|\Gamma(X_1)| + |\Gamma(X_2) \setminus Z|}{|X_1| + |X_2|} > \frac{|\Gamma(X_1)|}{|X_1|},$$

a contradiction. □

Corollary 2.1. *If X is a realizing set for $\text{bind}(G)$ of a non-Hallian graph G, then for each vertex $v \in X$, $\Gamma(v) \subseteq \Gamma(X \setminus \{v\})$.*

Proof. Let $X_1 = X \setminus \{v\}$, $X_2 = \{v\}$. By Lemma 2, $|\Gamma(v) \setminus Z|/|\{v\}| \leqslant |\Gamma(X_1)|/|X_1|$. Since $\text{bind}(G) < 1$, either $|\Gamma(X_1)|/|X_1| < 1$ or $|\Gamma(v) \setminus Z|/|\{v\}| < 1$. In the both cases we have $\Gamma(v) \setminus Z = \emptyset$ and the proof is completed. □

The following property is obvious.

Lemma 3. *If X is a realizing set for $\mathrm{bind}(G)$ of a non-Hallian graph G then there is no vertex v such that $X \cup \{v\}$ is an independent set and $\Gamma(v) \subseteq \Gamma(X)$.*

If S is a largest independent set of G, i.e., $|S| = \mathcal{B}_0(G)$ then S is called a $\mathcal{B}_0$-set of G. In [4], Mitchell, Hedetniemi and Goodman have presented a linear time algorithm for finding a $\mathcal{B}_0$-set of a tree. We can easily extend this result to k-trees.

Lemma 4. *Let G be a k-tree on $p \geqslant k+2$ vertices. There exists a $\mathcal{B}_0$-set which contains all endvertices of G.*

Proof. Let S be a $\mathcal{B}_0$-set of G. Assume there exists an endvertex v and $v \notin S$. Let $w_1, \ldots, w_k$ be the vertices adjacent to v. It is obvious these vertices induce the complete graph. We have two possibilities:
(i) One of the vertices $w_1, \ldots, w_k$ belongs to S, say w_i. In this case $(S \setminus \{w_i\}) \cup \{v\}$ is also a $\mathcal{B}_0$-set of G.
(ii) The set S does not contain w_i, $1 \leqslant i \leqslant k$. In this case $S \cup \{v\}$ is an independent set, a contradiction. □

The next lemma is obvious.

Lemma 5. *Let G be a graph and G_1, G_2 be components of G. If G_1 is the complete graph K_p, $p \geqslant 1$ then $\mathcal{B}_0(G) = 1 + \mathcal{B}_0(G_2)$.*

Similarly as in [4] using lemmas 1 and 2 we can inductively prove the BETA-ZERO algorithm. Notice that our algorithm is almost the same algorithm as in [4].

Algorithm BETA-ZERO

Given a k-tree T on $p \geqslant k+2$ vertices. $\mathcal{B}_0$-set is denoted BOSET.
STEP 0. BOSET $:= \emptyset$, $G := T$ and label all vertices "independent".
STEP 1. If G has "independent" endvertex v adjacent to vertices $w_1, \ldots, w_k$ then BOSET := BOSET $\cup \{v\}$; $G := G - v$; label $w_1, \ldots, w_k$ "dependent"; and go to STEP 1.
STEP 2. If G has "dependent" endvertex u then $G := G - u$; and go to STEP 1.
STEP 3. At this point, G consists of a complete graph K_k. If G contains "independent" endvertex v then BOSET := BOSET $\cup \{v\}$ and STOP otherwise STOP.

Theorem 6. *Let $G \in \mathcal{H}$. Then each realizing set for $\mathrm{bind}(G)$ is a subset of a $\mathcal{B}_0$-set of G.*

Proof. Let X be a realizing set for $\mathrm{bind}(G)$. If X is a maximal independent set then it is easy to check that X is a $\mathcal{B}_0$-set of G. Hence suppose X is a proper subset of a maximal independent set and X is not a subset of a $\mathcal{B}_0$-set of G. By Corollary 1.1, X contains endvertices. We denote the set of all endvertices of X, by X_1. Let X' be the maximal subset of X being a subset of a $\mathcal{B}_0$-set of G. Since X_1 is a subset of a $\mathcal{B}_0$-set then $X' \neq \emptyset$. This implies that there exists a vertex v of $X \setminus X'$ such that v is adjacent to at least two non-adjacent vertices w, z and these vertices are not adjacent to the vertices of X'. Notice that each vertex u, $u \neq v$

of $X \setminus X'$ has at least one adjacent vertex which is not adjacent to vertices of X'. Putting $Z = \Gamma(X') \cap \Gamma(X \setminus X')$, we obtain $|\Gamma(X \setminus X') \setminus Z| \geq |X \setminus X'| + 1$.

Consequently, $|\Gamma(X \setminus X') \setminus Z|/|X \setminus X'| > 1$. Applying Lemma 2 to the partition $X = (X', X \setminus X')$ we obtain that X is not a realizing set, a contradiction. □

Theorem 7. *Let $G \in \mathcal{H}$. If X is a realizing set for* bind(G) *then there exists a $\mathcal{B}_0$-set containing all endvertices such that X is a subset of the $\mathcal{B}_0$-set.*

Proof. If X is a realizing set for bind(G) and X is a $\mathcal{B}_0$-set then by Lemma 4 the theorem is true. Assume X is a proper subset of a $\mathcal{B}_0$-set and suppose there is no $\mathcal{B}_0$-set containing all endvertices and such that X is a subset of this set. Hence, there exists an endvertex v such that $v \notin X$. By the properties of the largest independent set, there is a vertex $w \in X$ and adjacent to v. This implies $v \notin \Gamma(X \setminus \{w\})$.

On the other hand, we obtain $\Gamma(w) \subseteq \Gamma(X \setminus \{w\})$ (by Corollary 2.1), a contradiction. □

Let S be the $\mathcal{B}_0$-set of G determined by the BETA-ZERO algorithm. Let $v \in S$, $\deg(v) \geq k + 1$. We say that *vertex v has the exchange property* if there exists a vertex $w \in \Gamma(v)$ with the following property:

$$\text{for every } z \in \Gamma(w) \text{ and } z \neq v, z \notin S. \tag{$*$}$$

It is easy to notice that if v has the exchange property then there exists a $\mathcal{B}_0$-set S', $S' \neq S$ such that $w \in S'$ and obviously w has the exchange property.

Lemma 8. *Let $G \in \mathcal{H}$. If X is a realizing set for* bind(G) *then X contains no vertex with the exchange property.*

Proof. Assume that $v \in X$ and v has the exchange property. Let w be the vertex adjacent to v satisfying the property $(*)$. This implies that $w \notin \Gamma(X \setminus \{v\})$ i.e., $\Gamma(v) \not\subseteq \Gamma(X \setminus \{v\})$. By Corollary 2.1 X is not the realizing set, a contradiction. □

By the theorems 6, 7 and Lemma 8 we obtain the next property of realizing sets for graphs of the class $\mathcal{H}$.

Theorem 9. *Let $G \in \mathcal{H}$. If X is a realizing set for* bind(G) *then X is a subset of a $\mathcal{B}_0$-set determined by the BETA-ZERO algorithm and X contains no vertex with the exchange property.*

By all described above properties of realizing sets for bind(G) where $G \in \mathcal{H}$, we can calculate bind(G) in the following way.

I. Determine $\mathcal{B}_0$-set S.
II. Find all vertices of S with the exchange property and denote the set of these vertices by A.
III. Find all subsets X of $S \setminus A$ satisfying the following properties:
    X contains endvertices and for any $v \in X$, $\Gamma(v) \subseteq \Gamma(X \setminus \{v\})$ and there is no vertex $w \in ((S \setminus A) \setminus X)$ such that $\Gamma(w) \subseteq \Gamma(X)$.
IV. Calculate $|\Gamma(X)|/|X|$ for each set X found in step III.

References

[1] A. Faragó, *F-Independence Number of Graphs*, Proc. Combinatorics, 7th Hung. Colloq. Eger, 1987, pp. 221–226.

[2] F. Harary, E. M. Palmer, *On Acyclic Simplicial Complexes*, Mathematika **15** (1968), 115–122.

[3] V. G. Kane, S. P. Mohanty, E. G. Straus, *Which Rational Numbers are Binding Numbers?*, Graph Theory **5** (1981), 379–384.

[4] S. Mitchell, S. Hedetniemi, S. Goodman, *Some Linear Algorithms on Trees*, Proc. 6th S–E Conf. Combinatorics, Graph and Computing, pp. 467–483.

[5] D. J. Rose, *Triangulated Graphs and The Elimination Process*, Math. Anal. Appl. **32** (1970), 597–609.

[6] D. R. Woodall, *The Binding Number of a Graph and Its Anderson Number*, J. Comb. Theory (B) **15** (1973), 225–255.

Danuta Michalak
Institute of Mathematics,
Higher College of Engineering,
Podgorna 50, 65–246 Zielona Góra,
Poland

An Extension of Brook's Theorem

Peter Mihók

0. The well-known Brooks' theorem can be formulated as follows

Theorem 1. (Brooks 1941 [1]) *If a connected graph G does not contain $K_{1,k+1}$, $k \geq 3$, then G is k-colorable unless $G = K_{k+1}$.*

The aim of this note is to prove the following extension of Brooks' theorem

Theorem 2. *Let T be a tree on $k+2$ vertices, $k \geq 3$. If a connected graph G does not contain T, then G is k-colorable unless $G = K_{k+1}$.*

1. We consider finite undirected graphs without loops and multiple edges. In general, we follow the notation and terminology of [2]. The vertex set of a graph G is denoted by $V(G)$, the set of vertices adjacent to a vertex v by $N(v)$ and the symbol $\bar{N}(v)$ denotes the set $N(v) \cup \{v\}$. We say that G *contains* H if G has a subgraph isomorphic to H. By a *path* P *in* G we mean also the subgraph of G consisting of vertices and edges of a path P in G. Let $P: x_0 x_1 \ldots x_t$, $t \geq 1$, be a path in a tree T, we say that *a path* $Q: y_0 y_1 \ldots y_t$ *in a graph* G *is extendible to* T *in* G, if G has a subgraph S ($Q \subseteq S$) isomorphic to T such that there is an isomorphism between T and S in which x_i correspond to y_i, for $i = 0, 1, \ldots, t$. For $k \geq 0$, a graph G is called k-*degenerate* (see [3], [4]) if every subgraph H of G has the minimum degree $\delta(H) \leq k$. Evidently every k-degenerate graph G is $k+1$-colorable. It is easy to prove

Lemma 1. *If the minimum degree of G, $\delta(G) \geq k$, $k \geq 1$, then G contains every tree T on $k+1$ vertices.*

Moreover, it is also easy to see that

Lemma 2. *Let $P: x_0 x_1 \ldots x_t$ be a path in a tree T on $k+1$ vertices and $\delta(G) \geq k$. Then each path $Q: y_0 y_1 \ldots y_t$ of G is extendible to T in G.*

2. Our object in this note is to considerably strengthen Lemma 1. We are going to prove

Lemma 3. *Let G be a connected graph with $\delta(G) \geq k$, $k \geq 1$, other than K_{k+1}. Then G contains every tree T on $k+2$ vertices, except for $T = K_{1,k+1}$ if G is k-regular.*

235

Proof. Obviously, if G is not k-regular and $\delta(G) \geq k$, then G contains $K_{1,k+1}$. So, let $T \neq K_{1,k+1}$ be any tree on $k+2$ vertices. We have to prove that G contains T. We can suppose that G does not contain K_{k+1} (otherwise G obviously contains T). Let $P\colon x_0 x_1 \ldots x_{t-1} x_t$ be a longest path in T. Let us denote by T' the subtree of T obtained by removing the endvertices x_0 and x_t from T. By lemma 2 any path $Q\colon y_0 y_1 \ldots y_{t-1}$ in G is extendible to T' in G. Let us consider under which conditions is Q extendible to T in G. If H (isomorphic to T') is the extension of Q to T' in G, then to obtain T (extending H in G) it is sufficient to find two different vertices y_0 and y_t, adjacent to y_1 and y_{t-1} in G respectively, such that both $y_0, y_t \notin V(H)$. It is easy to see that such vertices y_0 and y_t do not exist in G only if $\deg y_1 = k$ and $\overline{N(y_1)} = \overline{N(y_{t-1})}$. Let us denote by S the set $\overline{N(y_1)}$. It holds $|S| = k+1$, $S \supsetneq V(H) \supseteq V(Q)$ and since G does not contain K_{k+1}, in the subgraph of G induced by S there exist two nonadjacent vertices z_1 and z_2. Since $\deg z_1 \geq k$ in G, z_1 is adjacent to at least one vertex w of G such that $w \notin \overline{N(y_1)}$. If $z_1 \in V(Q)$ i.e. $z_1 = y_i$ for some $i = 2, 3, \ldots, t-2$ let us consider the path $Q'\colon z_1 = y_i y_{i+1} \ldots y_{t-1} y_{i-1} \ldots y_2 y_1$. (If $z_2 \in V(Q)$ we can proceed the same way as for z_1). Finally, if both $z_1, z_2 \notin V(Q)$, then let us consider the path $Q''\colon z_1 y_{t-1} \ldots y_3 y_1$. In both cases, since $\overline{N(z_1)} \neq \overline{N(y_1)}$ the paths Q' and Q'' are extendible to T in G. The lemma is thus proved. □

3. Proof of Theorem 2

Let T be a tree on $k+2$ vertices and assume the connected graph $G \neq K_{k+1}$ does not contain T. If $T = K_{1,k+1}$, then according to Brooks' theorem G is k-colorable. If $T \neq K_{1,k+1}$, then by Lemma 3, G must be $(k-1)$-degenerate (otherwise G contains T) and thus G is k-colorable, too. □

References

[1] R. L. Brooks, *On colouring the nodes of the network*, Proc. Cambridge Phil. Soc. **37** (1941), 194–197.

[2] F. Harary, *Graph theory*, Addison-Wesley, Reading, Mass., 1969.

[3] D. R. Lick, A. T. White, *k-degenerate graphs*, Canad. J. Math. **22** (1970), 1082–1096.

[4] J. Mitchem, *An extension of Brooks' theorem to n-degenerate graphs*, Discrete Math. **17** (1977), 291–198.

Peter Mihók
P. J. Šafárik University,
Košice
Czechoslovakia

On Sectors in a Connected Graph

LADISLAV NEBESKÝ

Roughly speaking, the present note concerns travelling through a shortest path, namely in the case when the aim was unexpectedly removed.

0. Let G be a connected graph (in the sense of [1]) with vertex set $V(G)$ and edge set $E(G)$. If $u_1, u_2 \in V(G)$, then we denote by $d_G(u_1, u_2)$ the distance between u_1 and u_2 in G. Let $v_1, v_2 \in V(G)$, and let P be a $v_1 - v_2$ path in G; we say that P is a shortest $v_1 - v_2$ path in G if the length of P equals $d_G(v_1, v_2)$. Let $w_1, w_2 \in V(G)$; by the interval $I_G(w_1, w_2)$ we mean the set of all $w \in V(G)$ with the property that w belongs to a shortest $w_1 - w_2$ path in G; moreover, we denote by $N_G(w_1, w_2)$ the set of all vertices w' such that w' is adjacent to w_1 and belongs to $I_G(w_1, w_2)$.

Let G be a connected graph, and let $u, v \in V(G)$. By the sector $S_G(u, v)$ we shall mean the set

$$\{u\} \cup \{t \in V(G - u);\, N_G(t, u) - N_G(t, v) \neq \emptyset\}.$$

It is not difficult to show that the sector $S_G(u, v)$ is the set of all $t \in V(G)$ with the property that there exists a shortest $t - u$ path P in G such that t is the only vertex on P', for each shortest $t - v$ path P' in G. Clearly, $I_G(u, v) \subseteq S_G(u, v)$. It is obvious that if $u = v$, then $|S_G(u, v)| = 1$. Instead of $I_G(u, v)$ and $S_G(u, v)$ we shall write $I(u, v)$ and $S(u, v)$, respectively.

Intervals in connected graphs were intensively studied by Mulder [2]. Sectors were introduced by the author in [4], but not under this name; in that paper the functions $S(u, v) \cap S(v, u)$ and $S(u, v) \cup S(v, u)$, for $u, v \in V(G)$, were studied. Note that the author's interest in sectors has its origin in his research in semiotics; cf. [3].

1. We shall now compare intervals and sectors.

Lemma. *Let G be a connected graph with an odd cycle. Then there exist distinct $u, v, w \in V(G)$ such that*

$$I(u, w) \subseteq S(v, w) \cap S(w, v) \text{ and } I(v, w) \subseteq S(u, w) \cap S(w, u). \qquad (*)$$

Proof (outlined). We denote by m the minimum integer n with the property that there exists a cycle of length $2n + 1$ in G. There exist distinct $u, v, w \in V(G)$ such that $uv \in E(G)$, $d_G(u,w) = m = d_G(v,w)$ and $I(u,w) \cap I(v,w) = \{w\}$. It is not difficult to show that (∗) holds. □

Proposition 1. *Let G be a connected graph. Then the following statements are equivalent:*

(i) *G is bipartite;*

(ii) *$I(u,v) \subseteq S(u,w)$ and $I(u,w) \subseteq S(u,v)$ if and only if $v = w$, for any $u, v, w \in V(G)$.*

Proof (outlined). Let (i) hold. Consider arbitrary $u, v, w \in V(G)$. Obviously, if $v = w$, then $I(u,v) \subseteq S(u,w)$ and $I(u,w) \subseteq S(u,v)$. Let $v \neq w$. Then either $v \notin I(u,w)$ or $w \notin I(u,v)$. Without loss of generality we assume that $v \notin I(u,w)$.

We first assume that for every $v' \in I(u,v) - \{u\}$, there exists $u' \in N_G(v', u)$ such that $v' \in N_G(u', w)$. Then we can conclude that $v \in I(u, w)$, which is a contradiction.

We now assume that there exists $v_0 \in I(u,v) - \{u\}$ such that $v_0 \notin N_G(u_0, w)$, for each $u_0 \in N_G(v_0, u)$. Since G is bipartite, it is not difficult to show that $N_G(v_0, u) \subseteq N_G(v_0, w)$, and thus $v_0 \notin S(u, w)$. Therefore, (ii) holds.

Conversely, according to our lemma, (ii) ⇒ (i). □

Proposition 2. *Let G be a connected graph. Then the following statements are equivalent:*

(i) *G is a tree;*

(ii) *$I(u,w) \subseteq S(v,w)$ and $I(v,w) \subseteq S(u,w)$ if and only if $u = v$, for any $u, v, w \in V(G)$.*

Proof (outlined). Clearly, (i) ⇒(ii). As follows from our lemma, if (ii) holds then G contains no odd cycle. Assume that G is bipartite but it is not a tree. Let m denote the minimum integer n with the property that there exists a cycle of length $2n + 2$ in G. There exist $u, v, w \in V(G)$ such that $d_G(u, w) = m = d_G(v, w)$, $d_G(u,v) = 2$, $I(u,w) \cap I(v,w) = \{w\}$ and if $t \in I(u,v) - \{w\}$, then $t \notin I(u,w) \cup I(v,w)$. It is not difficult to prove that $I(u,w) \subseteq S(v,w)$ and $I(v,w) \subseteq S(u,w)$. Thus, (ii) ⇒ (i), which completes the proof. □

2. Let G be a connected graph. We say that G is distinguishing if $v \neq w \Rightarrow S(u,v) \neq S(u,w)$, for any $u, v, w \in V(G)$. Similarly, we say that G is distinguishing from the left if $\bar{u} \neq \bar{v} \Rightarrow S(\bar{u}, \bar{w}) \neq S(\bar{v}, \bar{w})$, for any $\bar{u}, \bar{v}, \bar{w} \in V(G)$. As follows from Proposition 1, every connected bipartite graph is distingushing. As follows from Proposition 2, every tree is distinguishing from the left. But $K(m, n)$, $2 \leqslant m \leqslant n$, is not distinguishing from the left. Petersen graph is both distinguishing and distinguishing from the left.

Let F and H be vertex-disjoint graphs. We denote by $F + H$ the join in the sense of [1]. If H is trivial, then instead of $G + H$ we shall write $G + o$. Especially, if G is a cycle of length n, then we say that $G + o$ is an n-wheel. It is not difficult to show that n-wheel is distinguishing if and only if it is distinguishing from the left if and only if $\geqslant 8$. If G is a Petersen graph, then $G + o$ is distinguishing but not distinguishing from the left.

If G is a graph and $U \subseteq V(G)$, then $\langle U \rangle_G$ denotes the subgraph of G induced by U.

Proposition 3. *Let G be a connected graph. Assume that there exists a partition Π of $V(G)$ such that*
  (i) $|\Pi| \geqslant 2$;
  (ii) $E(\langle U \cup W \rangle_G) = E(\langle U \rangle_G) \cup E(\langle W \rangle_G)$ *or* $E(\langle U \rangle_G + \langle W \rangle_G)$, *for any distinct U, $W \in \Pi$*;
  (iii) $|V| \geqslant 2$ *and $\langle V \rangle_G + o$ is distinguishing, for each $V \in \Pi$. Then G is distinguishing.*

Proof (outlined). Consider arbitrary $u, v, w \in V(G)$ such that $v \neq w$. We denote by U_v and U_w the elements of Π such that $v \in U_v$ and $w \in U_w$. We distinguish two cases:

1. $U_v = U_w$. Denote $H = \langle U_v \rangle_G + o$. Let t denote the only vertex of H not belonging to U_v. We put $u' = u$ if $u \in U_v$ and $u' = t$ if $u \notin U_v$. It is not difficult to see that $S(u,v) \cap U_v = S_H(u',v) \cap U_v$ and $S(u,w) \cap U_v = S_H(u',w) \cap U_v$. Since H is distinguishing, we have $S(u,v) \neq S(u,w)$.

2. $U_v \neq U_w$. Without loss of generality, let $u \notin U_w$. Then

$$S(u,v) \cap U_w = \emptyset \text{ or } U_w,$$
$$S(u,w) \cap U_w = \{s \in U_w;\ d_G(s,w) \leqslant 1\}.$$

If $U_w \subseteq S(u,w)$, then it is not difficult to see that $\langle U_w \rangle_G + o$ is not distinguishing, which is a contradiction. If $U_w - S(u,w) \neq \emptyset$, then we can see that $S(u,v) \neq S(u,w)$.

Therefore, G is distinguishing. □

If F and H are vertex-disjoint graphs, than we denote by $F \times H$ their cartesian product.

Proposition 4. *Let F and H be connected vertex-disjoint graphs. If both F and H are distinguishing, then $F \times H$ is also distinguishing.*

Proof (outlined). Assume that both F and H are distinguishing. Denote $G = F \times H$. The case when G is trivial is obvious. Let $|V(G)| \geqslant 2$. Consider arbitrary $u, v, w \in V(F)$ and arbitrary $\bar{u}, \bar{v}, \bar{w} \in V(H)$ such that $(v, \bar{v}) \neq (w, \bar{w})$. Without loss of generality we assume that $\bar{v} \neq \bar{w}$. We denote by H_u the subgraph of G induced by $\{(u,t);\ t \in V(H)\}$. Obviously, H_u is a copy of H. It is not difficult to prove that

$$S\big((u,s),(r,t)\big) \cap V(H_u) = S_{H_u}\big((u,s),(u,t)\big)$$

for any $r \in V(F)$ and any $s, t \in V(H)$. Since H_u is distinguishing, $S_{H_u}\big((u,\bar{u}), (u,\bar{v})\big) \neq S_{H_u}\big((u,\bar{u}),(u,\bar{w})\big)$. Thus G is also distinguishing. □

Note that neither K_3 nor $K_3 \times K_2$ are distinguishing.

Question. Let G be a connected graph. Is it true or not that if G is distinguishing from the left, then it is distinguishing?

References

[1] M. Behzad, G. Chartrand, L. Lesniak-Foster, *Graphs & Digraphs*, Prindle, Weber & Schmidt, Boston, 1979.
[2] H. M. Mulder, *The Interval Function of a graph*, Mathematisch Centrum, Amsterdam, 1980.
[3] L. Nebeský, *Signs and environment*, Slovo a slovesnost **50** (1989), 109–113. (Czech)
[4] _____, *On certain extensions of intervals in graphs*, Čas. pěst. mat. **115** (1990), 171–177.

Ladislav Nebeský
Faculty of Philosophy,
Charles University,
nám. J. Palacha 2,
116 38 Praha 1,
Czechoslovakia

Fourth Czechoslovakian Symposium on
Combinatorics, Graphs and Complexity
J. Nešetřil and M. Fiedler (Editors)
© 1992 Elsevier Science Publishers B.V. All rights reserved.

Irreconstructability of Finite Undirected Graphs from Large Subgraphs

Václav Nýdl

All graphs considered are finite, simple and undirected. For a real q and an integer n we denote by n_q the integral part of the product $q \cdot n$.

Müller in [2] proved that for every real q, $\frac{1}{2} < q < 1$, there exists an integer N such that for every $n > N$ most graphs with n vertices can be uniquely reconstructed from n_q-vertex subgraphs. We show that in Müller's result, the word "most" cannot be replaced by "all." More precisely, for every real $q < 1$ and every integer N we exhibit two non-isomorphic graphs with $n > N$ vertices having the same collections of n_q-vertex subgraphs. The main construction of this paper is a more transparent modification of the proof given by the author in [4].

1. Preliminaries

If $G = (V, E)$ is a finite, simple, undirected graph with the set of vertices V and the set of edges E then for every $W \subset V$ we denote by G/W the induced subgraph with the set of vertices W. We use the symbol $\simeq$ for the isomorphism of graphs. The set $\{1, 2, \ldots, r\}$ will be denoted by I_r. We denote by $\mathrm{ind}_m(G)$ the set of all induced subgraphs of G having exactly m vertices.

Definition 1.1. Two graphs G_1, G_2 are called m-indistinguishable if there exists a bijection $f\colon \mathrm{ind}_m(G_1) \to \mathrm{ind}_m(G_2)$ such that $H \simeq f(H)$ for every $H \in \mathrm{ind}_m(G_1)$.

In [3] we proved the following theorem.

Theorem 1.2. Let m be an integer. For $u = 1, 2$ let $G_u = (V_u, E_u)$ be a graph and let us denote by P_u^m the set of all subsets Y of V_u such that Y has $\leqslant m$ elements and G_u/Y is connected. Then G_1, G_2 are m-indistinguishable iff there exists a bijection $g\colon P_1^m \to P_2^m$ such that $G_1/Y \simeq G_2/g(Y)$ for every $Y \in P_1^m$.

Further, we use a special construction described in [1], [5].

Construction 1.3. Let r be an integer and $\mathfrak{A}_r$ be the group of all even permutations on I_r. We find a graph $A_r = (X_r, Q_r)$ (called an extension of $\mathfrak{A}_r$) such that $I_r \subset X_r$ and
(I) for every $\alpha \in \mathfrak{A}_r$ there exists exactly one automorphism α^+ of the graph A_r satisfying $\alpha^+(x) = \alpha(x)$ for every $x \in I_r$,
(II) A_r has no other automorphisms than those of the form α^+,

(III) for every $x \in X_r$ there is $\deg(x) > 1$ in the graph A_r.

2. Main construction

We describe graphs $G(r, k, \pi)$ derived from the graph A_r by adding r disjoint branches emanating from the vertices $1, 2, \ldots, r$ and ending by clusters of pendant vertices (see Fig. 1).

Construction 2.1. Let $r > 1$, k be integers and let $\pi = (\pi_i)_{i=1}^r$ be a sequence of integers. Let $T_i = \{T_{i,j}; j = 1, \ldots, \pi_i\}$ be a set with π_i elements. Let $T_1, T_2, \ldots, T_r$ be pairwise disjoint and denote by T their union. Let $Z = I_r \times I_k = \{(i, j); i \in I_r, j \in I_k\}$, let X_r, Z, T be pairwise disjoint sets. Put $V = X_r \cup Z \cup T$.

Further, we define for all $i \in I_r$ the sets of edges $S_i = \{((i, k), t_{i,j}); j = 1, \ldots, \pi_i\}$ and denote by S their union. Also for all $i \in I_r$ let

$$R_i = \{(i, (i, 1))\} \cup \{((i, j), (i, j+1)); 1 \leqslant j < k\}$$

and denote by R their union. Put $E = Q_r \cup S \cup R$.

Finally, put $G(r, k, \pi) = (V, E)$ (see Fig. 1).

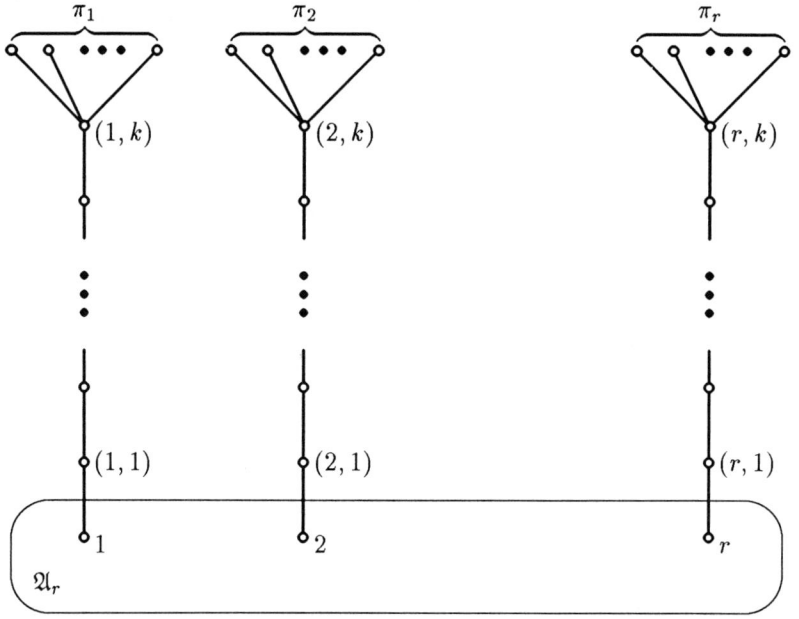

FIGURE 1

We will mention some properties of the graphs $G(r, k, \pi)$.

Lemma 2.2. Let r, k be integers, π, π' be sequences of r integers. Let $\varepsilon\colon G(r, k, \pi) \to G(r, k, \pi')$ be an isomorphism. Then
(a) $\varepsilon(I_r) = I_r$ and $\varepsilon\colon I_r \to I_r$ is an even permutation,

(b) $\pi_i = \pi'_{\varepsilon(i)}$ for every $i \in I_r$.

Proof. We successively get $\varepsilon(T) = T'$ (since T and T' contain exactly all the vertices of degree 1), $\varepsilon(Z) = Z$, $\varepsilon(X_r) = X_r$ and (use (II) from 1.3.) $\varepsilon(I_r) = I_r$. Then $\varepsilon(T_i) = T'_{\varepsilon(i)}$ for $i \in I_r$. □

Definition 2.3. Let r, k be integers, π, π' sequences of r integers and α an even permutation on I_r such that $\pi_i = \pi'_{\alpha(I)}$ for every $i \in I_r$. Using the sets V, V' described for r, k, π, π' in 2.1. we define the mapping $\hat{\alpha} : V \to V'$ as follows: $\hat{\alpha}(x) = \alpha^+(x)$ for $x \in X_r$; $\hat{\alpha}((i,j)) = (\alpha(i), j)$ for $i \in I_r$ and $j \in I_k$; $\hat{\alpha}(t_{i,j}) = t_{\alpha(i),j}$ for $i \in I_r$ and $1 \leq j \leq \pi_i$.

Lemma 2.4. *Every mapping $\hat{\alpha}$ defined in 2.3. is an isomorphism from $G(r, k, \pi)$ onto $G(r, k, \pi')$.*

3. Main result

Throughout this part we will use two special sequences π, π' of $r > 3$ integers defined as $\pi_i = i$ for $i \in I_r$, $\pi'_1 = 2$, $\pi'_2 = 1$, $\pi'_i = i$ for $i > 2$ and two special graphs $G = (V, E) = G(r, k, \pi)$, $G' = (V', E') = G(r, k, \pi')$.

Definition 3.1. The set $J \subset I_r$ is called d-set if it contains at least two integers greater than 2 and p_j, q_j will denote the first two such integers in J. For every d-set J on I_r we define an even permutation $\alpha_J : I_r \to I_r$ as $\alpha_J(1) = 2$, $\alpha_J(2) = 1$, $\alpha_J(p_J) = q_J$, $\alpha_J(q_J) = p_J$, $\alpha_J(i) = i$ otherwise.

Definition 3.2. For a d-set $J \subset I_r$ we define reduced sequences $\sigma = \text{red}_J \pi$, $\sigma' = \text{red}_J \pi'$ as $\sigma_{p_J} = \sigma_{q_J} = \sigma'_{p_J} = \sigma'_{q_J} = 1$, $\sigma_i = \pi_i$, $\sigma'_i = \pi'_i$ otherwise. According to 2.3. we define two isomorphisms β_J, β'_J as $\beta_J = \hat{\alpha}_J : G(r, k, \sigma) \to G(r, k, \sigma')$, $\beta'_J = \hat{\alpha}_J : G(r, k, \sigma') \to G(r, k, \sigma)$.

Definition 3.3. For every $Y \subset V$ we define $\text{dis}(Y) = \{i \in I_r ; Y \cap T_i = \emptyset\}$.

Lemma 3.4. *Let $Y \subset V$ be such that $J = \text{dis}(Y)$ is a d-set. Using 3.2. we define the set $g(Y) = \beta_J(Y) \subset V'$. Analogously, for $Y' \subset V'$ and $J = \text{dis}(Y')$ a d-set, we define $g'(Y') = \beta'_J(Y') \subset V$. Then*
(a) $G/Y \simeq G'/g(Y)$, $G'/Y' \simeq G/g'(Y')$,
(b) $g'(g(Y)) = Y$, $g(g'(Y')) = Y'$.
Proof. The properties (a), (b) are direct consequences of the definitions. □

Main theorem. *Let N be an integer, q a real, $0 < q < 1$. Then there exist two non-isomorphic graphs on $n > N$ vertices which are m-indistinguishable for every $m \leq qn$.*

Proof. Take an integer $r \geq \dfrac{5}{1-q}$, put $s = \dfrac{r(r+1)}{2}$, then $k = \max(s, N, a_r)$, where a_r is the number of vertices of A_r. Now, $G_1 = G = G(r, k, \pi)$, $G_2 = G' = G(r, k, \pi')$ are non-isomorphic (according to 2.2.) and have $n = a_r + rk + s$ elements with $N < n \leq (r+2)k$. To show m-indistinguishability of G_1, G_2 we apply theorem 1.2. for g from lemma 3.4. with property (a) (and g' from 3.4. and property (b) to prove that g is a bijection). It only remains to prove that for every $Y \in P_1^m$ ($Y' \in P_2^m$, respectively) dis(Y) (dis(Y'), respectively) has to be a d-set.

Suppose $Y \in P_1^m$ and $J = \mathrm{dis}(Y)$ is not a d-set. Then J has less than 4 elements and for every two $i, j \notin J$ all the elements (i, h), (j, h) for $h \in I_k$ are contained in Y (since G_1/Y is connected). Finally, $\dfrac{m}{n} > \dfrac{(r-3)k}{(r+2)k} = 1 - \dfrac{5}{r+2} > q$, which is a contradiction. $\square$

References

[1] Z. Hedrlín, A. Pultr, *Symmetric relations (undirected graphs) with given semigroups*, Monatsh. Math. **69** (1965), 318–322.

[2] V. Müller, *Probabilistic reconstruction from subgraphs*, Comment. Math. Univ. Carolinae **17** (1976), 709–719.

[3] V. Nýdl, *Some results concerning reconstruction conjecture*, Proceedings of the 12th Winter School on Abstract Analysis, Suppl. ai Rendiconti del Circolo Math. di Palermo, 1984, pp. 234–246.

[4] _____, *Finite undirected graphs which are not reconstructible from their large cardinality subgraphs*, to appear.

[5] A. Pultr, V. Trnková, *Combinatorial, Algebraic and Topological Representations of Groups, Semigroups and Categories*, North-Holland, 1980.

Václav Nýdl
AF VŠZ, Sinkuleho 13,
370 05 Č. Budějovice, Czechoslovakia

Fourth Czechoslovakian Symposium on
Combinatorics, Graphs and Complexity
J. Nešetřil and M. Fiedler (Editors)
© 1992 Elsevier Science Publishers B.V. All rights reserved.

On Inefficient Proofs of Existence and Complexity Classes

CHRISTOS H. PAPADIMITRIOU

Recently a complexity theory of non-constructive existence proofs has been developed by the author and others, categorizing and understanding problems for which a mathematical proof guarantees that a solution exists, but no efficient algorithm for finding it is known. This paper is basically a review of this active field, although the last part on probabilistic methods does not appear anywhere else.

In Complexity Theory we usually study *languages*, or equivalently computational problems whose answer is either "yes" or "no." In this paper we focus on *search problems*, that is, problems in which an output more elaborate than "yes" or "no" is sought. Such problems seek the computation of partial multivalued functions, that is, relations $R \subset \Sigma^* \times \Sigma^*$. (We assume throughout that relation R is polynomially computable and balanced, that is, if $(x, y) \in R$ then $|y| \leq p(|x|)$ for some polynomial p depending on R.) FP is the class of all search problems that can be solved in polynomial time; that is, there is a polynomial algorithm which, given x either finds a y such that $R(x, y)$ (or decides that no such y exists. In the latter case we say that $(x, \mathbf{no}) \in R$ where $\mathbf{no}$ is a special symbol; thus, we shall henceforth assume that all partial functions in FP are *total*). On the other hand, FNP is the class of search problems R such that there is a polynomial algorithm deciding whether $(x, y) \in R$. Here partiality is an important ingredient of the complexity of this class.

FP and FNP are traditionally studied in terms of their surrogates P and NP. For example, whether FP = FNP is exactly the famous P = NP question. There are certain aspects of the issue, however, that cannot be easily captured by languages and P or NP. Suppose for example that we wish to study the class of *total* multivalued functions R in FNP, that is, all R in FNP such that for all x there is a y with $(x, y) \in R$. We call the class of all such search problems that always have a solution TFNP [MP]. Obviously, FP $\subseteq$ TFNP $\subseteq$ FNP, and no proof of a proper inclusion is in sight since it would establish that P $\neq$ NP. Is TFNP = FP? Is it

---

Research supported by the ESPRIT Basic Research Action No. 3075 ALCOM, a grant from the Volkswagen Foundation to the Universities of Patras and Bonn, and by the NSF.

true, that is, that it is easy to find a solution when you know that there is always one? We feel this is an important question for Complexity Theory and Mathematics alike.

TFNP is of interest to mathematicians, because every natural problem in TFNP must come with a proof of existence.[1] For example, the fundamental theorem of arithmetic together with Pratt's certification scheme for primes [Pr] establish that factoring (into certified prime divisors) is a problem in TFNP—this may bother those who hoped for a quick proof that TFNP = FP.

But there are far simpler problems in TFNP, whose membership in FP is unclear. In [JPY] we defined a broad class of such problems, namely PLS (for *polynomial local search*). A typical problem in this class is the following: Given n cities and a distance matrix, find a traveling salesman tour that cannot be improved by replacing two edges. In general, a problem A in PLS is defined in terms of two polynomial algorithms computing the functions c (assigning a cost to a solution), and f (mapping a solution either to itself, or to one with a better cost) for an input x, where $S = \Sigma^{|x|}$ is the set of all solutions. We wish to find a solution mapped to itself by f, which must exist. Notice that totality for functions in PLS is established by invoking the following "lemma:"

Every finite directed acyclic graph has a sink.

The dag for invoking the "lemma" is the graph of f (without the self-loops). It is a dag because of the cost-improving property. In other words, we can view f and c as an *implicit syntactic way* for specifying an exponentially large dag. Class PLS contains a host of problems that are not known to be in FP. Several important examples are known to be PLS-complete [JPY, PSY, Kre].[2]

There are several other interesting subclasses of TFNP, besides PLS. Like PLS, we can think of these classes as containing total functions whose proof of totality is based on a simple "graph-theoretic lemma" like the sink-in-a-dag one above. For example, one class called PLF (for *polynomial leaf*) is based on the "parity argument:"

Any finite graph has an even number of nodes of odd degree.

Perhaps the most famous such problem is finding a *Brouwer fixpoint* of a continuous function from the unit simplex to itself [HV]; it is in PLF because the modern proof of Brouwer's Theorem employs Sperner's Lemma, and thus, indirectly, the parity argument. Other typical problems in this class, for which no polynomial algorithm is known, are these: Given an odd-degree graph, and a Hamilton cycle in it, find another (it must exist by Smith's Theorem [Th]); given two disjoint Hamilton cycles, find another decomposition of their union into two Hamilton cycles (it exists by a theorem in [Th]); given a Hamilton path in a digraph, find either another one, or

---

[1] In fact, S. Poljak, P. Pudlák, and D. Turzik [PPT] many years ago defined the notion of *purely constructive* problems, which captures exactly our present concept of "rigorously total functions," and includes all problems and classes defined below.

[2] Recall that a problem is complete for complexity class C if it is in C, and all other problems in C reduce to it; NP-completeness is perhaps the best-known example.

one in the complement of the digraph; and, given a root of a set of polynomial equations with n variables in the field modulo p, where the sum of the degrees is less than n, find another (it must exist by Chevalley's Theorem in Number Theory).

A problem A in PLF is defined in terms of a polynomial-time algorithm M. Let x be an input for A. The *configuration space* $C(x)$ is $\Sigma^{[p(|x|)]}$, the set of all strings of length at most $p(n)$, a polynomial. Given a configuration c, M outputs in time $p(|x|)$ a set $M(x,c)$ of at most two configurations. We say that two configurations c, c' are *neighbors*, written $N(c, c')$, if $c \in M(x, c')$ and $c' \in M(x, c)$. Obviously, N is a symmetric graph of degree at most two. M is such that $M(x, 0\ldots 0) = \{1\ldots 1\}$, and $0\ldots 0 \in M(x, 1\ldots 1)$, so that $0\ldots 0$ is always a leaf (the *standard leaf*). Problem A is the following search problem: "Given x, find a leaf of $C(x)$ other than $0\ldots 0$." PLF is the class of all problems A defined as above.

Strictly speaking, PLF is based on an even more trivial lemma, stating that *any graph with degrees two or less has an even number of leaves*, but we can show that the class based on the more general parity argment coincides with this weaker one. The proof uses the *chessplayer's algorithm* [Pa], explained next. Suppose that everybody remembers all games of chess they have played in their life. You have played an odd number of games, and you must find another odd player (known to exist by the parity argument). Here is a solution: We require that each player has paired up his/her past games so that game $2i - 1$ is the "mate" of game $2i$. The algorithm is this: Ask your last opponent if he is odd; if so, you are done. If not, you ask the address of his playmate in the game that is the mate, in his game history, of the game with you, and visit her. If she is odd you are done, but otherwise you ask for the address of her opponent in the game that is the mate (in her history) of her game with the previous player. And so on. You may come back to yourself many times (in which case you disregard parity), but the algorithm is guaranteed to terminate at another odd player. Alas, this algorithm may take time proportional to the number of all games of chess ever played! But it does reduce the original graph of chess games to one of degree two or less (the one traversed by the algorithm). Similar arguments serve to show that other classes coincide with PLF, such as the class based on the following variant of the parity argument: *If in a bipartite graph there is a node with degree not divisible by p, then there is another such node* (the proof of Chevalley's Theorem is based on this lemma).

Other variants of the parity argument seem to yield different classes. The most conspicuous is the one concerning *directed graphs* (that is, the two nodes in $M(x, c)$ are now ordered, designated "possible predecessor" and "possible successor"). Solution $00\ldots 0$ is always a source, and we are asking for another source *or a sink*. The resulting class, called PDLF, is trivially a subset of PLF, but we cannot prove that it is equal to it (or, of course, distinct). This is especially interesting, because of the following result:

Theorem 1. [Pa] *Finding a Brouwer fixpoint in a 3-dimensional continuous map is PDLF-complete.*

The input to this problem is an algorithm that computes a piecewise linear continuous map from the unit tetrahedron to itself; the intricate details of the definition are omitted.

By the way, tampering with the definition of PLF can have devastating effects. For example, suppose that we had defined PLF in the following way: We insist that the output be not *any* other leaf, but *the particular* other leaf connected to 0...0. This is a natural enough variant, since this is the leaf returned by the obvious algorithm, the leaf we *know* exists. Call this class PLF′.

Theorem 2. [Pa] *PLF′ = PSPACE.*

That is, the freedom to return another leaf (which is not guaranteed to exist!), strangely enough makes the problem much simpler.

What other natural problems in TFNP are there? In other words, what other *inefficiently constructive* existence proofs are there in Mathematics, that give rise to nontrivial computational problems? The Borsuk-Ulam Theorem comes to mind: Every continuous map from the n-sphere to $\mathbb{R}^n$ maps two antipodal points to the same point. In fact, the following is a computational problem immediately related to it: Given $2n^2$ points in n dimensions, divided into n classes with $2n$ points each, find a hyperplane that leaves n points of each class on each side (this can be called the "discrete generalized ham sandwich problem"). As a combinatorial proof of Tucker's Lemma (from which the Borsuk-Ulam Theorem follows) reveals [FT], this problem is also in PLF, as the parity argument lies at the roots of [FT]'s proof. Many more examples, no doubt, exist that are not known to me.

Another important source of inefficient proofs of existence is Lovász's Local Lemma [Sp]. We can define a class PLL (for *polynomial local lemma*).[3] A problem A in PLL is defined in terms of a polynomial algorithm M, much like PLS and PLF. Given input x, the set of possible solutions is again $\{0,1\}^{[p(|x|)]}$. M takes as inputs triples of the form string-integer-string; on input (x, j, λ) (where x is the original input, λ is the empty string, and $j \leq p(|x|))$ M generates a set $D_j \subset \{1, 2, \ldots, p(|x|)\}$, with $|D_j| \leq \log |x|$ for all j. Intuitively, the D_j's are the domains of the local conditions in the Local Lemma. On input (x, j, y) where $1 \leq j \leq q$ and $y \in \{0,1\}^{|D_j|}$, M outputs "yes" or "no" (these are the local conditions).

What is the desired output for input x? To define it, let $N(x, j)$ be $\{i \neq j : D_i \cap D_j \neq \emptyset\}$, and let $a(x, j)$ be the number of y's for which $M(x, j, y) =$ "yes" (notice that these numbers are polynomial-time computable). If it so happens that for some j

$$\sum_{i \in N(x,j)} e \frac{a(x,j)}{2^{|D_j|}} \geq 1,$$

then the correct output is the string "*local lemma does not apply*"; otherwise, the correct output is any string z in $\{0,1\}^{[p(|x|)]}$ such that $M(x, j, z|_{D_j}) =$ "no" for all $j \leq q$. The existence of at least one such string z is guaranteed by the Local Lemma [Sp]. It is easy to see that FP ⊆ PLL ⊆ TFNP.

Class PLL contains several interesting problems in Combinatorics, for which no polynomial algorithm is known. Examples: Given a directed graph with indegree

---

[3] This definition builds on ideas of J. Kratochvíl and M. Fellows [Kra], who defined a sequence of classes PLL[k], one for each size of the D_j's in our definition.

and outdegree ten, find an even cycle [Al]; or, given a set of clauses with eight literals in each, and with at most ten appearences of each variable, find a satisfying truth assignment [KST]. In fact, it has some (unfortunately, rather unnatural) complete problems.

If we rely on the crude probabilistic method, the situation is much less interesting. Define the class PPM (for *polynomial probabilistic method*) to have the same definition with PLL, except that the condition now becomes weaker:

$$\sum_{j=1}^{q} \frac{a(x,j)}{2^{|D_j|}} \geq 1.$$

If this condition does not hold, a simple argument establishes that at least one string satisfying $M(x, j, z|_{D_j}) = $ "no" for all $j \leq q$ exists [Sp]. In fact, a calculation shows that a *constant fraction of all strings* satisfy this condition, and hence finding one is solvable by a randomized algorithm: PPM $\subseteq$ FZPP (the class of search problems that can be solved by randomized algorithms). A little more care establishes a stronger result:

Theorem 3. [Kou] *PPM = FP.*

The algorithm in the proof of Theorem 3 finds the bits of z one-by-one, always choosing the bit that maximizes the probability that a string satisfying all conditions on the remaining bits exists. Unfortunately, there seems to be no obvious way of extending this strategy to PLL. A general technique for making the local lemma constructive remains an important open question.

Are there other natural examples of inefficiently constructive proofs of existence, or have we exhausted all such naturally occurring problems? Let me end by introducing another subset of TFNP which seems to be distinct from FP and the classes defined above: The class PPP (for polynomial pigeonhole principle) is the set of all problems reducible to the following one: Given a Boolean circuit C with n inputs and outputs such that $C(x) \neq 0^n$ for all inputs $x \in \{0,1\}^n$, find two inputs $x \neq y$ such that $C(x) = C(y)$; such inputs exist by the pigeonhole principle. It is open whether PPP = FP. For example, the following problem is in PPP, but is not known to be in FP: Given n positive integers whose sum is less than $2^n - 1$, find two subsets with the same sum. It is perhaps amusing to notice that the same class *only with monotone circuits* coincides with FP.

Acknowledgments

While trying to formulate the questions posed in this paper I benefitted a lot from discussions with Noga Alon, Manuel Blum, Elias Koutsoupias, Jan Kratochvíl, Nimrod Megiddo, Svatopluk Poljak, Mike Saks, Steve Vavasis, Emo Welzl, and Mihalis Yannakakis.

References

[Al] N. Alon, private communication, May 1990.

[FT] R. M. Freund, M. J. Todd, *A Constructive Proof of Tucker's Combinatorial Lemma*, J. of Combinatorial Theory A **30** (1981.), 321–325.

[HV] M. Hirsch, S. Vavasis, *Exponential Lower Bounds for Finding Brouwer Fixpoints*, J. of Complexity 5 (1989), 379–416, Proc. 1987 FOCS, see also M. Hirsch, C. H. Papadimitriou, S. Vavasis.

[JPY] D. S. Johnson, C. H. Papadimitriou, M. Yannakakis, *How Easy is Local Search?*, Proc. 26th Annual Symp. Foundations Comp. Sci., 1985, pp. 39–42; J. Comp. Syst. Sci. **37** (1988), 79–100.

[Kra] J. Kratochvíl, private communication, July 1990.

[KST] J. Kratochvíl, P. Savický, Z. Tuza, *One more occurrence of variables makes satisfiability jump from trivial to NP-complete*, (submitted).

[Kou] E. Koutsoupias, private communication, August 1990.

[Kre] M. W. Krentel, *Structure of Locally Optimal Solutions*, Proc. 30th Annual Symp. Foundations Comp. Sci., 1989, pp. 216–221.

[MP] N. Megiddo, C. H. Papadimitriou, *A Note on Total Functions, Existence Theorems, and Computational Complexity*, IBM Research Report RJ 7091, Theoretical Computer Science, 1989, to appear.

[Pa] C. H. Papadimitriou, *On Graph-Theoretic Lemmata and Complexity Classes*, Proc. 1990 FOCS.

[PSY] C. H. Papadimitriou, A. S. Schäffer, M. Yannakakis, *On the Complexity of Local Search*, Proc. 1990 STOC.

[PPT] S. Poljak, D. Turzik, P. Pudlák, *Extensions of k-subsets to $k+1$-subsets: Existence versus Constructability*, Commentationes Mathematicae Universitatis Carolinae **23** no. 2 (1982), 337–349.

[Pr] V. R. Pratt, *Every Prime has a Succinct Certificate*, SIAM J. on Computing **5** (1974).

[Sp] J. Spencer, *Ten Lectures on the Probabilistic Method*, SIAM, CBMS-NSF Regional Conference Series No. 52, 1987.

[Th] A. Thomason, *Hamiltonian Cycles and Uniquely Edge Colourable Graphs*, Annals of Discrete Math. **3** (1978), 259–268.

Christos H. Papadimitriou
Department of Computer Science and Engineering,
University of California at San Diego
USA

Fourth Czechoslovakian Symposium on
Combinatorics, Graphs and Complexity
J. Nešetřil and M. Fiedler (Editors)
© 1992 Elsevier Science Publishers B.V. All rights reserved.

Optimal Coteries on a Network

CHRISTOS H. PAPADIMITRIOU AND MARTHA SIDERI

In a computer network nodes may fail, and the failures may in fact partition the network into two or more connected components. (We assume that only nodes, and not edges, fail; however, our results easily extend to edge failures.) In this case, it is important that the network continue its operation, but no two connected components continue operating simultaneously and independently (thus reaching incompatible decisions). In [BM] a protocol was proposed, whereby at most one component continues. This protocol involves the notion of a *coterie*. A coterie for a graph $G = (V, E)$ is a family of subsets of V, with the property that any two sets in it intersect. The protocol is simply this: A connected component can operate if and only if it contains a set in the coterie. Obviously no two connected components can operate simultaneously. A common way of implementing coteries is by *voting*, whereby each node in V is assigned a real number, and a set of nodes is in the coterie if it is a majority, that is, the numbers in it add up to more than the numbers outside it. However, it was observed by Yannakakis that the vast majority of coteries cannot be implemented by voting (interestingly, all coteries considered and proved optimal in this paper can be implemented by voting).

Our interest lies in evaluating coteries in terms of their performance, and calculate the optimum coterie in each situation. We are given a graph, a probability of failure p_i for each node i, and a coterie. Consider an experiment in which each nodes fails with probability p_i, independently. The outcome of the experiment is the *size of the component that contains a set in the coterie* if such a set exists, and zero otherwise. The performance of the coterie is the expectation of this outcome. We seek to design a coterie for G and the p_i's that has the maximum such performance.

One first observation is that we can assume that the coterie in consideration $\mathcal{C} = \{C_1, \ldots, C_n\}$, besides

(1) *Any two sets in it intersect*

has the following properties:

Research supported by the ESPRIT Basic Research Action No. 3075 ALCOM, a grant from the Volkswagen Foundation to the Universities of Patras and Bonn, and by the NSF.

(2) Any C_i induces a connected subgraph of G (if not, we can replace C_i with the collection of all minimal supersets of C_i that have this property, without changing the performance).

(3) Each C_i is minimal, in that it cannot be replaced by a subset and still satisfy (1) and (2) (such C_i's can be replaced by the subset in hand, improving the performance).

(4) C is maximal, in that no new set can be added to it and satisfy (1) through (3) (if such a set exists, adding it improves the performance).

Our first result is that, computationally speaking, this problem is quite difficult:

Computing the optimum coterie can be done in exponential nondeterministic time, and we can show that it is #P-hard, that is, at least as difficult as computing the permanent of a matrix.

In view of this result, we may want to examine some easy special cases. For example, trees. It turns out that this is easy:

The optimum coterie in a tree is a monarchy, that is, the coterie consisting of one set, $\{c\}$; c is called the center, and can be computed in quadratic time. Even if all probabilities of failure are all equal to p, the optimal location of the center changes according to the precise value of p. For large p, it is the node of largest degree; for small p it is the root that minimizes average distance to the nodes; and so on.

We are thus encouraged to look at the cycle. We can show the following useful lemma:

If $\mathcal{C} = \{C_1, \ldots, C_n\}$ is the optimum coterie in the cycle, then each C_i is an arc, and for any arc longer than one node there is exactly one other arc that intersects it in each of its endpoints. It can be computed in quadratic time.

It is interesting to look at the case of the cycle with n nodes and equal probabilities p. Assume that n is very large. What is the optimal coterie here? Perhaps surprisingly, the answer depends on the product pn, the expected number of failures:

If pn is less than 2, then the optimal coterie is the democracy, *where each node has one vote (and one node has a tie-breaking second vote, if n is even). As pn increases beyond 2, then the optimum coterie is the* canonical odd oligarchy, *that is, all nodes have zero votes, except for $2k+1$ nodes (for some integer $k \geqslant 0$) spread as evenly as possible around the cycle, which get one vote each. In particular, $2k+1$ is optimum when pn is near $pn = 2 + \frac{1}{8k^2}$. For $pn > 2.63\ldots$ the monarchy is optimal.*

The cases of trees and cycles can be combined. Define a *cactus* to be a graph such that no two cycles intersect in an edge (that is, all biconnected components are either edges or cycles).

There is a polynomial-time algorithm that finds the optimum coterie for any cactus and any probabilities of failure.

In fact, our algorithm can also handle *weights* on the nodes of the cactus.

References

[BM] D. Barbara, H. Garcia-Molina, *The Vulnerability of Vote Assignments*, ACM Transactions on Computer Systems **4** no. 3 (1986), 187–213.

Christos H. Papadimitriou
University of Patras,
Visiting from the Department of
Computer Science and Engineering,
University of California at San Diego, USA

Martha Sideri
University of Patras,
Computer Technology Institute,
Koloktroni 3,
26221 Patras, Greece

Fourth Czechoslovakian Symposium on
Combinatorics, Graphs and Complexity
J. Nešetřil and M. Fiedler (Editors)
© 1992 Elsevier Science Publishers B.V. All rights reserved.

On Some Heuristics for the Steiner Problem in Graphs

JÁN PLESNÍK

The Steiner problem in graphs (networks) is to find a minimum cost tree T spanning a prescribed subset Z of vertices in a given graph G with positive edge costs. There exists a vast bibliography of the Steiner problem and thus the reader is referred to surveys (see e.g. Winter [13] and Hwang and Richards [4]). Note that the Steiner problem is solvable in polynomial time for any fixed cardinality of Z but in general it is an NP-hard problem even for special cases [13]. Therefore several polynomial time heuristics have been proposed to provide at least an approximate solution. No heuristic (published before this Symposium) is known to have a worst-case error ratio less than $2 - \varepsilon$ for $\varepsilon > 0$ (i.e. for any $\varepsilon > 0$ there is an instance of the Steiner problem with $c(T_H)/c^* > 2 - \varepsilon$, where $c(T_H)$ is the cost of the tree T_H produced by the heuristic H and c^* is the cost of an optimal solution). Note that during this Symposium Zelikovsky [14] has presented an $\frac{11}{6}$-approximation heuristic. His heuristic means an important step in this area but, of course, it could not be involved in our contribution. The purpose of this paper is to show that many heuristics have the following property: none of them is superior to any other in terms of the quality of the approximate solution. In fact we involve all the main heuristics from the literature and show an extreme incomparability. We denote: $n = |V(G)|$, $m = |E(G)|$, $p = |Z|$.

The following seven heuristics are considered:

$H1$: The minimum spanning tree heuristic. This heuristic is usually attributed to Kou, Markowsky and Berman [5] but it was developed by El-Arbi [2] and then several times rediscovered by others (cf. [4, 13]). In fact the core of this heuristic is due to E. F. Moore (see [3]).

$H2$: The minimum path heuristic; Takahashi and Matsuyama [8].

$H3$: The contraction heuristic; Plesník [6]. We will use the version described in Winter [13].

$H4$: The average distance heuristic; proposed by Rayward-Smith [7] and analysed by Waxman and Imase [11].

$H5$: The first heuristic of Chen [1]; analysed by Widmayer [12].

$H6$: The second heuristic of Chen [1]; analysed by Widmayer [12].

$H7$: The heuristic of Wang [9]; analysed by Widmayer [12].

All these heuristics run in polynomial time. More precisely, in time $O(n^2p)$: $H1$, $H2$ and $H7$; in time $O(n^3)$: $H3$ and $H4$; the complexity of $H5$ is $O(mnp\log n)$ and that of $H6$ is $O(mp^2 \log n)$. (For faster versions of $H1$ see references in [4]).

As to the quality of approximation each of these heuristics has the property that for any output T cost $c(T) \leqslant (2 - 2/p)c^*$.

We say that a heuristic H' wins over a heuristic H'' with ratio $\varrho > 1$ if there is an instance of the Steiner problem such that for any outputs the corresponding costs $c_{H'}$ and $c_{H''}$ fulfil

$$\varrho c_{H'} \leqslant c_{H''}.$$

For example, for any $\varepsilon > 0$, $H4$ wins over $H1$ with ratio $2 - \varepsilon$ as one can see in the following figure where δ is positive and sufficiently small; the Z-vertices are black and p is sufficiently large.

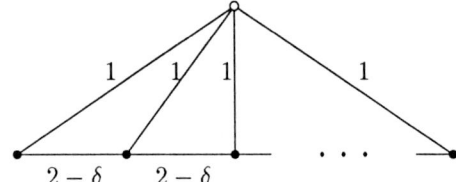

Such examples have appeared in the literature rather sporadically. A systematic research has been done by Widmayer [12] who proved that for any pair (H', H'') of distinct heuristics from the set $\{H1, H2, H5, H6, H7\}$, H' wins over H'' with ratio $\varrho > 1$, but his ϱ-numbers are small ($\frac{17}{15}$ or so). We have:

Theorem. *For any pair (H', H'') of distinct heuristics from the set $\{H1, H2, \ldots, H7\}$, except for $(H1, H3)$ and $(H3, H1)$, H' wins over H'' with ratio $2 - \varepsilon$ for any given $\varepsilon > 0$.*

The proof consists of several constructions. Some of them are simple (cf. our figure) while others are more sophisticated and lengthy. Therefore the proof cannot be presented here and will be published elsewhere.

Remark 1. The two exceptions in the Theorem are open questions. In fact, we cannot establish any ratio $\varrho > 1$.

Remark 2. Very recently, we have proposed and analysed modifications of heuristics $H3$ and $H4$ and accordingly extended the Theorem.

References

[1] N. P. Chen, *New algorithms for Steiner tree on graphs*, Proc. IEEE Intl. Symp. Circuits and Systems (1983), 1217–1219.

[2] C. El-Arbi, *Une heuristique pour le probleme de l'arbre de Steiner*, RAIRO Operations Research **12** (1978), 207–212.

[3] E. N. Gilbert and H. O. Pollak, *Steiner minimal trees*, SIAM J. Appl. Math. **16** (1968), 1–29.

[4] F. K. Hwang and Dana S. Richards, *Steiner tree problems*, preprint.

[5] L. Kou, G. Markowsky and L. Berman, *A fast algorithm for Steiner trees*, Acta Informatica **15** (1981), 141–145.
[6] J. Plesník, *A bound for the Steiner tree problem in graphs*, Math. Slovaca **31** (1981), 155–163.
[7] V. J. Rayward-Smith, *The computation of nearly minimal Steiner trees in graphs*, Int. J. Math. Educ. Sci. Technol. **14** (1983), 15–23.
[8] H. Takahashi and A. Matsuyama, *An approximate solution for the Steiner problem in graphs*, Math. Japonica **24** (1980), 573–577.
[9] S. M. Wang, *A multiple source algorithm for suboptimum Steiner trees in graphs*, Proc. Intl. Workshop Graphtheor. Concepts in Comp. Science (H. Noltemeier, ed.), Würzburg, 1985, pp. 387–396.
[10] B. M. Waxman, *Routing of multipoint connections*, IEEE J. Select. Areas Comm. **6** (1988), 1617–1622.
[11] B. M. Waxman and M. Imase, *Worst-case performace of Rayward-Smith's Steiner tree heuristic*, Inf. Process. Lett. **29** (1988), 283–287.
[12] P. Widmayer, *Fast approximation algorithms for Steiner's problem in graphs*, Dissertation, Universität Karlsruhe, 1987.
[13] P. Winter, *Steiner problem in networks: a survey*, Networks **17** (1987), 129–167.
[14] A. Z. Zelikovsky, *An $\frac{11}{6}$-approximation algorithm for the Steiner problem on graphs*, these Proceedings.

J. Plesník
KNOM MFF UK,
Mlynská dolina,
842 15 Bratislava,
Czechoslovakia

Cycle Covers of Graphs with a Nowhere-Zero 4-Flow
(Abstract)

ANDRÉ RASPAUD

A *cycle* in a graph G is a connected 2-regular subgraph. The *length* of a cycle is the number of its edges. A *cycle cover* $\mathcal{C}$ of G is a set of cycles such that each edge of G belongs to at least one cycle of $\mathcal{C}$. The *length* of $\mathcal{C}$ is the sum of the lengths of the cycles in $\mathcal{C}$ and it is denoted by $l(\mathcal{C})$.

It was conjectured by Itai and Rodeh that every bridgeless graph G has a cycle cover $\mathcal{C}$ of length at most $|E(G)|+|V(G)|-1$. This conjecture is still open. G. Fan has proved that, if G is a simple graph with a nowhere-zero 4-flow, G has a cycle cover $\mathcal{C}$ of length at most $|E(g)| + |V(G)| - 2$ and has conjectured that, if G is a simple graph with a nowhere-zero 3-flow, then G has a cycle cover $\mathcal{C}$ such that $l(\mathcal{C}) \leqslant |E(G)| + |V(G)| - 3$.

It is easy to see that the complete bipartite graph $G = K_{3,3m}$ has a nowhere-zero 3-flow and that any cycle cover of G has length at least $|E(G)| + |V(G)| - 3$. This shows that this conjecture, if true, is best possible.

We prove that, if G is a simple graph, different from K_4, with a nowhere-zero 4-flow, then G has a cycle cover $\mathcal{C}$ with $l(\mathcal{C}) \leqslant |E(G)| + |V(G)| - 3$.

If a graph has a nowhere-zero 3-flow it has a nowhere-zero 4-flow, this result implies the conjecture of G. Fan.

Corresponding paper to be published in *J. of Graph Th.*

André Raspaud
La.B.R.I.,
Université Bordeaux I,
351, cours de la Libération,
33405 Talence Cedex,
France

Fourth Czechoslovakian Symposium on
Combinatorics, Graphs and Complexity
J. Nešetřil and M. Fiedler (Editors)
© 1992 Elsevier Science Publishers B.V. All rights reserved.

Minimax Results and Polynomial Algorithms in VLSI Routing

ANDRÁS RECSKI

This survey is an informal introduction to some results in VLSI routing. There are hundreds or perhaps thousands of papers on this broad subject, including heuristic algorithms with sometimes very good practical performance as well as some very deep theoretical results. The present article tries to give an introduction for interested mathematicians, concentrating on some simple special subproblems which can be used as illustrations of the applicability of combinatorial optimization.

Introduction

We suppose that the devices of the electric equipment to be designed are already placed into their final positions on the board and we wish to interconnect some terminals by wires. Technological constraints require that the wires must use grid edges only (to keep sufficient distance etc.) and, obviously, if wires interconnect terminals belonging to different nets then these wires must not intersect. There are several layers for these wires and a wire can leave a layer for another one using a "via hole" in a grid point.

For example, Fig. 1 shows a part of a board, the shaded areas are occupied by devices, the dots are the terminals (those with identical numbers belong to the same net) and the thin lines show the grid. From the graph-theoretical viewpoint, the routing is a collection of vertex-disjoint Steiner trees in the graph of Fig. 2 (if there are three layers). In this particular case a solution is shown in Fig. 3 or in a simplified way in Fig. 4. Via holes are denoted by heavy vertical lines and by squares, respectively.

There is an additional constraint in most technologies that if a via hole is used for one Steiner tree then its intersection with any third layer must not be used by another Steiner tree. However, we restrict ourselves mostly to two layers.

Moreover, we mainly concentrate on channel routing (where the terminals are situated on two parallel lines, see the formal definitions below); and in most cases a net will consist of two terminals only, hence the Steiner trees will simply be paths.

The solution in Fig. 3 is called unconstrained since there is no special constraint on the location of the paths or Steiner trees within the graph except that they are pairwise vertex disjoint. There are some constrained models as well (the paths must

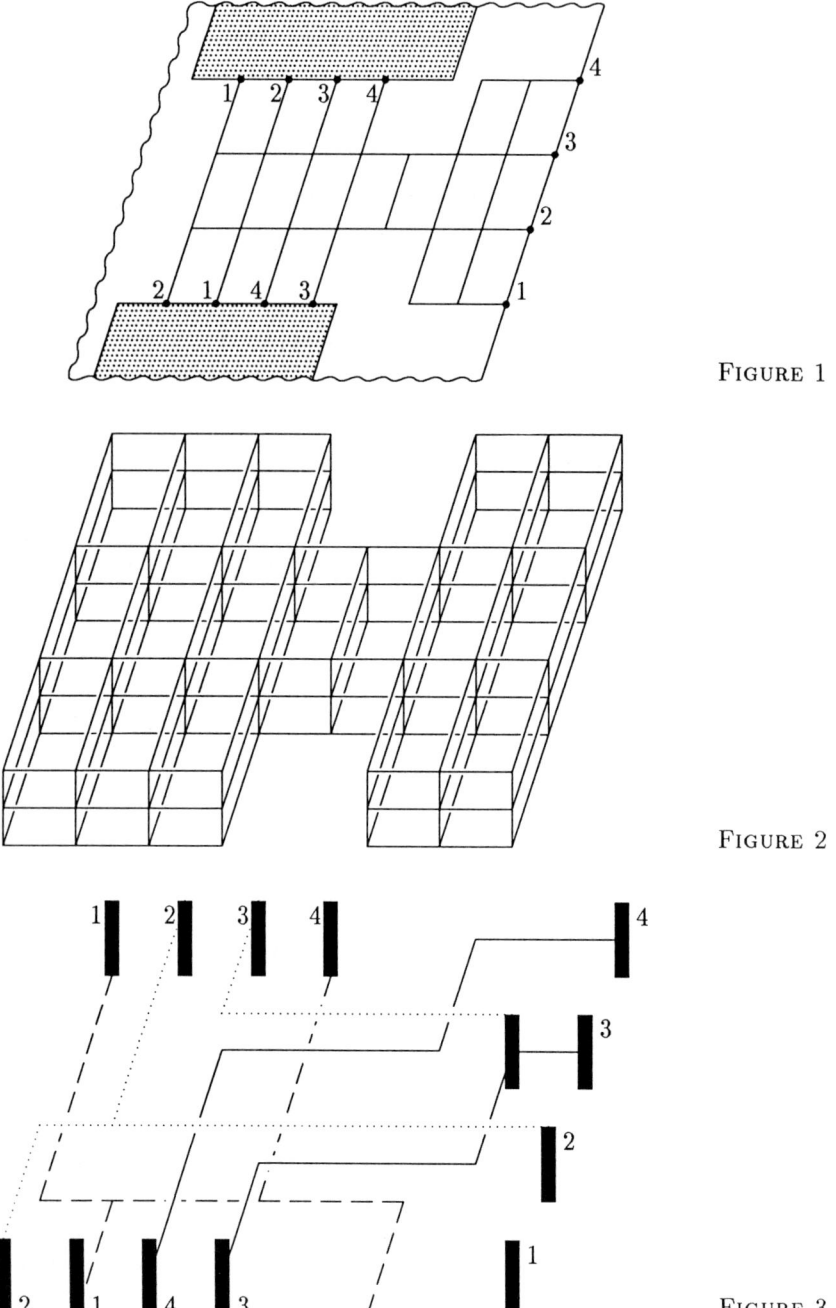

FIGURE 1

FIGURE 2

FIGURE 3

be edge disjoint, or some layers must contain horizontal or vertical segments only etc). These are also explained in the first section.

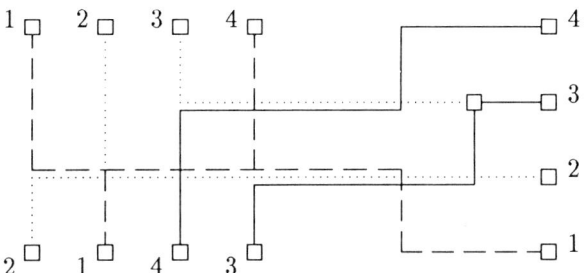

FIGURE 4

I. Basic definitions

A *channel* is a rectangular grid of *tracks* (numbered from 0 to $w + 1$) and *columns* (numbered from 1 to m), where w is the *width* and m the *length* of the channel. A *net* $N = (\{t_1, \ldots, t_u\}, \{b_1, \ldots, b_l\})$ is a collection of terminals, where the lower terminal b_j is located at the grid point $(b_j, 0)$ on track 0 (at the *lower boundary*) and the upper terminal t_j at the grid point $(t_j, w+1)$ on the track $w+1$ (at the *upper boundary*). N is called *2-terminal* if $u = l = 1$, and *multiterminal* otherwise. A 2-terminal net $(\{t\}, \{b\})$ or shortly (t, b) is of shape / if $t \geqslant b$ and of shape \ if $t < b$.

A *channel routing problem (CRP)* is a set of pairwise disjoint nets $\mathcal{N} = \{N_1, \ldots, N_n\}$. A CRP is *dense* if every boundary point belongs to some net. The *solution* (also called *layout*) of a CRP is a set $\mathcal{G} = \{G_1, \ldots, G_n\}$ of subgraphs (typically Steiner trees) G_i of G (also called *wires*) such that G_i connects the terminals of N_i, for $i = 1, \ldots, n$ under the conditions of the corresponding wiring model.

A very special case of the CRP is the *single row routing problem (SRRP)* where every terminal is located at the upper boundary. (Warning: some authors define SRRP in a different way, allowing wires on "both sides" of the row of the terminals.)

The wiring models are formulated as restrictions on the mutual relations of the subgraphs. The simplest approach is the *single layer model (SLVM)* where the wires are vertex-disjoint subgraphs of the grid graph G. For some reasons (see Section IV) the *edge-disjoint case* on a single layer ($SLEM$) is also of considerable interest.

The *unrestricted 2-layer model (TLUM)* requires vertex-disjoint wires again, however, the initial graph G is not a planar rectangular grid but an $m \times (w+2) \times 2$ cubic grid. In this case a lower or upper terminal is a vertex-pair $\{(b_j, 0, 0), (b_j, 0, 1)\}$ or $\{(t_j, w+1, 0), (t_j, w+1, 1)\}$, respectively, together with the edge connecting the two vertices.

As a special case of the 2-layer problem, many people studied the so called *Manhattan model (TLMM)* where one layer is reserved for the horizontal and one for the vertical edges.

The two basic problems of routing are as follows:
(P1) Decide whether a problem is solvable.

(P2) If yes, determine the layout with minimum width.

Some other objectives can also be of importance (like minimizing total wiring length or total number of via holes) but are disregarded in the present paper.

II. On the mutual relation of the models

Obviously, if a problem can be solved in SLVM then so it can in TLMM. If a problem can be solved in TLMM then its projection leads to a SLEM-solution where the wires are edge disjoint and crossings of shape + are permitted but using a common vertex with a shape ⊥ (also called *knock-knee*) is prohibited. See also p. 153 of [6].

Proposition 1. *The relation of the four models is shown in Fig. 5.*

Proof. Since the inclusions are clear, all we need are Examples A through E in the indicated positions of the chart of the figure. These examples are given in Fig. 6. Examples A, B, C and D are obvious. In case of E the figure clearly shows that the specification can be realized in SLEM. Its unrealizability in TLUM will follow from the combination of Lemma 10 and Proposition 11 below since the nets 3, 4 and 5 pairwise cross each other. □

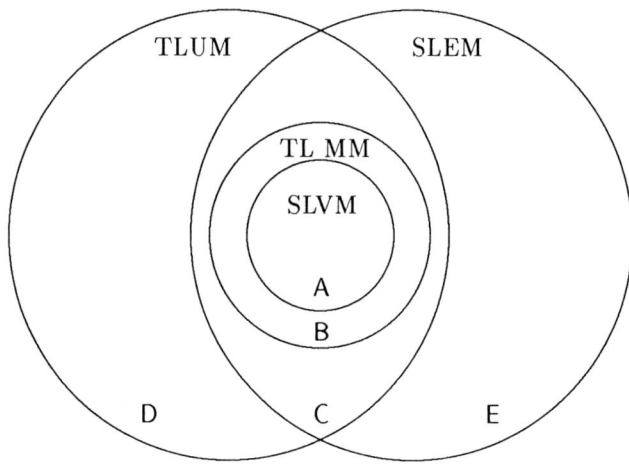

FIGURE 5

III. Some results in the single row routing problem

We call two nets $\{t_1, t_2, \ldots, t_k\}$ and $\{t'_1, t'_2, \ldots, t'_l\}$ (with $t_1 < t_2 < \ldots < t_k$ and $t'_1 < t'_2 < \ldots < t'_l$) *crossing* if $t_i < t'_{i'} < t_j < t'_{j'}$ or if $t'_{j'} < t_j < t'_{i'} < t_i$ for some $1 \leq i < j \leq k$ and $1 \leq i' \leq j' \leq l$; and we call them *intersecting* if the intervals $[t_1, t_k]$ and $[t'_1, t'_l]$ are non-disjoint. Every crossing pair is intersecting but not vice versa.

It is very natural to define two graphs. The vertices of both graphs correspond to the nets of a given specification. Two vertices are adjacent in the *crossing graph* G_C or in the *intersection graph* G_I if the corresponding nets are crossing or intersecting, respectively. For example, Fig. 6 below shows G_I for the specification 1, 2, 2, 1, 3, 4, 4, 3, 5, 5, 1.

Clearly, G_C is a subgraph of G_I. For example, if the terminals are 1, 2, 3, 3, 2, 4, 1, 4 then G_C consists of a single edge $\{1,4\}$ while G_T contains every possible edge except $\{2,4\}$ and $\{3,4\}$.

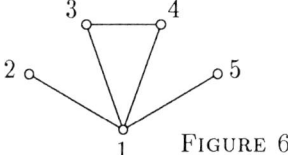

FIGURE 6

Proposition 2. *A specification is solvable in SLVM if and only if no two nets are crossing. If this condition is met then the minimum width equals the maximum number of pairwise intersecting nets, that is, the size $\omega(G_I)$ of the maximum clique in G_I.*

The statement is trivial, see Fig. 7.

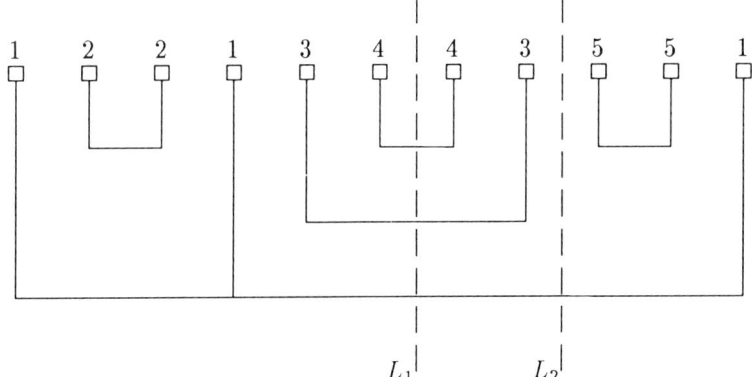

FIGURE 7

Remark 1. If a net consists of two adjacent terminals only (like nets 2 and 6 in Fig. 8) and they can be connected using the boundary edges then the minimum width may sometimes be reduced by one. Hence nets of form $\{t, t+1, \ldots, t+k\}$ should be disregarded in the definition of the graphs G_I and G_C.

Proposition 3. *Every specification is solvable in TLMM and the minimum width equals $\omega(G_I)$.*

Proof. Let L be a vertical line cutting the board in two and let us define the congestion $c(L)$ as the number of nets which are separated by L. For example, $c(L_1) = 3$ and $c(L_2) = 1$ for lines L_1, L_2 in Fig. 7 since nets 1, 3 and 4 each have terminals on both sides of L_1 but only net net 1 has them on both sides of L_2. By the definition of G_I clearly $\omega(G_I) = \max\{c(L)$ for every vertical line $L\}$.

If we consider the Manhattan model TLMM then every horizontal wire segment uses the same side of the board, hence the maximum congestion is clearly an upper bound for the minimum width. Equality follows from the simple observation that G_I is an interval graph, hence it is perfect (see [4] for example), thus $\omega(G_I)$ equals the chromatic number $\chi(G_I)$ and proper colourings of G_I are just the proper layouts (since horizontal wire segments can share the same track if and only if the

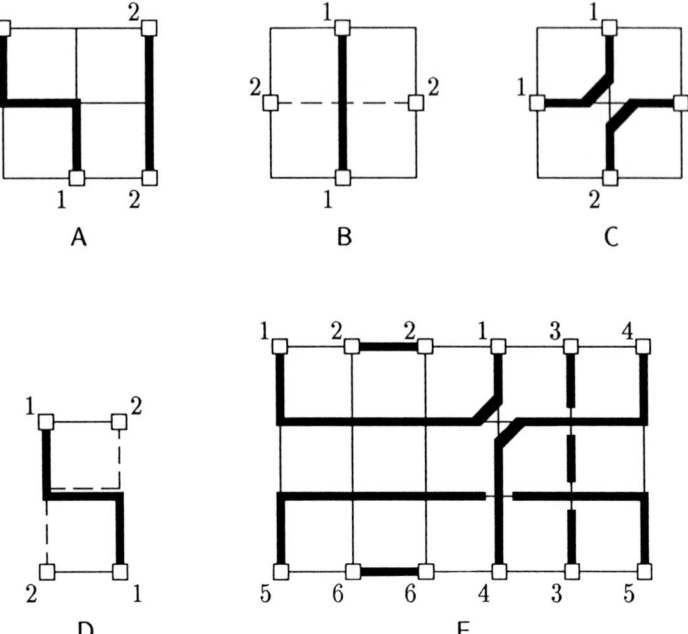

FIGURE 8

corresponding vertices of G_I are pairwise non-adjacent). Fig. 9 shows an example with $\omega = \chi = 4$. □

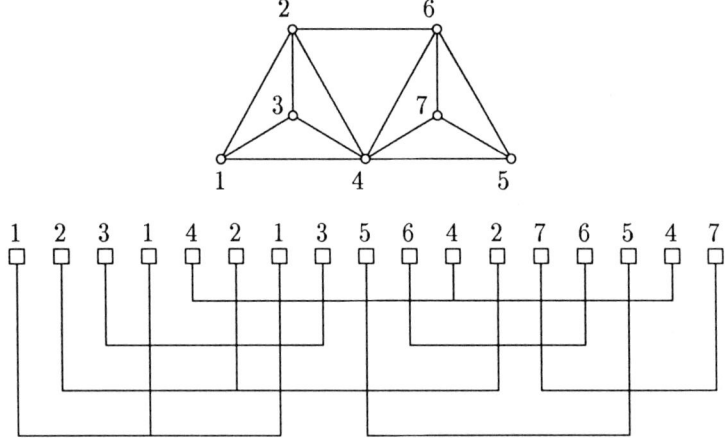

FIGURE 9

Alternatively, one can prove this statement by verifying the following algorithm which places the horizontal wire segments of the nets (shortly, the intervals) on the tracks.

Step 1. If the list $\mathcal{L}$ of the intervals to be arranged is empty, stop. Otherwise consider the interval with minimum left end coordinate, place it to a new track, denote its right end coordinate by x and delete it from $\mathcal{L}$.

Step 2. Consider those intervals whose left and coordinate is greater than x. If there are none, go to Step 1. Otherwise choose the one with minimum left end coordinate (i.e. closest to x), place it to the actual track, denote its right end coordinate by x and delete it from $\mathcal{L}$. Go to Step 2.

The simplicity of this result was probably a major motivation for the extensive research in routing in TLMM.

Let us change our model from TLMM to SLEM or TLUM. Clearly, every specification can be realized (this was already the case in TLMM). The width cannot be reduced in SLEM; the upper bound arising from $\max\{c(L)\}$ is still sharp (cf. Remark 1). However, appearently little is known about the minimum width in TLUM except that it is between $\lfloor \omega(G_I)/2 \rfloor$ and $\omega(G_I)$ and both bounds may be attained by suitable specifications. If the specification is dense, the lower bound is the upper integer part $\lceil \omega(G_I)/2 \rceil$ but one less can sometimes be obtained for non-dense problems, see Fig. 13 below.

The following small observation seems to be new. Recall that a tree is called a *caterpillar* if and only if deleting all the vertices of degree one the result is a path.

Proposition 4. *A dense specification can be realized with width one in TLUM if and only if each component of its crossing graph G_C is a caterpillar.*

Proof. The sufficiency can essentially be seen from the example shown in Fig. 10. For the necessity we need two observations.

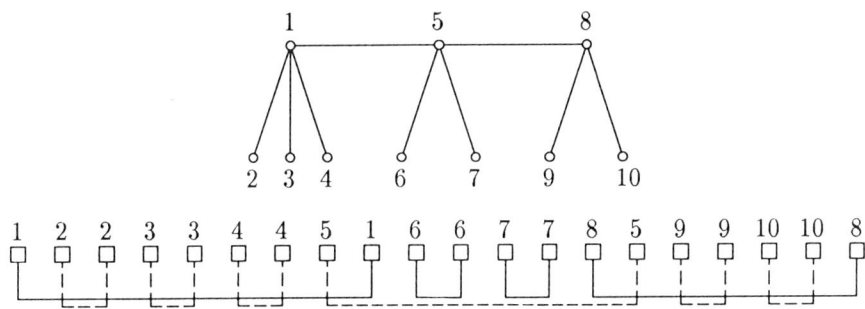

FIGURE 10

Lemma 5. *The intersection graph G_I is chordal (every circuit with length $\geqslant 4$ of it contains at least one diagonal).*

Lemma 6. *If a specification is dense and realizable with width one then there exists a layout without using via holes (other than the terminals).*

Lemma 5 is well known, see [4] (recall that the graphs arising as G_I for some specification are just the interval graphs). For proving Lemma 6 suppose that a layout contains a via hole V next to terminal node N. The line NV cuts the board in two parts. Interchange the role of the two sides of the board (i.e. the continuous and the dotted lines) on, say, the left hand side of this line: V is not needed any more (see Fig. 11).

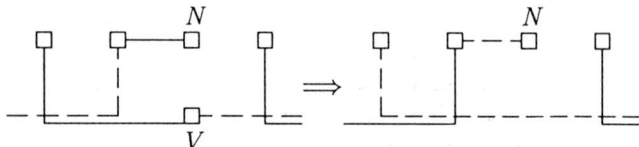

FIGURE 11

In order to finish the proof of Proposition 4 first observe that each component of G_C must be a tree, for otherwise its supergraph G_I must contain a circuit of length 3 leading to minimum width $\geq \lceil \omega(G_I)/2 \rceil > 1$. Suppose now that such a component is a tree but not a caterpillar. Then it should contain three paths, each of length at least two, starting from a common vertex 1, see Fig. 12. Since $\{1,3\} \notin E(G_C)$, the specification must be of form ..3..2..3..1..2..1.. or ..1..2..1..3..2..3.., that is, net 2 must "leave" the interval of net 1 towards "left" or "right". The same holds for net 4 in the other direction and then 6 and 7 cannot be placed anywhere. □

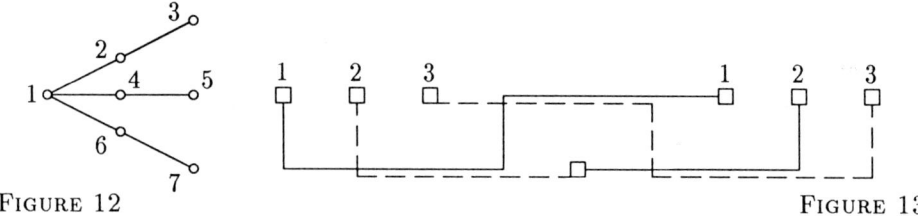

FIGURE 12

FIGURE 13

Remark 2. We have used the density several times: Otherwise Lemma 6 and the stronger upper bound $\lceil \omega(G_I)/2 \rceil$ would not necessarily be true and G_C could contain a circuit, see Fig. 13.

IV. Some results in the channel routing problem

The intersection graph and the crossing graph must be redefined first. Their vertex sets will still be the set of nets. Two nets $N = (\{t_1, \ldots, t_u\}, \{b_1, \ldots, b_l\})$ and $N' = (\{t'_1, \ldots, t'_{u'}\}, \{b'_1, \ldots, b'_{l'}\})$ with $t_1 < t_2 < \ldots < t_u$ etc. are *intersecting* if the intervals

$$[\min\{t_1, b_1\}, \max\{t_u, b_l\}] \quad \text{and} \quad [\min\{t'_1, b'_1\}, \max\{t'_{u'}, b'_{l'}\}]$$

are non-disjoint and they are *crossing* if there are some subscripts $1 \leq i \leq u$, $1 \leq i' \leq u'$, $1 \leq j \leq l$, $1 \leq j' \leq l'$ so that

$$(t_i - t'_{i'})(b_j - b'_{j'}) < 0$$

The channel routing problem (CRP) is solvable in the SLVM if and only if no two nets are crossing. A general formula for the minimum width can be found in [8]. Deciding the solvability of a CRP is NP-complete in the TLMM ([9], [13]) but it is polynomial in the SLEM, even if terminals are allowed on all the four sides of the rectangle ([1]). Every CRP is solvable in the TLUM ([10]; see also a linear time algorithm in [12]) but the complexity of finding the minimum width seems to be an open problem.

In what follows, some special cases of the above results are mentioned for the CRP with 2-terminal nets only.

Proposition 7. *The CRP with 2-terminal nets only is solvable in the SLVM if and only if no two nets are crossing. In this case (also called river routing, see [2]) the minimum width of the layout (the answer to (P2)) is the smallest number k having the property*

$$k \geqslant n \quad \text{or} \quad t_p \geqslant b_{p-k} + k \quad \text{for} \quad p = k+1, \ldots, n.$$

Proposition 8 [1]. *Suppose that at least one of the four corners of the grid does not belong to any net. Then the 2-terminal CRP is solvable in the SLEM and the minimum width equals the maximum congestion $\omega(G_I)$.*

Fig. 14 illustrates the proposition. If all the four corners belong to some nets, the problem may be unsolvable. The solvable cases are characterized in the same paper [1] but they are more complicated and not included here. Example D of Fig. 6 is the simplest unrealizable specification.

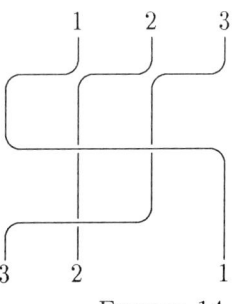

FIGURE 14

This example also shows the reason for the unrealizability in the TLMM. Let us define a constraint digraph so that the vertices are the 2-terminal nets and there is a directed edge (N_1, N_2) from the net $N_1 = (t_1, b_1)$ to the net $N_2 = (t_2, b_2)$ if and only if $t_1 = b_2$. Such an edge (N_1, N_2) indicates that if the wires of the layout should contain only a single horizontal line segment for each net then the segment corresponding to N_1 must lay "higher" than that corresponding to N_2, see Fig. 15. If the horizontal connections within a net can be realized by several line segments (also called *doglegs*) then the vertex set of the constraint digraph must contain several copies of this net, see Fig. 16. In particular, we obtain

Proposition 9. *A 2-terminal CRP with no doglegs permitted is solvable in the TLMM if and only if the constraint digraph contains no directed circuits. If this condition is met then the length of the longest directed path is a lower bound for the minimum width.*

A. Frank conjectures that the length of the longest directed path plus the maximum congestion (plus perhaps 1 or 2) is an upper bound for the minimum width.

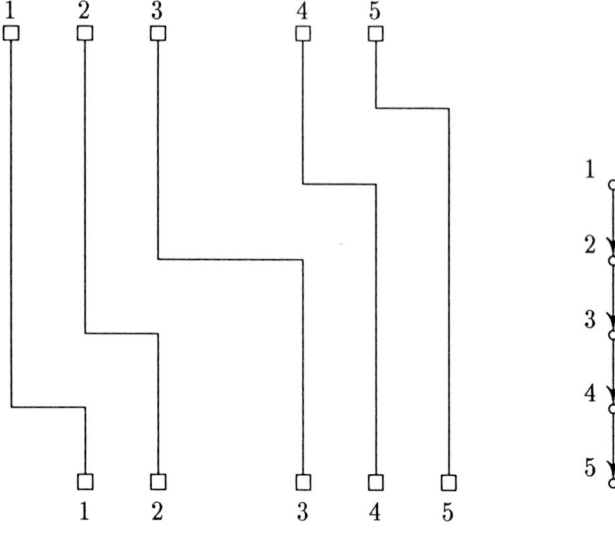

FIGURE 15

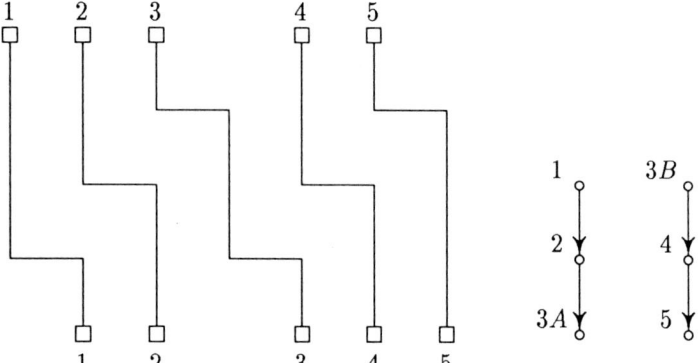

FIGURE 16

If doglegs are permitted, the cyclic constrains may be eliminated by realizing some nets with several horizontal line segments. These nets must correspond to a *vertex feedback set* of the constraint digraph (i.e. a set of vertices which together cover every directed circuit). Deciding whether there exists a feedback set with given cardinality is known to be NP-complete.

We saw that every CRP is solvable in the TLUM. Fig. 17 shows such a realization, it indicates the general principle as well: Nets of the shape \ are realized first, in decreasing order of their lower coordinate, then nets of the shape / in increasing order of their upper coordinate. The width is clearly not minimum, see a compactification in Fig. 18.

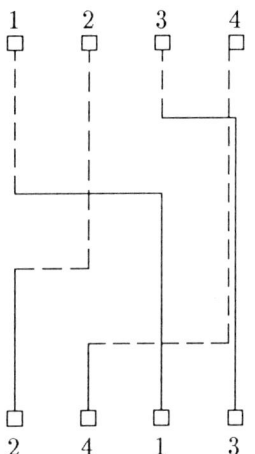
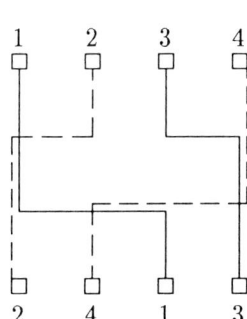

FIGURE 17 FIGURE 18

Lemma 10. *If a dense 2-terminal CRP is solvable in the TLUM with width 1 or 2 then no via hole is required.*

Proof. Suppose that a layout contains a via hole V next to a terminal node N. By the density N must be connected to V on, say, the first side of the board and then a line leaves V to some direction on the second side. Delete V and relocate the line section NV from the first side to the second one. □

This observation motivates our study in the via-free layouts of dense 2-terminal CRP's.

Proposition 11 (Recski and Strzyzewski, 1990). *A dense 2-terminal CRP is solvable in the TLUM without via holes if and only if its crossing graph G_C is bipartite. The minimal width w equals the maximum degree in G_C.*

The first statement is trivial: Bipartiteness is necessary for via-free routing on two layers and the two bipartition classes determine two separate river routing problems.

Remark 3. *Crossing graphs arising from 2-terminal CRP are just the permutation graphs, see [4], for example. They may contain chordless circuits but of length 4 only, see Lemma 1.3 in [11].*

As a corollary of Proposition 11 one can clearly obtain

Proposition 12. *A dense 2-terminal CRP with crossing graph G_C is solvable in the TLUM with width 1 if and only if G_C consists of isolated vertices and vertex-disjoint edges only, and it is solvable with width 2 if and only if G_C consists of vertex disjoint paths (possible isolated vertices included) and even circuits only.*

Even the characterization of the CRP's which are solvable with width 3 is apparently open. Some partial results are presented in [12]. We only mention the following conjecture here, which is proved for $w = 3$. Let $n(L)$ be the number of via holes adjacent to a vertical line L. Clearly, $n(L) \leqslant 2(w-2)$ can be obtained for a dense 2-terminal CRP, by Lemma 10. We conjecture that if a dense 2-terminal CRP is solvable in TLUM with width w then there exists a layout with

$$c(L) + 2n(L) \leqslant 2w \quad \text{for every } L.$$

Acknowledgements. Part of the research was performed during a stay at the Research Institute of Discrete Mathematics, Bonn. The work was partially supported by the Hungarian Academy of Sciences (Contract No. OTKA 1059) and by the Alexander-von-Humboldt Foundation. Useful conversations with I. Abos, G. Bacsó, V. Chvátal, A. Frank, B. Korte, M. Middendorf, A. Sebö and F. Strzyzewski are gratefully acknowledged.

References

[1] A. Frank, *Disjoint paths in a rectilinear grid*, Combinatorica **2** (1982), 361–371.

[2] D. Dolev, K. Karplus, A. Siegel, A. Strong and J. D. Ullman, *Optimal wiring between rectangles*, Proc. *13th STOC Symp.*, 1981, pp. 312–317.

[3] S. Gao and M. Kaufmann, *Channel routing of multiterminal nets*, Proc. *28th FOCS Symp.*, 1987, pp. 316–325.

[4] M. C. Golumbic, *Algorithmic graph theory and perfect graphs*, Academic Press, New York, 1980.

[5] D. S. Johnson, *The NP-Completeness column: an ongoing guide*, Journal of Algorithms **3** (1982), 381–395.

[6] D. S. Johnson, *The NP-Completeness column: an ongoing guide*, Journal of Algorithms **5** (1984), 147–160.

[7] B. W. Kernighan, D. G. Schweikert and G. Persky, *An optimum channel-routing algorithm for polycell layouts of integrated circuits*, Proc. *10th Design Automation Workshop*, 1973, pp. 50–59.

[8] M. Koebe and P. Dupont, *Single-layer channel routing*, J. Inf. Process. Cybern. EIK (formerly Elektron. Inf.-verarb. Kybern.) **24**, 7/8 (1988), 339–354.

[9] A. S. LaPaugh, *A polynomial time algorithm for optimal routing around a rectangle*, Proc. *21st FOCS Symp.*, 1980, pp. 282–293.

[10] M. Marek-Sadowska and E. Kuh, *General channel-routing algorithm*, Proc. *IEE (GB), 130, G, 3*, 1983, pp. 83–88.

[11] A. Recski, *2-layer routing of dense bipartite specifications with vertex-disjoint paths and via-holes*, Working paper 88550, University of Bonn, Research Institute of Discrete Mathematics.

[12] A. Recski and F. Strzyzewski, *Vertex disjoint channel routing on two layers*, Integer Programming and Combinatorial Optimization (R. Kannan and W. Pulleyblank, eds.), Waterloo, 1990, pp. 397–405.

[13] T. G. Szymanski, *Dogleg channel routing is NP-complete*, unpublished manuscript, quoted in [5].
[14] T. Yoshimura and E. Kuh, *Efficient algorithms for channel routing*, IEEE Trans. CAD-1 (1982), 25–35.

András Recski
Dept. Math., Fac. Electr. Eng.,
Technical Univ. Budapest,
H–1521 Budapest, Hungary

Critical Perfect Systems of Difference Sets

D. G. ROGERS

1. Definitions, Notation and an Illustrative Example

We call a vector $\mathbf{a} = (a_0, a_1, \ldots, a_n)$ of integers a_i, $0 \leq i \leq n$, arranged such that $0 = a_0 < a_1 < \cdots < a_n$, a *component*, and the set $D(\mathbf{a}) = \{a_i - a_j : 0 \leq j < i \leq n\}$ the *difference set* of the component $\mathbf{a}$. Both $\mathbf{a}$ and $D(\mathbf{a})$ are said to have *valency* n; and $D(\mathbf{a})$ is *full* when all the $\frac{1}{2}n(n+1)$ differences $a_i - a_j$ are distinct.

We depict components and their difference sets as in Figure 1, which shows the example $\mathbf{a} = (0, 8, 18, 40, 51)$: $\mathbf{a}$ appears as the leading diagonal above the bar and the iterated differences appear as successive diagonals below the bar to form a *difference triangle*. A key equality for difference triangles is that the sum of entries in the kth row from the apex is equal to the sum of entries in the kth row from the bottom (this goes back to [9]); like other equalities for difference triangles, this may be proved by expressing all entries as sums of entries in the bottom row.

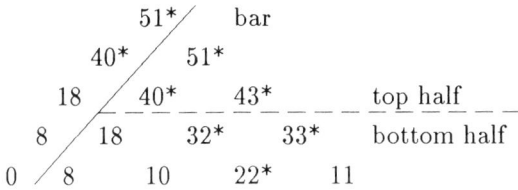

FIGURE 1: A Difference Triangle

Now the example in Figure 1 fits together with the difference triangles shown in Figure 2 to give a partition of the consecutive *run* of integers from 6 to 51 (inclusive) into full difference sets of valency at least 2 (as noted in [2]). A partition of this sort is called a *perfect system of difference sets*, the least difference is called the *threshold* (in our example, 6).

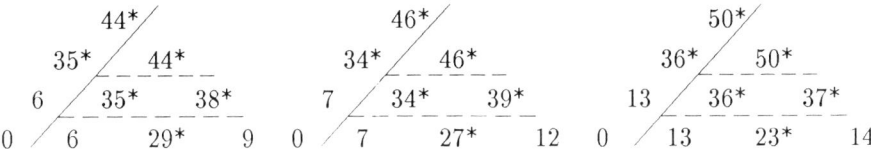

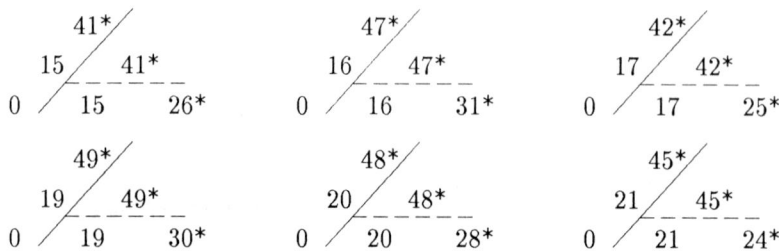

FIGURE 2: Further Difference Triangles

Perfect systems of difference sets have applications in convolutional coding, missile guidance and the layout of radio telescopes; they also give rise to certain cyclic block designs or, equivalently, in the guise of graceful labellings, to decompositions of certain complete graphs into edge-disjoint copies of the complete graph. For a survey, see [1]. We continue this survey here using an example to illustrate more recent results and open problems.

The perfect system shown in Figures 1 and 2 partitions the run of integers from 6 to 51 in two further ways. The top halves of the difference triangles taken together form a subrun (from 41 to 51), as do the bottom halves (from 6 to 33). A perfect system with this property is called *critical*. Notice that if we sum the key equality over the top and bottom halves of the difference triangles of a perfect system and bound the resulting sums above and below by sums of runs of consecutive integers, then we obtain an inequality, known as the *BKT Inequality* (see [1, 3, 14]), which holds with equality precisely when the system is critical.

Next, if the entries in the components marked by an asterisk are all increased by some positive integer p then the differences from 6 to 21 (unmarked) remain unchanged, but the differences from 22 to 51 (marked with an asterisk) are all increased by p. A perfect system in which the run of integers splits apart at x (in this case, 22) is said to have a *split* at x; the splitting point x is like a second threshold and splits allow perfect systems to be pulled apart and reassembled into larger systems (*addition*; see [11]).

2. Results

If there is a critical perfect system with m difference sets of valency n and threshold c then $n = 2$ or 3 and $m = 2c - 1$; for $n = 2$, such *regular* systems of *valency* n are known for all c, but, for $n = 3$, the existence problem remains unsettled. However, the following recent result shows that these latter systems, where they exist, have special significance:

Theorem. *There are no critical perfect systems with threshold c without difference sets of valency 2, other than those consisting of $2c - 1$ difference sets of valency 3.*

A proof appears in [16], based on summing over the following equality which holds in any difference triangle of valency n, say, at least 3: for $u = [\frac{1}{2}n]$ and $1 \leqslant k \leqslant [\frac{1}{4}(n + 1)]$, the sum of the entries in the $(u + k)$th row from the bottom minus the sum of the entries in the kth row from the apex is equal to the sum of the

$[\frac{1}{2}(u+1)] + 1 - 2k$ centrally placed entries in the uth row from the bottom, where, as usual, $[x]$ denotes the integer part of x. (In fact, this equality holds for more general values of u, but the choice $u = [\frac{1}{2}n]$ facilitates the proof of the theorem; as before, we sum first over k for each difference set of valency at least 3 in the system and then over these difference sets.)

The range of possibilities when there *are* difference sets of valency 2 is, as yet, unknown; Figures 1 and 2 show that difference sets of valency greater than 3 can then appear. An interesting case occurs when difference sets are of mixed valency 2 or 5: in this case, if there is a critical perfect system with threshold c with m_n difference sets of valency n for $n = 2$ or 5, then, by the BKT Inequality, $m_2 = 2c - 1$ without any apparent restriction on m_5. But this is only apparent, since summing over a further equality gives $m_5 \leqslant 2c - 1$, which may be strengthened to $m_5 \leqslant c - 1$ if the system has a split at $3c + 6m_5 - 1$, the lowest value for which a split may occur.

For these systems, a structual analysis can be given (see [5, 12]) leading to our second recent result.

Theorem. *If there is a critical perfect system with threshold c, with m_n difference sets of valency n for $n = 2$ or 5, and with a split at $3c + 6m_5 - 1$ for some c, then there is such a system for all sufficiently large c. Moreover, there are such systems at least when $m_5 = 1, 5, 6$ and 7, but not for $m_5 = 2, 3$ or 4.*

For the cases $m_5 = 2, 3$ and 4, partial results appear in [6].

3. Complete Permutations

The theory of perfect systems involves at several points the allied notion of a *complete permutation*; a permutation π of the set N_c of integers in modulus less than c is *complete* when the set of differences $\pi(i) - i$, for $|i| < c$, is again the set N_c. If π is a complete permutation of N_c, then the difference sets of the components $(0, 4c + i - 2, 6c + \pi(i) - 3)$, for $|i| < c$, form a critical perfect system with threshold c (and split at $3c - 1$). Since there are complete permutations of N_c for all positive integers c (indeed, their number increases rapidly with c), this ensures the existence of regular critical perfect systems of valency 2.

More generally, we say that an n-tuple $(\pi_1, \ldots, \pi_n)$ of permutations π_r, $1 \leqslant r \leqslant n$, of N_c is a *completely compatible set* of valency n for N_c when π_1 is the identity permutation and $\pi_r \pi_s^{-1}$, $1 \leqslant s < r \leqslant n$, are complete permutations of N_c. The result just mentioned is then the first instance of a multiplication theorem for perfect systems (see [1, 8, 10]): if there is a (critical) perfect system of m difference sets, all of valency at most n, with threshold c and a completely compatible set of valency n for N_d, then there is a (critical) perfect system of $(2d-1)m$ difference sets with threshold $2cd - c - d + 1$. Thus it follows that, since the difference set of $(0, 1, 4, 6)$ is a regular critical perfect system of valency 3 and there are completely compatible sets of valency 3 for N_c for infinitely many values of c, there are regular critical perfect systems of valency 3 for N_c for infinitely many values of c. While there are *no* regular critical perfect systems of valency 4 (or greater), completely compatible sets of valency 4 for N_c are known, again for infinitely many values of c, so that our illustrative example in Section 1 may be multiplied to give infinite family of examples showing that the condition that there are no difference sets of valency 2 in our first theorem cannot be relaxed. But nothing is known of completely compatible sets of valency n for N_c when $n \geqslant 5$ and the precise conditions under which such sets exist is a challenging open problem (see, further, [7, 10]).

Because of this ignorance, our proof of the second theorem does not depend on multiplication. However, the structual analysis of the critical perfect systems considered there reveals that the difference sets of valency 2 arise from complete permutations where now the permutation is partially specified. An arithmetic of complete permutations subject to such *constraints* is developed in [4, 13, 15]. For example, underlying the second theorem is the result that if partial specification which is independent of c is satisfied by a complete permutation of N_c for *some* c, then it is satisfied by a complete permutation of N_c for *all* sufficiently large c.

Acknowledgements

It is a rare privilege and pleasure to speak on the subject of perfect systems of difference sets in Czechoslovakia since the two Republics are united by the joint researches in this area of Anton Kotzig and Jaromir Abrham. When I met Professor Abrham last, in Singapore in 1986, we little thought that we should be able to meet again in Czechoslovakia, certainly not so soon. I am sorry that Professor Kotzig was unable to attend, but he is well represented by his work.

I am glad to acknowledge the support and hospitality of the University of Aberdeen.

References

[1] J. Abrham, *Perfect systems of difference sets—a survey*, Ars Combin. **17A** (1986), 5–36.

[2] J. Abrham, A. Kotzig and P. J. Laufer, *Remarks on the minimum number of components in perfect systems of difference sets*, Congress. Numer. **52** (1986), 7–19.

[3] J. C. Bermond, A. Kotzig and J. Turgeon, *On a combinatorial problem of antennas in radio astronomy*, Proc. Fifth Hungarian Combinatorial Colloquium, Keszthely, 1976, North Holland, Amsterdam, 1976, pp. 135–149.

[4] D. J. Crampin and D. G. Rogers, *An arithmetic of complete permutations with constraints, II: Case studies*, Discrete Math. **70** (1988), 241–256.

[5] G. M. Hamilton and D. G. Rogers, *Further results on irregular, critical perfect systems of difference sets, I: Split systems*, Discrete Math., to appear.

[6] T. Hayasaka, S. Saito and D. G. Rogers, *Further results on irregular, critical perfect systems of difference sets, II: Systems without splits*, Discrete Math., to appear.

[7] P. D. Johnson and D. G. Rogers, *Compatible additive permutations on finite integral bases*, Graphs and Combinatorics **2** (1986), 43–53.

[8] A. Kotzig and J. Turgeon, *Perfect systems of difference sets and additive sequences of permutations*, Proc. Tenth SE Conf. on Combinatorics, Graph Theory and Computing, Boca Raton, 1979, Vol 2 (Utilitas Mathematica Winnipeg 1979), pp. 629–636.

[9] B. Lindström, *An inequality for B_2-sequences*, J. Combinatorial Theory **6** (1969), 211–212.

[10] D. G. Rogers, *A multiplication theorem for perfect systems of difference sets*, Discrete Math., to appear.

[11] _____, *Addition theorems for perfect systems of difference sets*, J. London Math. Soc. **(2) 23** (1981), 385–395.

[12] _____, *Irregular, extremal perfect systems of difference sets*, J. London Math. Soc. **(2) 34** (1986), 19–211.

[13] _____, *An arithmetic of complete permutations with constraints, I: An exposition of the general theory*, Discrete Math. **70** (1988), 219–240.

[14] _____, *Critical perfect systems of difference sets with components of odd size*, Quart. J. Math (Oxford) **(2) 39** (1988), 501–512.

[15] _____, *On the general Erdős conjecture for perfect systems of difference sets and embedding partial complete permutations*, to appear.

[16] _____, *Critical perfect systems of difference sets without components of size two*, to appear.

D. G. Rogers
Department of Mathematical Sciences,
The University, Aberdeen, UK, AB9 2TY

Fernley House, The Green,
Croxley Green, UK, WD3 3HT

Fourth Czechoslovakian Symposium on
Combinatorics, Graphs and Complexity
J. Nešetřil and M. Fiedler (Editors)
© 1992 Elsevier Science Publishers B.V. All rights reserved.

Some Operations (Not) Preserving the Integer Rounding Property

ANNA RYCERZ

Operations that preserve the total dual integrality of linear systems of inequalities $Ax \leqslant b$ have been investigated by Cook and Schrijver. A generalization of this problem consists in considering linear systems with the integer rounding property and the behaviour of this property under operations. This problem is studied in the present note.

1. Introduction

Let A be a rational $m \times n$-matrix, and let b be a rational vector of length m. Several polyhedra occurring in combinatorics are described by linear systems of inequalities $Ax \leqslant b$ which have the integer rounding property (cf. [2], [4]).

Baum and Trotter [2] defined a linear system $Ax \leqslant b$ to have the *integer rounding property* if

$$\min\{yb \colon yA = w,\ y \geqslant 0,\ y \text{ integral}\} = \lceil \min\{yb \colon yA = w,\ y \geqslant 0\} \rceil \qquad (1.1)$$

for each integral vector w for which $\min\{yb \colon yA = w,\ y \geqslant 0\}$ is finite. (If q is a rational number, then $\lceil q \rceil$ is the least integer greater than or equal to q.)

This property is considered in [4] as some generalization of the total dual integrality. The system of inequalities $Ax \leqslant b$ is called *totally dual integral* if the minimum in the linear programming duality equation

$$\max\{wx \colon Ax \leqslant b\} = \min\{yb \colon yA = w,\ y \geqslant 0\} \qquad (1.2)$$

has an integral optimal solution for each integral vector w for which the optima exist.

In general the total dual integrality is also defined for non-integral matrices A, but here we are only interested in integral matrices.

Cook [3] and Schrijver [7] have considered operations that preserve the total dual integrality of linear inequality systems. These operations are useful because

---

Support of KBN of Poland in 1990, Grant P/05/003/90–2, is acknowledged.

they allow certain assumptions to be made on the form of a linear system without losing any generality. Some of these operations also preserve the integer rounding property of systems of linear inequalities (cf. [6]). In this note we present further results of the behaviour of the integer rounding property under operations.

2. Operations preserving the integer rounding property

Giles and Orlin [5] (cf. also [7, Thm. 22. 18]) make clear that the integer rounding property can be characterized in terms of Hilbert bases. A set of integral vectors $a_1, \ldots, a_k$ forms a Hilbert basis if each integral vector in the convex cone spanned by $a_1, \ldots, a_k$ is a nonnegative integer combination of $a_1, \ldots, a_k$.

Theorem 2.1 [5]. *The system $Ax \leqslant b$, with b integral, has the integer rounding property if and only if the rows of the matrix*

$$\begin{pmatrix} A & b \\ 0 & 1 \end{pmatrix} \tag{2.1}$$

form a Hilbert basis or equivalently the system

$$\begin{pmatrix} A & b \\ 0 & 1 \end{pmatrix} X \leqslant 0, \tag{2.2}$$

where $X = (x, x_0)^T \in \mathbb{R}^{n+1}$ and $x = (x_1, \ldots, x_n)$, is totally dual integral.

This theorem allows us to prove some propositions given below. The following three propositions are given in [6], but for completeness they are included here.

Throughout this paper we assume that A and C are integral matrices with at least one non-zero column and b, d are integral vectors.

τ-homogenization

For any polyhedron $P = \{x \in \mathbb{R}^n : Ax \leqslant b\}$ and $\tau \in \{-1, 1\}$ we define the τ-homogenization of P to be

$$\tau\text{-hog}(P) := \{(x, \tau)^T \in \mathbb{R}^{n+1} : x \in P\}^{00}, \tag{2.3}$$

where for $T \in \mathbb{R}^{n+1}$, $T^0 = \{y \in \mathbb{R}^{n+1} : xy \leqslant 0 \; \forall x \in T\}$ denotes the polar cone of T. It is known (cf. [1]) that

$$\tau\text{-hog}(P) = \{X = (x, x_0)^T \in \mathbb{R}^{n+1} : Ax - \tau b x_0 \leqslant 0, \; \tau x_0 \geqslant 0\} =$$
$$= \{X : B_\tau X \leqslant 0\}, \tag{2.4}$$

where

$$B_\tau = \begin{pmatrix} A & -\tau b \\ 0 & -\tau \end{pmatrix}. \tag{2.5}$$

Proposition 2.2. *If the system $Ax \leqslant b$ has the integer rounding property, then the system $B_\tau X \leqslant 0$ also has the integer rounding property.*

Translation

Proposition 2.3. *If $Ax \leqslant b$ has the integer rounding property and c is an integral vector, then the system $Ax \leqslant b + Ac$ has the integer rounding property.*

Nonnegative integer combination

Proposition 2.4. *If $Ax \leqslant b$ and $Cx \leqslant d$ define the same polyhedron, and each inequality in $Ax \leqslant b$ is a nonnegative integer combination of inequalities in $Cx \leqslant d$, then $Ax \leqslant b$ having the integer rounding property implies $Cx \leqslant d$ has the integer rounding property.*

Unimodular transformations

Let U be an integral $n \times n$-matrix such that $\det(U) = \pm 1$.

Proposition 2.5. *If $Ax \leqslant b$ has the integer rounding property, then $AUx \leqslant b$ also has the integer rounding property.*

Proof. Let $Ax \leqslant b$ have the integer rounding property. By Theorem 2.1 it suffices to show that the rows of the matrix

$$\begin{pmatrix} AU & b \\ 0 & 1 \end{pmatrix} \qquad (2.6)$$

form a Hilbert basis. Note that

$$\begin{pmatrix} AU & b \\ 0 & 1 \end{pmatrix} = \begin{pmatrix} A & b \\ 0 & 1 \end{pmatrix} \begin{pmatrix} U & 0 \\ 0 & 1 \end{pmatrix}.$$

Using three facts: 1^0 with U also the matrix $\begin{pmatrix} U & 0 \\ 0 & 1 \end{pmatrix}$ is unimodular, 2^0 the system $\begin{pmatrix} A & b \\ 0 & 1 \end{pmatrix} X \leqslant 0$ is totally dual integral and 3^0 the result of Cook [3] which says that total dual integrality is maintained under unimodular transformation, we have that the system $\begin{pmatrix} AU & b \\ 0 & 1 \end{pmatrix} X \leqslant 0$ is totally dual integral or equivalently the rows of the matrix (2.6) form a Hilbert basis. □

Repeating variables

The integer rounding property is maintained also under the following operation.

Proposition 2.6. *If $Ax \leqslant b$ has the integer rounding property, then the system $Ax + A_n x_{n+1} \leqslant b$, where A_n is the n-th column of A and x_{n+1} is a new variable, has the integer rounding property also.*

Proof. To see that the system $Ax + A_n x_{n+1} \leqslant b$ has the integer rounding property it suffices to show that the rows of the matrix

$$\begin{pmatrix} A & A_n & b \\ 0 & 0 & 1 \end{pmatrix} \qquad (2.7)$$

form a Hilbert basis. To see this, let (w, w_{n+1}, w_0) be an integral vector in the cone generated by the rows of the matrix (2.7). So there exist $y \geqslant 0$, $y_0 \geqslant 0$ with

$$w = yA, \quad w_{n+1} = yA_n, \quad w_0 = yb + y_0. \tag{2.8}$$

(2.8) implies $w_{n+1} = w_n$. Moreover (w, w_0) is an integral vector in the cone generated by the rows of $\begin{pmatrix} A & b \\ 0 & 1 \end{pmatrix}$. As $Ax \leqslant b$ has the integer rounding property, there exist an integral vector $\bar{y} \geqslant 0$ and an integer $\bar{y}_0 \geqslant 0$ such that $w = \bar{y}A$, $w_0 = \bar{y}b + \bar{y}_0$. This gives $w = \bar{y}A$, $w_{n+1} = \bar{y}A_n$, $w_0 = \bar{y}b + \bar{y}_0$. So (w, w_{n+1}, w_0) is a nonnegative integer combination of the rows of (2.7), i.e. the rows of the matrix (2.7) form a Hilbert basis. □

3. Operations not preserving the integer rounding property

The following two transformations generally do not preserve the integer rounding property.

Removing variables

Proposition 3.1. *If $Ax + ax_0 \leqslant b$, where a is an integral column vector, has the integer rounding property, then the system $Ax \leqslant b$ does not necessarily have the integer rounding property.*

Example. Let $A = 2$, $a = 3$ and $b = 1$, $2x + 3x_0 \leqslant 1$ has the integer rounding property because the rows of the matrix $\begin{pmatrix} 2 & 3 & 1 \\ 0 & 0 & 1 \end{pmatrix}$ form a Hilbert basis. But $2x \leqslant 1$ does not have the integer rounding property as the rows of the matrix $\begin{pmatrix} 2 & 1 \\ 0 & 1 \end{pmatrix}$ do not form a Hilbert basis.

Finally, we state another negative result which can be shown immediately.

Adding right-hand sides

Proposition 3.2. *If $Ax \leqslant b$ and $Ax \leqslant d$ have the integer rounding property, then $Ax \leqslant b + d$ has not always the integer rounding property.*

References

[1] A. Bachem and M. Grötschel, *New Aspects of Polyhedral Theory*, Modern Applied Mathematics-Optimization and Operations Research (B. Korte, eds.), North-Holland, 1982, pp. 51–106.

[2] S. Baum and L. Trotter, *Finite Checkability for Integer Rounding Properties in Combinatorial Programming Problems*, Mathematical Programming **22** (1982), 141–147.

[3] W. Cook, *Operations that Preserve Total Dual Integrality*, Operations Research Letters **2** (1983), 31–35.

[4] W. Cook, J. Fonlupt and A. Schrijver, *An Integer Analogue of Carathéodory's Theorem*, Journal of Combinatorial Theory B **40** (1986), 63–70.
[5] R. Giles and J. Orlin, *Verifying Total Dual Integrality*, Manuscript.
[6] A. Rycerz, *Operations that Preserve the Integer Rounding Property of Linear Inequality Systems*, Graphs, Hypergraphs and Matroids III (M. Borowiecki and Z. Skupień, eds.), Higher College of Engineering in Zielona Góra, 1989, pp. 111–116.
[7] A. Schrijver, *Theory of Linear and Integer Programming*, Wiley-Chichester, 1986.

Anna Rycerz
Institute of Mathematics,
University of Mining and Metallurgy,
Kraków, Poland

Fourth Czechoslovakian Symposium on
Combinatorics, Graphs and Complexity
J. Nešetřil and M. Fiedler (Editors)
© 1992 Elsevier Science Publishers B.V. All rights reserved.

Optimal Embedding of a Tree into an Interval Graph in Linear Time

PETRA SCHEFFLER

1. Introduction

An interval supergraph for a given graph is called optimal if its clique number is smallest possible. The problem to find this embedding is NP-hard in general. We present a linear-time algorithm for the case of trees. It is shown that every tree T is a subgraph of an interval graph with clique number less than $\log_3(2|V(T)|+1)$. The considered problem is equivalent to finding an optimal *path-decomposition*, an optimal *node search* strategy or a linear layout with minimum *vertex separation* (see [S]). It has practical applications in VLSI-design (the so called *Gate-Matrix-Layout-Problem*) and in linguistics.

The graphs considered here are finite, undirected and simple. The *interval thickness* $\theta(G)$ of a graph G is the minimum over the clique numbers of all interval graphs having a subgraph G.

Since the problem to determine the interval thickness of a graph is NP-hard [KF], an efficient algorithm for this problem in the general case hardly exists. Nevertheless, we proved:

Theorem 1. *For any fixed integers k and d there is an algorithm deciding in time $O(n^k)$ for a given graph with maximal degree at most d, whether its interval thickness is at most k (see [S]).*

2. The interval thickness of trees

Let T be a tree and t any of its nodes. Every connected component of $T \setminus \{t\}$ is called a *branch of T at t*, it is denoted by T_{ts} if it contains the node s (usually a neighbour of t). If the tree has root w, for every edge (t,s) the node t is called the father of s if t lies on the path from s to w. The maximal subtree of T rooted at t is denoted by T^t. The following crucial theorem is the basis for the proposed algorithm.

Theorem 2. *For any integer $k \geqslant 2$ and any tree T holds: $\theta(T) \geqslant k+1$ iff there exists a node $t \in V(T)$ with at least three branches T_{tv}, T_{tu} and T_{tw} at t such that $\theta(T_{tv}) \geqslant k$, $\theta(T_{tu}) \geqslant k$ and $\theta(T_{tw}) \geqslant k$.*

Proof. ($\Leftarrow$) Assume there is a node $t \in V(T)$ with three branches of interval thickness k. Then in any optimal interval model of T there must be a place for each thickness-k-branch containing only its intervals. But then the middle place intervals separate the other two branches in contradiction to their connectedness in T.

($\Rightarrow$) Let T have interval thickness $k+1$. Then every node t has at least one branch of thickness k or more, since otherwise for T an interval model of thickness k would exist.

Consider the edge set $\mathsf{F} := \{(s,t) \colon \theta(T_{st}) \geqslant k \text{ and } \theta(T_{ts}) \geqslant k\}$ in T. It is not empty since a tree has one more edge than nodes. If there is a node with 3 incident edges in F we are done. So assume F to consist of some paths. Actually, F is connected: Otherwise, there would exist an edge (s,t) in $T \setminus \mathsf{F}$ on a path between two components of F, but then both branches T_{st} and T_{ts} would contain a subtree of thickness at least k, consequently the edge (s,t) would belong to F in contradiction to its choice.

Let the path F consist of the nodes $t_1, \ldots, t_l$ in this order. Then every node t_i (for $2 \leqslant i \leqslant l-1$) has exactly two branches of thickness k or more, namely $T_{t_i t_{i-1}}$ and $T_{t_i t_{i+1}}$. The nodes t_i and t_l have exactly one branch of thickness k or more each, namely $T_{t_1 t_2}$ resp. $T_{t_l t_{l-1}}$. All other branches at a node t_i ($i = 1, \ldots, l$) having thickness smaller than k combine with t_i to a subtree. Its optimal interval supergraph has an interval model of thickness at most k with the interval corresponding to t_i containing all other intervals. By joining these interval models in the order of F (and adding intersections of t_i- with t_{i+1}-intervals) we get an interval embedding of the whole tree T of thickness k. This contradicts the assumption. $\square$

As a corollary we get by induction that the interval thickness of a tree with n nodes is always less than $\log_3(2n+1) + 1$.

Moreover, we may determine the interval thickness for special classes of trees using this theorem. For example, the complete binary tree of height h has interval thickness $\lceil \frac{h}{2} \rceil + 1$.

3. The algorithm

We present the ideas of the linear algorithm proving the following theorem here, all details are contained in [S].

Theorem 3. *There is a linear-time algorithm to determine the interval thickness $\theta(T)$ of a given tree T.*

The algorithm starts with the leaves of the tree and computes for increasing k all pending subtrees with interval thickness k. The root of such a subtree is stored in a FIFO-list $QUEUE(k)$ for $k = 1, \ldots, \log_3(2n+1)$ while it has not been merged to its father. The orientation of the tree is computed by the algorithm. The kernel of the algorithm is a recursive procedure MERGE calculating the interval thickness of a tree obtained joining two subtrees by an edge. Unfortunately, it is not possible to determine the interval thickness of the join correctly given only the interval thickness of the two subtrees. Denote by $\theta_1(t)$ the interval thickness of a subtree T^t rooted at node t. Define in addition $\theta_2(t)$ to be equal to zero if there is an optimal interval model of T^t with the root t as endinterval and let be $\theta_2(T) = 1$ otherwise. Moreover, we need the number of branches at t having maximal interval thickness. Store this in a parameter $\theta_3(t) = |\{s : t = FATHER(s)$ and $\theta_1(s) = \theta_1(t)\}|$. Obviously $0 \leq \theta_3(t) \leq 2$. The triple $\theta(t) = (\theta_1(t), \theta_2(t), \theta_3(t))$ is called the *interval-decompositon-vector* of T^t. We use the following facts proved applying theorem 2 (see [S]):

Lemma 1. *Let the tree T be obtained from the subtrees T^t and T^s joined by the edge (s, t). Then the interval thickness $\theta(T)$ satisfies:* $\max\{\theta_1(t), \theta_1(s)\} \leq \theta(T) \leq \max\{\theta_1(t), \theta_1(s)\} + 1$.

Lemma 2. *Let T^t be an oriented tree with the root t and with the interval-decomposition-vector $\Theta = (\theta_1, \theta_2, \theta_3)$. Then:*
(i) *If $\theta_3 = 0$, then also $\theta_2 = 0$ and T^t has an interval model of width θ_1 with the interval for t containing all other intervals.*
(ii) *If $\theta_2 = 0$, then there is no node ν in the tree T^t possessing two sons of interval thickness as large as θ_1.*
(iii) *If $\theta_3 = 2$, then $\theta_2 = 1$.*
(iv) *If $\theta_2 = 1$, then there is exactly one node ν in the tree T^t possessing two sons of interval thickness θ_1.*

The node ν in the last case (iv) plays an important role: A third branch of it can get interval thickness θ_1 later, when a new subtree T^s is added. Then the interval thickness of T increases by one. This branch is $T_{\nu FATHER(\nu)} = T \setminus T^\nu$. To take account of this fact we store also the interval-decomposition-vector of this subtree and link it by a pointer $REST(t)$ to the vector of T^t. A chain of pointers may occur there, but it is not longer than $\theta_1(t)$ since the rest subtrees must have decreasing interval thickness. According to the following rules the procedure MERGE calculates the interval-decomposition-vector of a join of two trees in time $O(k)$, where k is the interval thickness of the larger subtree:

Lemma 3. *Let a tree T be obtained joining the subtrees T^t and T^s. Let $\theta(t)$ and $\theta(s)$ be their interval-decomposition-vectors. Then the tree T has the following interval-decomposition-vector $\varrho(t)$:*

1. case: $\theta_1(s) > \theta_1(t)$

1a) $\theta_2(s) = 0 \Rightarrow \varrho(t) := (\theta_1(s), 0, 1)$
1b) $\theta_3(s) = 2 \Rightarrow \varrho(t) := (\theta_1(s), 1, 1)$
1c) $\theta_2(s) = \theta_3(s) = 1 \Rightarrow \varrho(t) := \begin{cases} (\theta_1(s) + 1, 0, 0), & \text{if } \theta_1(T \setminus T_\nu) = \theta_1(s) \\ \theta(s) & , \text{otherwise} \end{cases}$

2. case: $\theta_1(s) = \theta_1(t)$

2a) $\theta_2(t) = 1$ or $\theta_2(t) = 1 \Rightarrow \varrho(t) := (\theta_1(s) + 1, 0, 0)$
2b) $\theta_2(s) = \theta_2(t) = 0$ and $\theta_3(t) = 0 \Rightarrow \varrho(t) := (\theta_1(s), 0, 1)$

2c) $\theta_2(s) = \theta_2(t) = 0$ and $\theta_3(t) = 1 \Rightarrow \varrho(t) := (\theta_1(s), 1, 2)$

3. case: $\theta_1(s) < \theta_1(t)$

3a) $\theta_2(t) = 0 \qquad \Rightarrow \varrho(t) := \theta(t)$
3b) $\theta_3(t) = 2 \qquad \Rightarrow \varrho(t) := \theta(t)$
3c) $\theta_2(t) = \theta_3(t) = 1 \Rightarrow \varrho(t) := \begin{cases} (\theta_1(t) + 1, 0, 0), & \text{if } \theta_1(T \setminus T_v) = \theta_1(t) \\ \theta(t) & , \text{otherwise.} \end{cases}$

It should be clear, how these rules are applied for an algorithm MERGE (with recursion in cases 1c) and 3c)) having worst case time complexity $O(\theta(T))$. In addition the main program MIN_THICKNESS is used, organizing the calls of the procedure MERGE for all edges $(s,t) \in E(T)$ in the needed order. The interval-decomposition-vector of every node is initialized to $\theta := (1,1,2)$. For the leaves this is the final vector. Their fathers can also get an easy treatment. Obviously, for all nodes t all sons of which are leaves we have $\theta(t) = (2,0,0)$. This will be used in step (1). Notice that the procedure MERGE is called at most once for any node s, so we get at all time complexity $\sum_s O(\theta(T_s)) = \sum_s O(\log |V(T_s)|) = O(n)$.

For the correctness proof observe yet, that at least two nodes are contained in the QUEUEs until the last step. During the last step (when $i = n-1$) the interval-decomposition-vector of the root is updated by merging the last of its branches to T_w.

```
PROCEDURE MINTHICKNESS (T)
//Input:  tree T with n = |V(T)|; output:  interval thickness θ(T)//
  (0) FOR all t ∈ V(T) DO θ(t) := (1,1,2); i := 0 ENDFOR
      //Preparation:  leaves are trivial branches//
  (1) FOR all leaves s DO i := i + 1;
          IF i < n THEN label s ENDIF
          the only node with (s,t) ∈ E(T) denote by t
          FATHER(s) := t;  θ(t) := (2,0,0)
          IF t has only one unlabeled neighbour v THEN
              FATHER(t) := v;  Put t into QUEUE(2)
          ENDIF
      ENDFOR
      //Treatment of the inner nodes of the tree//
  (2) WHILE i < n DO k := the smallest number of a nonempty QUEUE
          take s out of QUEUE(k)
          i := i + 1; label s; t := FATHER(s)
          CALL MERGE (T^t, T^s, T^t)
          IF t has only one unlabeled neighbour v THEN
              FATHER(t) := v;  Put t into QUEUE(θ_1(t))
          ENDIF
      ENDWHILE
```

```
                //Reading the interval thickness of the root//
 (3)   w := the only unlabeled node; FATHER(w) := NIL;  θ := θ₁(w)
       RETURN ("The interval thickness of T is", θ)
       END
```

Recently, we were informed that Ellis, Sudborough and Turner obtained a similar result considering the vertex separation of graphs. Another similar algorithm is contained in the paper of Möhring [M]. Our linear algorithm can be modified easily so that an optimal embedding and an interval model is also constructed.

References

[KF] T. Kashiwabara, T. Fujisawa, *NP-completness of the problem of finding a minimum-clique-number interval graph containing a given graph as a subgraph*, Proc. ISCAS 1979, pp. 657–660.

[M] R. H. Möhring, *Graph problems related to Gate Matrix Layout and PLA folding*, TU Berlin, FB 3, Report 233/1989.

[S] P. Scheffler, *Die Baumweite von Graphen als ein Mass für die Kompliziertheit algorithmischer Probleme*, K.-Weierstraß-Institut für Mathematik, Berlin (R-MATH-04/89), 1989.

[Y] M. Yannakakis, *A polynomial algorithm for the min-cut linear arrangement of trees*, J. Assoc. Comput. Mach. **32** no. 4 (1985), 950–988.

Petra Scheffler
Karl-Weierstraß-Institut für Mathematik,
Berlin, Germany

Construction of Polytopal Graphs

W. SCHÖNE

Let be $P_2^6 = C(6,4)$ the cyclic 4-polytope with 6 vertices. The graph $G = (X, W)$ of P_2^6 is the complete graph K_6 with the vertex set $X = \{1, 2, \ldots, 6\}$. We denote the facets of P_2^6 by F_k and their vertex sets by I_k where $k = 1, 2, \ldots, 9$. We unite the sets to a set $S = \{I_k : I_k \subset X, \ k = 1, 2, \ldots, 9\}$ called the facet set of G. For $k = 1, 2, \ldots, 9$ we name by G_k the subgraph of G induced by I_k. We can assign to G another graph G^* which is the graph of a 4-polytope P_2^{6*} dual to P_2^6. We label the vertices of P_2^{6*} and G^* by $I_1, I_2, \ldots, I_9$. Two vertices I_j, I_k of G^* are joined by an edge if and only if $I_j \cap I_k = U_\alpha$, $\alpha \in \{1, 2, \ldots, 18\}$ where U_α is the vertex set of a subfacet F_α^5 of P_2^6. In this case we label the edge $u_\alpha = (I_j, I_k)$ of G^* by U_α. To G^* belongs a facet set

$$S^* = \{I_1^*, I_2^*, \ldots, I_7^*\} \quad \text{with } I_l^* \subset S.$$

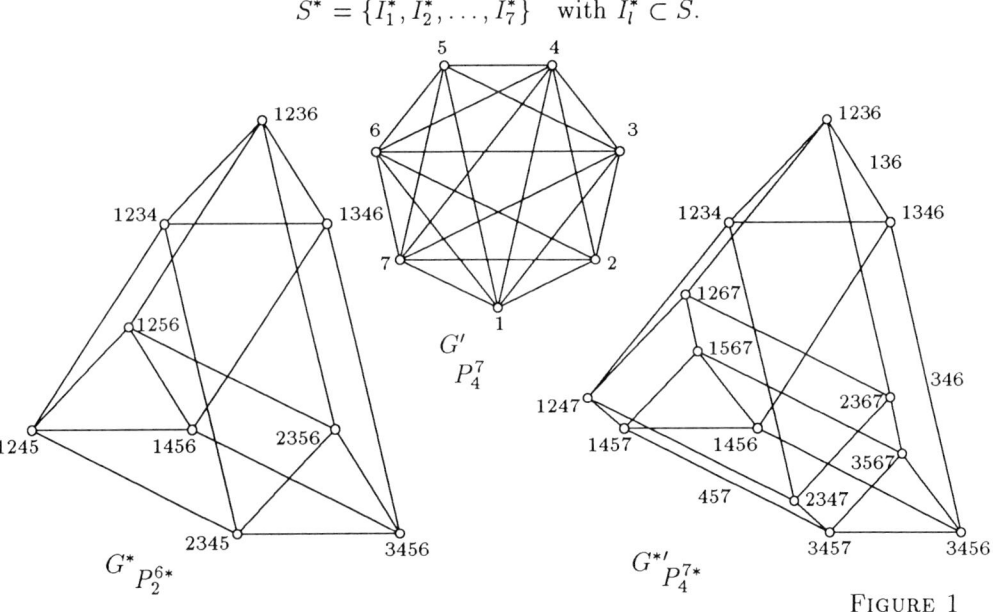

FIGURE 1

Fig. 1 shows the graph G^* with the set S. We take a k-face with the vertex set $I \subset X$ of P_2^6. For example we choose $I = \{2, 5\}$. So we have to put $k = 1$. Then we delete from S all facets containing I and we get so the set $F(I)$ of remaining facets. Further we consider the set

$$F^S(I) = \{U_\alpha : \text{It exists one and only one } I_k \in F(I) \text{ with } U_\alpha \subset I_k\}.$$

Now we can define an operation $\omega_{1,I}$ which transforms the graph G of P_2^6 in a polytopal graph $G' = (X', W')$ with 7 vertices.

We find $X' = X \cup \{7\}$ and $S' = F(I) \cup \{(U_\alpha \cup \{7\}): U_\alpha \in F^S(I)\}$ as a facet set of G'. In our example we have

$$S' = \{\{1,2,3,4\}, \{1,2,3,6\}, \{1,2,4,7\}, \{1,2,6,7\},$$
$$\{1,4,5,7\}, \{1,5,6,7\}, \{1,4,5,6\}, \{2,3,6,7\}, \{2,3,4,7\},$$
$$\{3,4,5,6\}, \{3,4,5,7\}, \{3,5,6,7\}, \{1,3,4,6\}\}.$$

The edge set W' of G' consists of all edges of the subgraphs G_k with $V(G_k) = I_k \in F(I)$ and of all edges which join the new vertex 7 with every vertex belonging to a set U_α of $F^S(I)$. The graph G' is the graph of the simplicial 4-polytope P_4^7 in the notation of B. Grünbaum ([1]). We write $P_4^7 = \omega_{1,I} P_2^6$. To $\omega_{1,I}$ belongs a dual operation $\omega_{1,I}^*$ which transforms the dual graph G^* of G with $S^* = \{S_1^*, S_2^*, \ldots, S_6^*\}$ in the dual graph $G^{*\prime}$ with $S^{*\prime} = \{S_1^{*\prime}, S_2^{*\prime}, \ldots, S_7^{*\prime}\}$. The operation $\omega_{1,I}^*$, can be realized geometrically by cutting off all and only all vertices of $S \setminus F(I)$ from the polytope P_2^{6*} by a hyperplane. All edges of P_2^{6*} which join a vertex of $F(I)$ with a vertex of $S \setminus F(I)$ are also cut by this hyperplane. We delete in G^* all vertices of $S \setminus F(I)$ and all edges joining these vertices and also all halves of cut edges incident with the deleted vertices. At the end of the remaining half of every cut edge labeled with U_α we put a new vertex and denote it by $U_\alpha \cup \{7\}$. Two new vertices U_α and $U_\beta \cup \{7\}$ are joined by a new edge with the label $(U_\alpha \cup \{7\}) \cap (U_\beta \cup \{7\})$ if and only if the sets $U_\alpha \cup \{7\}$ and $U_\beta \cup \{7\}$ have three common elements. We write $\omega_{1,I}^* P_2^{6*} = P_4^{7*}$. Analogously to $\omega_{1,I}$ and $\omega_{1,I}^*$ we can define two other operations $\omega_{1,U}$ and $\omega_{1,U}^*$. Here the set $U = \{U_{ik}: k = 1, 2, \ldots, r\}$ is a set of vertex sets of subfacets of P_2^6. To every U_{ik} there corresponds an edge u_{ik} ($k = 1, 2, \ldots, r$) with the label U_{ik} in G^*. The set U has the property that the subgraph G^r of G^* generated by the edges u_{ik} ($k = 1, 2, \ldots, r$) is connected and must be a proper subgraph of the 1-skeleton of a facet of P_2^{6*}. We delete from S all facets which contain an element U_{ik} of U and we denote by $F(U)$ the set of remaining facets in S. Then we put $F^S(U) = \{U_\alpha: \text{It exists one and only one } I_k \in F(U) \text{ with } U_\alpha \subset I_k\}$.

So we get a new polytopal graph $G'' = (X'', W'')$ with $X'' = X \cup \{7\}$ and $S'' = F(U) \cup \{(U_\alpha \cup \{7\}): U_\alpha \in F^S(U)\}$, where W'' consists of all edges of the subgraphs G_k with $I_k \in F(U)$ and of all edges joining the vertex 7 with every vertex belonging to a subfacet with a vertex set $U_\alpha \in F^S(U)$. The dual operation $\omega_{1,U}^*$ can be defined by cutting off all vertices and edges of the subgraph G^r by a hyperplane from the polytope P_2^{6*}. The operations $\omega_{1,U}$ and $\omega_{1,U}^*$ cannot be realized geometrically in every case. But in the case that to U corresponds a set of edges which is a path of G^* contained in the graph of a 2-face of P_2^{6*} then the operations $\omega_{1,U}$ and $\omega_{1,U}^*$ are geometrically realizable. This remains true for all simplicial d-polytopes. By the operations $\omega_{1,I}$ and $\omega^*_{1,U}$, where U corresponds to such a path we get all 37 combinatorial types of simplicial 4-polytopes with 8 vertices (or of simple 4-polytopes with 8 facets) from the 4-simplex.

For an example we choose $U = \{\{2,4,5\},\{2,3,5\}\}$ and we get from the polytope P_2^6 respectively P_2^{6*} the sets $F(U) = \{\{1,2,3,4\},\{1,2,3,6\},\{1,3,4,6\}\}$ and

$$F^S(U) = \{\{1,2,4\},\{1,2,6\},\{1,4,5\},\{1,5,6\},\{2,3,4\},$$
$$\{2,3,6\},\{3,4,5\},\{3,5,6\}\}.$$

We get by $\omega_{1,U}$ from P_2^6 the simplicial 4-polytope P_5^7 with 14 facets ([1]).

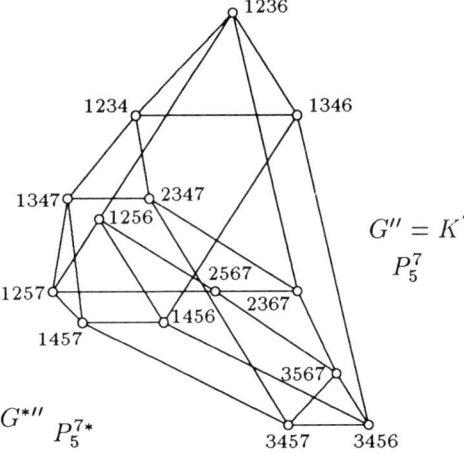

FIGURE 2

In Fig. 2 the graph G'' and the graph $G^{*''}$ of P_5^7 and P_5^{7*} are drawn. The facet set of G'' is given by

$$S'' = \{\{1,2,3,4\},\{1,2,3,6\},\{1,2,4,7\},\{1,2,5,6\},$$
$$\{1,2,5,7\},\{1,3,4,6\},\{1,4,5,6\},\{1,4,5,7\},\{2,3,4,7\},$$
$$\{2,3,6,7\},\{2,5,6,7\},\{3,4,5,6\},\{3,4,5,7\},\{3,5,6,7\}\}.$$

The graph G is the complete graph with 7 vertices.

References

[1] B. Grünbaum, V. P. Sreedharan, *An Enumeration of Simplicial 4-Polytopes with 8 Vertices*, Journal of Combinatorial Theory **2** no. 4 (1967).

[2] W. Schöne, *Special systems of linear equations and graphs of convex polytopes*, to be reprinted in the Ringel-Festschrift 1990.

Dr. W. Schöne
Technische Universität Chemnitz,
Sektion Mathematik,
PSF 964,
9010 Chemnitz, Germany

Fourth Czechoslovakian Symposium on
Combinatorics, Graphs and Complexity
J. Nešetřil and M. Fiedler (Editors)
© 1992 Elsevier Science Publishers B.V. All rights reserved.

More About Two-Graphs

J. J. Seidel

1. Introduction

The notion of a regular two-graph was proposed in 1970 by G. Higman as a setting for the 2-transitive representation for certain simple groups. This idea was worked out by Taylor [24]. It turned out that strong relations exist with switching of graphs [15]. At two occasions [20], [21] the present author has been involved in surveying the subject of two-graphs. In addition, the Tables of two-graphs [5] in part served the same purpose. Several new results during the past 10 years again justify a view back.

In one of our surveys [21] we stated that in the subject various parts of mathematics are interrelated, such as linear algebra and matrix methods, cohomology over $\mathbb{F}_2$, configurations in Euclidean space, finite geometries, and computational techniques. Apart from further illustrations, the present paper adds to this list presentation of groups and approximation theory.

The subsections are as follows:

1. Introduction
2. Definitions
3. Enumeration
4. Representation
5. Two-graphs from trees
6. Presentation of groups
7. Regular two-graphs
8. Symmetric Hadamard (36)
9. Cocliques
10. Approximation and projection
11. References

2. Definitions

A *two-graph* (V, Δ) consists of a set V (the vertices) and a collection Δ of triples of vertices (the odd triples), such that for each 4 vertices the number of odd triples is even. In a *regular two-graph* each pair of vertices is in a constant number of odd triples.

Any (simple) graph $\Gamma = (V, E)$ gives rise to a two-graph (V, Δ) as follows. The odd triples of Δ are the triples of vertices which carry an odd number of edges of Γ. One checks easily that each graph on 4 vertices has either 4 or 2 or no odd triples. We shall see that different graphs can give rise to the same two-graph.

Our first *example* deals with the pentagon graph on $\{2,3,4,5,6\}$ extended by the isolated vertex 1. This graph $\Gamma = (V, E)$ is determined by its (± 1)-adjacency matrix E_6 (in which -1 indicates adjacency). The odd triples are:

$$\Delta = \{123,\ 134,\ 145,\ 156,\ 162,\ 523,\ 634,\ 245,\ 356,\ 462\}.$$

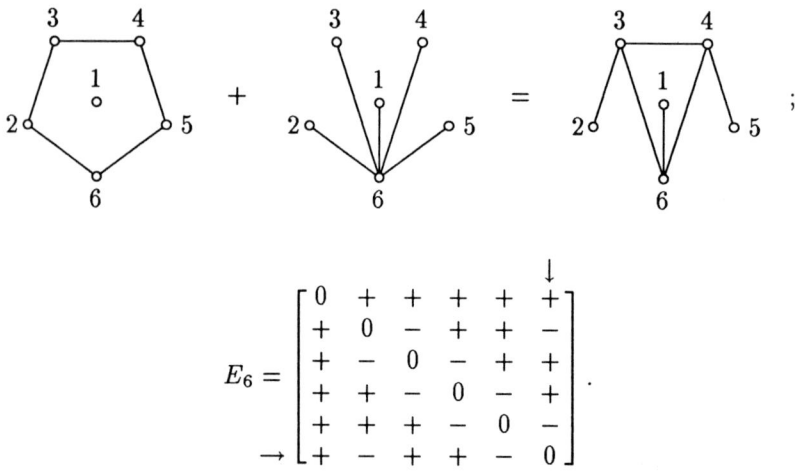

$$E_6 = \begin{bmatrix} 0 & + & + & + & + & + \\ + & 0 & - & + & + & - \\ + & - & 0 & - & + & + \\ + & + & - & 0 & - & + \\ + & + & + & - & 0 & - \\ + & - & + & + & - & 0 \end{bmatrix}.$$

We observe that the same set of odd triples holds for the graph $\Gamma' := \Gamma + $ star$_6$ (mod 2), whose (± 1)-adjacency matrix E'_6 agrees with E_6 apart from row 6 and column 6 which are multiplied by -1. We say that Γ' is obtained from Γ by *switching* with respect to vertex 6. Switching generates an equivalence relation, so there is a whole switching class which gives rise to the same two-graph.

In general, if a graph $\Gamma = (V, E)$ gives rise to the two-graph (V, Δ), then so does any graph in its *switching class*. This switching class is described by the mod 2 sums $\Gamma + B$, where B is any complete bipartite graph on V. This switching class is also described by the adjacency matrices DED, where E is the (± 1)-adjacency matrix of Γ and D is any diagonal matrix with diagonal entries ± 1. The eigenvalues of DED are those of E. Since there is a one-to-one correspondence between two-graphs and switching classes of graphs, we shall often transfer from one notion to the other. The *eigenvalues* of the two-graph will be those of the (± 1)-adjacency matrix of any graph in its switching class. We refer to the surveys for further details [5], [20], [21].

3. Enumeration

Let $N(n)$ denote the number of nonisomorphic two-graphs on n vertices. Then

$n =$	3	4	5	6	7	8	9
$N(n) =$	2	3	7	16	54	243	2038

The tables of [5] list all these two-graphs, and give several further details. Mallows and Sloane [16] found the equicardinality of two-graphs and Euler graphs, cf. [6] and [5] which also contain general Burnside-type formulae. The first tables [15], which include the eigenvalues of the (± 1)-adjacency matrices for $n \leqslant 7$, go back to 1966. Recently, they gained renewed interest in relation to presentations of groups, cf. Section 6. We quote two theorems from [22] which together determine all two-graphs whose (± 1)-adjacency matrix E has smallest eigenvalue $\geqslant -3$.

Theorem 3.1. *A graph (V, E) with positive definite $3I + E$ is switching equivalent to the void graph on n vertices, to the one-edge graph on n vertices, or to one of the following $2 + 3 + 5$ graphs on $5, 6, 7$ vertices, resp.*

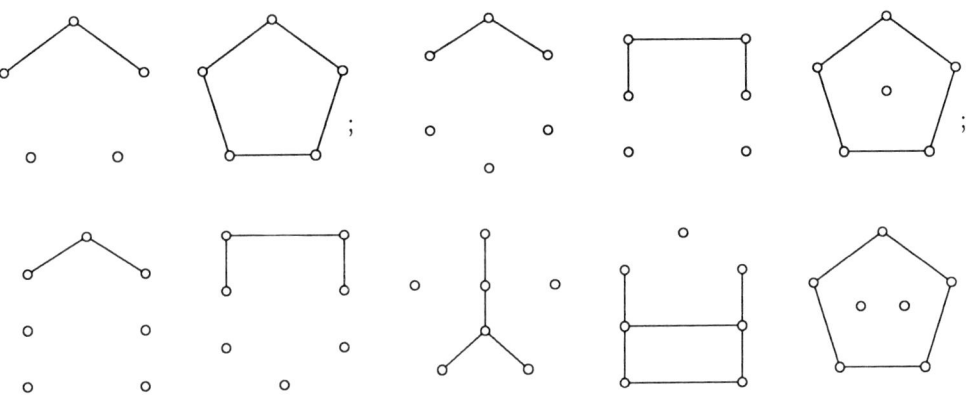

Theorem 3.2. *A graph (V, E) with positive semidefinite $3I + E$ is switching equivalent to a subgraph of $n\, K_2$ (disjoint union of n edges), or of the complement of the triangular graph on 8 symbols (the complement of the line graph of K_8).*

The first theorem follows from the tables, by induction for $n \geqslant 10$. The second theorem is a consequence of the classification of Euclidean root lattices, cf. [4], Thm. 3.13.1.

4. Representation

Any switching class of graphs (hence any two-graph) may be represented geometrically as a set of *equiangular lines* [15]. To see this, we consider the matrix $\varrho I + E$, where E is the (± 1)-adjacency matrix of a graph on n vertices, and $(-\varrho) = \eta_{\min}$ is the smallest eigenvalue of E, of multiplicity $n - d$ say. Then $\varrho I + E$ is positive semidefinite of rank d. By linear algebra, $\varrho I + E$ is the Gram matrix of the inner products of n vectors in Euclidean d-space. These vectors have equal norm ϱ, and mutual inner products ± 1. Hence the n lines spanned by these n vectors have pairwise the same angle with $\cos \varphi = 1/\varrho$. In the example in Section 2 these equiangular lines are the 6 diameters of the regular icosahedron in $\mathbb{R}^3$ at $\cos \varphi = 1/\sqrt{5}$.

If ϱ is taken to be any real number, then we obtain from $\varrho I + E$ n equiangular lines in Euclidean n-space if $\varrho > -\eta_{\min}$, or in indefinite space if $\varrho < -\eta_{\min}$. In

Section 3 we took $\varrho = 3$, and the matrix $3I + E$ to be positive (semi)definite of rank d, with $d \leqslant n$. This leads to a representation of the two-graph in terms of roots (vectors of norm 2), cf. [3], [8], [22].

To that purpose we let $\frac{1}{2}(3I + E)$ denote the Gram matrix of n vectors $\xi_1, \ldots, \xi_n$ in $\mathbf{R}^d$ with rank $(3I + E) = d \leqslant n$. We extend $\mathbf{R}^d$ to $\mathbf{R}^{d+1}$ by use of a vector $\sigma_0 \in \mathbf{R}^{d+1}$, $\sigma_0 \perp \mathbf{R}^d$, $(\sigma_0, \sigma_0) = 2$. Then we define

$$\sigma_i := \xi_i + \frac{1}{2}\sigma_0, \qquad \sigma'_i := -\xi_i + \frac{1}{2}\sigma_0, \qquad i = 1, \ldots, n.$$

The vectors thus obtained satisfy, for $i \neq j$,

$$(\sigma_i, \sigma_i) = (\sigma'_i, \sigma'_i) = 2, \qquad (\sigma_i, \sigma'_i) = (\sigma_i, -\sigma_0) = (\sigma'_i, -\sigma_0) = -1,$$
$$(\sigma_i, \sigma_j) = (\sigma'_i, \sigma'_j) = 1 - (\sigma_i, \sigma'_j) = 1 - (\sigma'_i, \sigma_j) \in \{0, 1\}.$$

For $\Sigma := \{\sigma_0, \sigma_0, \ldots, \sigma_n\}$ this yields

$$\text{Gram } \Sigma = \begin{bmatrix} 2 & j^t \\ j & \frac{1}{2}(3I + E + J) \end{bmatrix} = \begin{bmatrix} 2 & j^t \\ j & 2I + B \end{bmatrix},$$

where B is the (1,0) adjacency matrix of the complement of the graph $\Gamma = (V, E)$. Thus we have represented the two-graph (V, Δ) as a set of *roots* Σ on lines at $60°$ and $90°$ in $\mathbf{R}^{d+1}$. In the positive definite case these roots are contained in the well-known root systems of type A_{n+1}, D_{n+1}, E_{n+1}, cf. [8].

5. Two-graphs arising from trees

Cameron [7] constructs a two-graph (V, Δ) on n vertices from a given connected tree T on $n+1$ vertices as follows. Let V be the set of the n edges of T. The triples of edges of T are of two kinds: one edge is situated between the other two, or none is between the other two. Let $\Delta = \Delta(T)$ be the set of the triples of edges of T of the kind: none between the others. Considering the possibilities for quadruples of edges we see that $(V, \Delta(T))$ is a two-graph. We say that the two-graph $\Delta(T)$ *arises from the tree* T. The corresponding switching class of graphs is described as follows. Take any orientation of the edges of T. Construct the graph with vertex set V and adjacency of vertices iff the corresponding edges in T are oppositely directed on the path joining them. The graphs thus obtained form a switching class corresponding to the two-graph $(V, \Delta(T))$. Switching with respect to a set U of edges of T corresponds to reversing the orientation of the edges in U.

Examples. The switching classes of the graphs

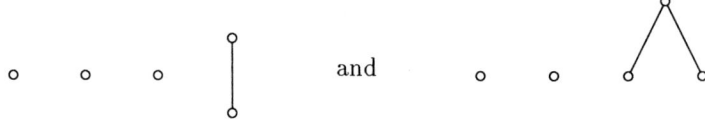

arise from the trees

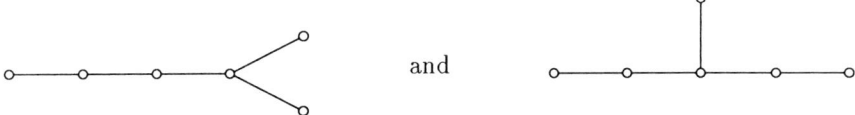

respectively.

The pentagon graph provides the smallest example of a two-graph which does not arise from a tree. For higher orders the two-graphs not arising from a tree provide a fast growing majority among the family of all two-graphs. Cameron [7] gave the following characterization in terms of forbidden substructures.

Theorem 5.1. *A two-graph Δ arises from a tree T iff it contains neither the pentagon nor the hexagon as an induced substructure. Moreover, if $\Delta(T_1)$ and $\Delta(T_2)$ are isomorphic, then so are T_1 and T_2.*

In the literature we found two further characterizations of special two-graphs in terms of forbidden substructures.

Theorem 5.2 is due to Frankl and Füredi [10]. They list two examples of two-graphs which contain no 4-cliques, and prove that these are all.

Example (i). As the example of Section 2 (the graph consisting of the pentagon and an isolated vertex), but each vertex replaced by any coclique.

Example (ii). The vertices of any regular $(2k+1)$-gon in the plane, the odd triples being those whose convex hull contains the center of the $(2k+1)$-gon.

Theorem 5.2. *The two-graphs containing no 4-cliques are those of examples (i) and (ii).*

Theorem 5.3 is due to Kratochvíl [14]. It characterizes 1-code-perfect two-graphs (each graph in the switching class has a 1-perfect code).

Theorem 5.3. *The two-graphs containing no pentagon and no quadrangle are the disjoint unions of three cliques.*

6. Groups represented by two-graphs

Let $\Gamma = (V, E)$ denote a graph with vertex set $V = \{1, 2, \ldots, n\}$ and (± 1)-adjacency matrix $E = [\varepsilon_{ij}]$ having smallest eigenvalue $\eta_{\min}$. We will be interested in the following *Tsaranov group*, cf. [22], defined in terms of generators and relations:

$$\mathrm{Ts}(\Gamma) := \left\langle x_1, x_2, \ldots, x_n : x_i^3 = 1, (x_i x_j^{\varepsilon_{ij}})^2 = 1, i \neq j = 1, \ldots, n \right\rangle.$$

Example. If Γ is the void graph, then the relations are $x_i^3 = 1$ and $(x_i x_j)^2 = 1$ for $i \neq j$; the group is the alternating group on $(n+2)$ symbols, cf. [9], p. 66.

Example. If Γ is the pentagon graph, then

$$\mathrm{Ts}(\Gamma) = \left\langle x_1, \ldots, x_5 : x_i^3 = 1 = (x_i x_{i+1}^{-1})^2 = (x_i x_{i+2})^2, \; i \bmod 5 \right\rangle$$

is the 27 lines group $U_4(2) = Sp_4(3) = O_6^-(2) = \mathrm{Cox}^+(E_6)$. The last group is the index 2 subgroup of the even elements of the Coxeter group of type E_6.

Replacing in $\mathrm{Ts}(\Gamma)$ any generator x_i by its inverse x_i^{-1} amounts to switching the graph Γ with respect to the vertex i. Hence we have

Theorem 6.1. *Switching equivalent graphs have isomorphic Tsaranov groups.*

Thus, the group $\text{Ts}(\Gamma)$ really depends on the two-graph Δ defined by Γ; we shall denote the group by $\text{Ts}(\Delta)$.

So far, results have been obtained for groups $\text{Ts}(\Delta)$ whose two-graphs are those occurring in Theorems 3.1, 3.2 and 5.1. We mention some of these from [22].

Theorem 6.2. *The group $\text{Ts}(\Delta)$ of the two-graph Δ is finite iff the (± 1)-adjacency matrix E has smallest eigenvalue > -3. If so, then $\text{Ts}(\Delta) = \text{Cox}^+(\Pi)$, where Π is a Coxeter-Dynkin graph of type A, D or E.*

Theorem 6.3. *If the two-graph $\Delta(T)$ arises from a tree T, then*

$$\text{Ts}(\Delta(T)) = \text{Cox}^+(T).$$

The *Coxeter groups* occurring above are those defined on a graph $\Pi = (\Pi, A)$ with vertex set Π and $(1,0)$-adjacency matrix A:

$$\text{Cox}(\Pi) := \langle \Pi \colon \pi_i^2 = 1,\ (\pi_i \pi_j)^3 = 1\ (i \sim j),\ (\pi_i \pi_j)^2 = 1\ (i \not\sim j) \rangle.$$

The Witt representation

$$P := \mathbf{Z} \cdot \Pi, \qquad \text{Gram(basis } p_\pi \colon \pi \in \Pi) = 2I - A,$$

connects this abstract Coxeter group with the

Weyl group $W(\Pi) := \langle w_\pi \colon P \to P \colon p \mapsto p - (p, p_\pi) p_\pi,\ \pi \in \Pi \rangle$,

Root system $\Phi(\Pi) := \{ w(p_\pi) \colon \pi \in \Pi,\ w \in W \}$.

Indeed, $\text{Cox}(\Pi) \cong W(\Pi)$ by Tits' theorem [3].

The Tsaranov group $\text{Ts}(\Delta)$ admits a similar representation [22]. Consider the index 2 extension $\text{Ts}^*(\Delta) = \text{Ts}(\Delta) \rtimes \langle x_0 \rangle$, where the automorphism x_0 maps each element of $\text{Ts}(\Delta)$ onto its inverse. Recalling from Section 4 the representation of Δ as a set of roots Σ, we define the lattice

$$L := \mathbf{Z} \cdot \Sigma, \qquad \text{Gram}(\sigma_0, \sigma_1, \ldots, \sigma_n) = \begin{bmatrix} 2 & j^t \\ j & \frac{1}{2}(3I + E + J) \end{bmatrix}.$$

Then the corresponding Weyl group is a quotient for $\text{Ts}^*(\Delta)$. However, it is not clear to us whether this representation of $\text{Ts}(\Delta)$ is faithful.

7. Regular two-graphs

For a graph (V, E) on n vertices, let the (± 1)-adjacency matrix E have the eigenvalues $\lambda_1, \ldots, \lambda_d$ and $(-\varrho)$ of multiplicity $(n - d)$. Then

$$\text{trace } E = 0 = \lambda_1 + \ldots + \lambda_d - (n - d)\varrho,$$
$$\text{trace } E^2 = n(n - 1) = \lambda_1^2 + \ldots + \lambda_d^2 + (n - d)\varrho^2.$$

These equations imply the inequality

$$\varrho^2(n - d) \leq d(n - 1), \qquad \text{equality iff} \qquad \lambda_1 = \ldots = \lambda_d.$$

The case of equality is of great interest for the two-graph which contains (V, E) in its switching class.

Definition 1. *A two-graph is regular if it has only two eigenvalues:*
$$(E + \varrho I)(E - \lambda I) = 0.$$

Definition 2. *A two-graph is regular if each pair of vertices is in a constant number of odd triples.*

The definitions are equivalent. The constant of definition 2 equals $\frac{1}{2}(\lambda + 1) \times (\varrho - 1)$, and $n = 1 + \lambda\varrho$. It follows that λ and ϱ are odd integers, unless $E^2 = \varrho^2 I$, the case of a conference matrix [20]. Below follows a state of affairs for regular two-graphs. The list is complete for $n \leqslant 30$, and not quite complete for $n \leqslant 50$. For $50 < n \leqslant 276$ we mention two open cases, and three spectacular regular two-graphs on 126, 176, 276 vertices. Their 2-transitive automorphism group are the unitary $U(3, 5^2)$, the Higman-Sims, and the Conway ·3 group, respectively, cf. [24], [20].

n	6	10	14	16	18	26	28	30	36
λ	$\sqrt{5}$	3	$\sqrt{13}$	5	$\sqrt{17}$	5	9	$\sqrt{29}$	7
ϱ	$\sqrt{5}$	3	$\sqrt{13}$	3	$\sqrt{17}$	5	3	$\sqrt{29}$	5
N	$\bar{1}$	$\bar{1}$	$\bar{1}$	$\bar{1}$	$\bar{1}$	$\bar{4}$	$\bar{1}$	$\bar{6}$	227

n	38	42	46	50	76	96	126	176	276
λ	$\sqrt{37}$	$\sqrt{41}$	$\sqrt{45}$	7	15	19	25	35	55
ϱ	$\sqrt{37}$	$\sqrt{41}$	$\sqrt{45}$	7	5	5	5	5	5
N	11	18	97	27	?	?	1	1	$\bar{1}$

Here n is the number of vertices, λ and $(-\varrho)$ are the eigenvalues, N is the number of known nonisomorphic solutions, and a bar denotes that all solutions have been established.

8. Symmetric Hadamard (36)

$$H^2 = 36I, \quad H = H^t = I + E, \quad (E - 5I)(E + 7I) = 0,$$

define a regular two-graph on 36 vertices. The following constructions are known:
  11 from Latin squares of order 6, cf. [5],
  80 from Steiner triple systems of order 15, cf. [5],
  136 new ones by Spence [23], by use of a computer.
Most of the new ones are of the following shape:

$$H = \begin{bmatrix} A & N \\ N^t & B \end{bmatrix} \begin{matrix} 20 \\ 16 \end{matrix}, \quad A = \begin{bmatrix} I+P & I-P \\ I-P & I+P \end{bmatrix} \begin{matrix} 10 \\ 10 \end{matrix}, \quad \begin{matrix} P^2 = 9I, \\ PJ = 3J, \end{matrix}$$

where P is the (± 1)-adjacency matrix of the Petersen graph. Via the defining equations of these hypotheses we calculate eigenvalues:

$$A^2 + NN^t = 36I, \quad AN + NB = 0, \quad B^2 + N^t N = 36I,$$

matrix: $\quad A^2 \quad\quad\quad A \quad\quad\quad NN^t \quad\quad N^t N \quad\quad B^2 \quad\quad\quad B$
spec: $\quad 36^{10}4^{10} \quad 6^5(-6)^5 2^{10} \quad 0^{10}32^{10} \quad 0^6 32^{10} \quad 4^{10}36^6 \quad (-2)^{10} 6^6.$

Hence $B^2 + 4B - 12I = 0$, $B - I$ is the matrix of a regular two-graph of size 16, and B is unique. Now we can introduce a 10×16 matrix M such that

$$NN^t = 16 \begin{bmatrix} I & I \\ I & I \end{bmatrix}, \quad N = \begin{bmatrix} M \\ M \end{bmatrix}, \quad \begin{matrix} MM^t = 16I \\ M^tM = 12I - 2B \end{matrix}, \quad MB + 2M = 0.$$

The 16 columns of M have $\cos\varphi = \pm\frac{1}{5}$, and constitute a eutactic star in 10-space. We make one column into the all-one column j, and put

$$M = D_1[C \; j]D_2, \quad D_i = \text{diag}(\pm 1),$$
$$C = BIBD(10, 4; 15, 6, 2), \quad \text{quasisymmetric}.$$

Gronau [11] showed that there are 3 types for C. Thus, the reasoning above identifies 100 new regular two-graphs of order 36, cf. [23].

9. Cliques

A two-graph is a *clique* if all triples of its vertices are odd triples. In regular two-graphs sub-cliques play an important role, cf. [24]. We mention two recent developments involving cliques. First we develop an upper bound for the size c of a clique in a regular two- graph, in terms of the smallest eigenvalue ϱ_2. From

$$A = \begin{bmatrix} I - J & N \\ N^t & B \end{bmatrix} \begin{matrix} c \\ n - c \end{matrix}, \quad (A - \varrho_1 I)(A - \varrho_2 I) = 0,$$

with $\varrho_1 > 0$ and $\varrho_2 < 0$, we infer

$$(I(1 - \varrho_1) - J)(I(1 - \varrho_2) - J) + NN^t = 0,$$
$$(c + \varrho_1 - 1)(c + \varrho_2 - 1) \leq 0, \quad c \leq 1 - \varrho_2.$$

We call *maximal cliques* those with $c = 1 - \varrho_2$.

The following definition was proposed by Haemers [12].

Definition 9.1. *A two-graph geometry is a set C of maximal cliques in a regular two-graph (V, Δ), such that each triple from Δ is in one clique from C.*

Haemers gives two examples. The second example yields an interesting structure, which provides a one-point extension of the sporadic partial geometry with parameters $s = 4$, $t = 17$, $\alpha = 2$, due to Haemers.

Example 9.2. $(n, \varrho_1, \varrho_2) = (10, 3, -3), (16, 5, -3), (28, 9, -3)$. Every odd triple is contained in $1 + \frac{1}{4}(\varrho_1 + 3)(\varrho_2 + 3)$ 4-cliques. For $\varrho_2 = -3$ this number equals 1. By taking C to be the set of all 4-cliques, we obtain three two-graph geometries.

Example 9.3. The Higman-Sims group acts 2-transitively on a regular two-graph with $(n, \varrho_1, \varrho_2) = (176, 35, -5)$. The subgroup M_{22} has an orbit C of size 18480 on the maximal 6-cliques of the two-graph, and every odd triple is contained in one 6-clique of C. This provides a two-graph geometry.

The following definition was proposed by Neumaier [18].

Definition 9.4. *A regular two-graph is completely regular if, for all k, every k-clique is in a constant number a_k of $(k+1)$-cliques.*

Theorem 9.5. *There are unique completely regular two-graphs on 10, 16, 28, 36 and 276 vertices. There are no others, except possibly on 1128 or on 3160 vertices.*

Neumaier [18] proved this theorem up to the existence of the cases 288, 96, 640. These were settled in the negative by Blokhuis and Brouwer [1], by Blokhuis and Wilbrink [2], and by Blokhuis and Haemers (cf. [2]), respectively.

10. Approximation and projection

Consider the problem to approximate, in the p-norm, the unit matrix I_n of size n by the class $\mathcal{A}_d$ of matrices of size $n \times n$ and rank d. The *approximation numbers* a_d are defined by

$$a_d = a_d(I_n; l_1^n, l_p^n) := \min_{A \in \mathcal{A}_d} \|I - A\|_{1,p} = \min_{A \in \mathcal{A}_d} \max_{\|x\|=1} \|x - Ax\|_p =$$
$$= \min_{A \in \mathcal{A}_d} \max_{1 \leqslant k \leqslant n} \|e_k - Ae_k\|_p = \min_{X_d < X_n} \max_{1 \leqslant k \leqslant n} \min_{x \in X_d} \|e_k - x\|_p.$$

Here $e_1, \ldots, e_n$ is an orthonormal basis of the n-space X_n, and X_d runs through all d-subspaces.

For $p = 2$ these numbers are known: $a_d = \sqrt{1 - d/n}$.

For $p \neq 2$ Pinkus [19] and Melkman [17] proved inequalities for a_d.

Theorem 10.1. *For $p = \infty$ the approximation numbers satisfy*

$$a_d \geqslant \left[1 + \sqrt{\frac{d(n-1)}{n-d}}\right]^{-1},$$

and equality holds iff there exists a regular two-graph with multiplicities d and $n - d$.

Similar inequalities, with similar iff conditions for equality, hold for projection constants.

Let X_n denote a Banach space, and let $X_d < X_n$ be a closed complemented subspace (there exists a projection $P \in L(X_n)$ with $P^2 = P$, $P(X_n) = X_d$). The *relative projection constant* of X_d in X_n is defined by

$$\lambda(X_d, X_n) := \inf\{\|P\|: \ P \text{ projection of } X_n \text{ onto } X_d\}.$$

The *absolute projection constant* of X_d is defined by

$$\lambda_d = \lambda(X_d) := \sup\{\lambda(X_d, X_n): \ X_n \geqslant X_d\}.$$

H. König [13] proved the following inequality for λ_d:

Theorem 10.2. *The projection constants satisfy*

$$\lambda_d \leq \frac{d}{n}\left(1 + \sqrt{\frac{(n-d)(n-1)}{d}}\right),$$

and equality holds iff there exists a regular two-graph with multiplicity d and $n-d$.

The connection between these theorems and the notion of a regular two-graph is as follows. Let the adjacency matrix E of a graph in the switching class have the spectrum $d \times \varrho$ and $(n-d) \times (-\sigma)$, then

$$\varrho\sigma = n - 1, \qquad \varrho d - \sigma(n-d) = 0,$$

hence

$$\varrho = \sqrt{\frac{(n-d)(n-1)}{d}}, \qquad \sigma = \sqrt{\frac{d(n-1)}{n-d}},$$

and

$$P := \frac{d}{n}\left(I + \frac{1}{\sigma}E\right) \qquad \text{has spectrum} \qquad d \times 1, \qquad (n-d) \times 0.$$

If this matrix exists (if the regular two-graph exists), then

$$\|P\|_{\infty,\infty} = \max_{0 \leq i \leq n} \sum_{k=1}^{n} |p_{ik}| = \frac{d}{n}\left(1 + \frac{n-1}{\sigma}\right) = \frac{d}{n}(1+\varrho),$$

hence the projection constant $\lambda_d \geq d(1+\varrho)/n$. But König proved $\lambda_d \leq d(1+\varrho)/n$, hence we have equality.

From the regular two-graph we also can make

$$A := \frac{n\sigma}{d(1+\sigma)}P, \qquad \text{with } |\delta_{ij} - a_{ij}| = \frac{1}{1+\sigma}, \qquad \text{for all } 1 \leq i,j \leq n.$$

The approximation number a_d in the case $p = \infty$ satisfies

$$a_d = \min_{A \in \mathcal{A}_d} \|I - A\|_{1,p} = \min_{A \in \mathcal{A}_d} \max_{i,j} |\delta_{ij} - a_{ij}| \leq \frac{1}{1+\sigma}.$$

But Pinkus proved $a_d \geq 1/(1+\sigma)$, hence we have equality.

The converse statements follow from a more complicated reasoning.

References

[1] A. Blokhuis, A. E. Brouwer, *Uniqueness of a Zara graph on 126 points and nonexistence of a completely regular two-graph on 288 points*, Papers dedicated to J. J. Seidel (P. J. de Doelder, J. de Graaf, J. H. van Lint, eds.), Techn. Univ. Eindhoven report 84-WSK-03, 1984, pp. 6–19.

[2] A. Blokhuis, H. Wilbrink, *Characterization theorems for Zara graphs*, Europ. J. Combin. **10** (1989), 57–68.

[3] N. Bourbaki, *Groupes et algèbres de Lie, Ch. IV, V, VI*, Hermann, 1968.
[4] A. E. Brouwer, A. M. Cohen, A. Neumaier, *Distance Regular Graphs*, Springer, 1989.
[5] F. C. Bussemaker, R. A. Mathon, J. J. Seidel, *Tables of two-graphs*, Combinatorics and graph theory, L.N.M. (S. B. Rao, ed.), vol. 885, Springer, 1981, pp. 70–112; *Report 79-WSK-05*, Techn. Univ. Eindhoven, 1979.
[6] P. J. Cameron, *Cohomological aspects of two-graphs*, Math. Zeitschr. **157** (1977), 101–119.
[7] _____, *Two-graphs and trees*, manuscript.
[8] P. J. Cameron, J. M. Goethals, J. J. Seidel, E. E. Shult, *Line graphs, root systems and elliptic geometry*, J. Algebra **43** (1976), 305–327.
[9] H. S. M. Coxeter, W. O. J. Moser, *Generators and relations for discrete groups*, Springer, 1965.
[10] P. Frankl, Z. Füredi, *An exact result for 3-graphs*, Discrete Mathem. **50** (1984), 323–328.
[11] H.-D. Gronau, *The 2-(10, 4, 2) designs*, Rostock Math. Kolloq. **16** (1981), 5–10.
[12] W. H. Haemers, *Regular two-graphs and extensions of partial geometries*, Europ. J. Combin., to appear.
[13] H. König, *Spaces with large projection constants*, Israel J. Mathem. **50** (1985), 181–188.
[14] J. Kratochvíl, *Perfect codes and two-graphs*, Comm. Math. U. Carol. **30** (1989), 755–760.
[15] J. H. van Lint, J. J. Seidel, *Equilateral point sets in elliptic geometry*, Proc. KNAW A **69**; Indag. Math. **28** (1966), 335–348.
[16] C. L. Mallows, N. J. A. Sloane, *Two-graphs, switching classes, and Euler graphs are equal in number*, SIAM J. Appl. Math. **28** (1975), 876–880.
[17] A. A. Melkman, *The distance of a subspace of $\mathbb{R}^m$ from its axes and n-widths of octahedra*, J. Approx. Theory **42** (1984), 245–256.
[18] A. Neumaier, *Completely regular two-graphs*, Arch. Math. **38** (1982), 378–384.
[19] A. Pinkus, *n-widths in approximation theory*, Chapter VI., Springer, 1985.
[20] J. J. Seidel, *A survey of two-graphs*, Coll. Intern. Teorie Combin., Atti dei convegni Lincei 17, Roma, 1976, pp. 481–511.
[21] J. J. Seidel, D. E. Taylor, *Two-graphs, a second survey*, Algebr. Methods in Graph Theory (L. Lovász, Vera T. Sós, eds.), Coll. Math. Soc. J. Bolyai, vol. 25, 1981, pp. 689–711.
[22] J. J. Seidel, S. V. Tsaranov, *Two-graphs, related groups, and root systems*, Bull. Soc. Math. Belgique, to appear.
[23] T. Spence, *New regular two-graphs of order 36*, in preparation.
[24] D. E. Taylor, *Regular two-graphs*, Proc. London Math. Soc. **35** (1977), 257–274.

J. J. Seidel
Vesaliuslaan 26,
5644 HK Eindhoven,
Netherlands

Fourth Czechoslovakian Symposium on
Combinatorics, Graphs and Complexity
J. Nešetřil and M. Fiedler (Editors)
© 1992 Elsevier Science Publishers B.V. All rights reserved.

These are the Two-free Trees

J. SHEEHAN AND C. R. J. CLAPHAM

Using the definition of k-free, a known result can be re-stated as follows: If G is not edge-reconstructible then G is k-free, for all even k. It is known that trees are edge-reconstructible; but an alternative proof of this can be obtained by combining the result above with the result outlined here that, apart from paths, all trees (except a finite number, which are determined) are not 2-free. The approach may be of use in obtaining further results on edge-reconstruction.

1. Introduction

The graphs in this paper are connected simple graphs with n vertices. Such a graph G will be considered as a spanning subgraph of K_n. The following is a definition of 'k-free', though the way that it is used will be seen more clearly below. Indeed, we recommend the reader to take, as the definition to work with, the property given later in Lemma 2.

Definition. *Suppose that G is a graph and that $1 \leqslant k \leqslant |E(G)|$. Then G is k-free if, for every subset A of $E(G)$ with $|A| = |E(G)| - k$, there exists an automorphism φ of K_n such that $E(G) \cap E(\varphi(G)) = A$. A graph is even-free if it is k-free, for all even k.*

Also the following result is known (see Nash-Williams [2]):

Lemma 1. *If G is not edge-reconstructible then, for every subset A of $E(G)$ such that $|A| \equiv |E(G)| \pmod{2}$, there exists an automorphism φ of K_n such that $E(G) \cap E(\varphi(G)) = A$.*

Now this lemma says that if a graph is not edge-reconstructible then it is even-free, and so a graph that is not k-free, for some even value of k, is edge-reconstructible. An investigation into which graphs are, or are not, k-free may therefore bring closer the settling of the Edge-Reconstruction Conjecture.

For example, it is known that trees are edge-reconstructible. An alternative method of establishing this is to show that, apart from paths, all trees (except a finite number, which are determined) are not 2-free. For then, paths and the finite number of other trees that are 2-free can clearly be shown to be edge-reconstructible; and the others, being not 2-free, are also edge-reconstructible. Here we shall outline a proof that the only 2-free trees are those shown in Figure 2. It has to be admitted that a complete account involves a lot of detail.

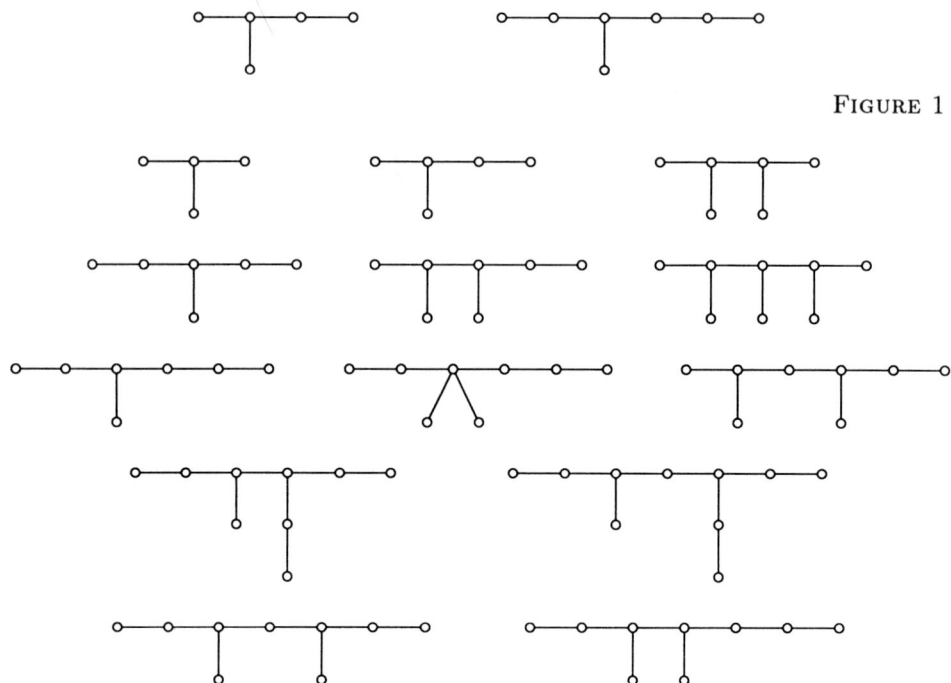

FIGURE 1

FIGURE 2

The definition above was presented in [4] and an investigation into which graphs are k-free, for all k, was begun in [1]. It was shown in [3] that, apart from paths, the only 1-free trees are the two shown in Figure 1.

The following method is used in practice to determine whether or not a graph is k-free. Let $\{e_1, e_2, \ldots, e_k\}$ be a set of edges of a graph G. A *replacing set* is any set of edges that can be added to $G - \{e_1, e_2, \ldots, e_k\}$ to form a graph isomorphic to G. The replacing set $\{f_1, f_2, \ldots, f_k\}$ is called a *disjoint replacing set* if $\{e_1, e_2, \ldots, e_k\} \cap \{f_1, f_2, \ldots, f_k\} = \emptyset$. If there is a disjoint replacing set for $\{e_1, e_2, \ldots, e_k\}$, we shall say that the set $\{e_1, e_2, \ldots, e_k\}$ is *replaceable*. The following is immediate.

Lemma 2. *The graph G is k-free if and only if every set of k edges is replaceable.*

So we shall outline a proof of the following:

Theorem. *Apart from paths, the only 2-free trees are those shown in Figure 2.*

The method will be to show that for all other trees it is possible to find a pair of edges $\{e_1, e_2\}$ that is not replaceable.

2. Sketch of the Proof

It is known that a tree T has either one or two centres. (We may also refer to a centre as a *central* vertex.) Let Z denote the induced subgraph of T whose vertex-set $V(Z)$ consists of the centre(s) of T. We define S, the *spine* of T, to be the induced subgraph whose vertex-set $V(S)$ consists of those vertices that belong to every longest path of T; thus, S is the intersection of all longest paths. Then S is a path and $V(S)$ contains the centre(s). If $v \in V(S)$, v is a *spinal* vertex.

Now let us define the *spinal distance* $\text{spd}(v)$ of a vertex v by

$$\text{spd}(v) = \min\{d(v,s) \mid s \in V(S)\}.$$

This is the distance of v from the spine. Then we shall call T a *spindly* tree if $\text{spd}(v) \leq 1$ for all v, except for a number of 'permissible' vertices with spinal distance 2. [The precise meaning of permissible is this: Suppose that $\text{spd}(v) = 2$. Let s be the spinal vertex distance 2 from v and let α be the centre nearest to s. Then v is *permissible* if either $s = \alpha$ or α and all the vertices between s and α have degree 2.]

The proof we are outlining can be broken into two parts. The first step about which more will be said below is to show that if T is not a spindly tree then a pair of edges that is not replaceable can be found. The second step consists of looking at spindly trees and showing that, again, a non-replaceable pair of edges can be found unless the tree is one of those in Figure 2. The details of this step will not be given here.

The first step, which proves that all 2-free trees are spindly, has to explain how, if the tree is not spindly, a non-replaceable pair of edges can be found. The first edge e_1 is a pendant edge (that is, incident with a vertex of degree 1) chosen in a very particular way; it is the end edge of a 'critical' path, whose definition is the subject of the next section. The second edge e_2 is a non-pendant edge, which can be chosen much more freely; all that is required in the choice of e_2, roughly, is that when e_1 and e_2 are removed, the new tree has the same centre(s) as the original.

3. Critical Path

The purpose then of this section is to arrive at the definition of critical path.

For any non-central vertex v, let us define the *predecessor* v^-: it is the unique vertex adjacent to v and belonging to the component of $T - v$ that contains Z. Thus, if α is a centre, $d(\alpha, v^-) = d(\alpha, v) - 1$. The definition of a successor will also be needed: For any non-central vertex v, let $N^+(v) = N(v) \setminus \{v^-\}$. If $v^+ \in N^+(v)$, v^+ is a *successor* of v. Thus, if α is a centre and v^+ is a successor of v, $d(\alpha, v^+) = d(\alpha, v) + 1$. For a central vertex α, let $N^+(\alpha) = N(\alpha) \setminus V(Z)$. In fact, the successors of v are precisely those vertices that have v as predecessor.

We'll also need this notation: For any non-central vertex v, let T_v be the component of $T - v^-$ that contains v. For central vertices, the appropriate definition is: If $V(Z) = \{\alpha, \beta\}$, let T_α be the component of $T - \beta$ that contains α, and T_β the component of $T - \alpha$ that contains β. If $V(Z) = \{\alpha\}$, let $T_\alpha = T$.

Define now the *weight* $w(v)$ of a vertex v by $w(v) = \min\{|V(T_u)|: u \in N^+(v) \setminus V(S)\}$, if $N^+(v) \not\subseteq V(S)$ (that is, if v has a non-spinal successor), and $w(v) = |V(T_{v^+})|$, if $N^+(v) = \{v^+\} \subseteq V(S)$ (which occurs only if v is spinal and has just one successor v^+).

Now let us call a path in T *centrifugal* (defined in the dictionary as 'tending away from the centre') if it is defined by vertices $v_0, v_1, \ldots, v_s$, where $s \geqslant 1$, $v_0 \in V(Z)$, each v_i is a successor of v_{i-1}, and $\deg(v_s) = 1$. For such a centrifugal path P, let $\lambda(P)$ be defined by

$$V(S) \cap \{v_0, v_1, \ldots, v_s\} = \{v_0, \ldots, v_{\lambda(P)}\}.$$

Thus $\lambda(P)$ is the length of the part that P has in common with the spine. Now let

$$\lambda_T = \min\{\lambda(P): P \text{ is a centrifugal path of } T\}.$$

It follows that λ_T is the minimum distance from a centre to a vertex of degree 1 or $\geqslant 3$. (If a centre has degree $\geqslant 3$, $\lambda_T = 0$.)

For any path P in T, defined by vertices $v_0, \ldots, v_r$, where $r \geqslant 1$, $v_0 \in V(Z)$ and each v_i is a successor of v_{i-1} (with v_r not necessarily of degree 1), define the *weight sequence* $\mathbf{w}(P)$ by

$$\mathbf{w}(P) = \bigl(w(v_0), \ldots, w(v_r), \ldots\bigr).$$

We shall want to compare weight sequences and this will be done by defining the following ordering. Let $\mathbf{w} = (w_0, w_1, w_2, \ldots)$ and $\mathbf{x} = (x_0, x_1, x_2, \ldots)$ be sequences of non-negative integers (with finitely many non-zero). Write $\mathbf{w} < \mathbf{x}$ if, for some i, $w_1 = x_1, \ldots, w_{i-1} = x_{i-1}$ and $w_i < x_i$.

So we arrive at the definition of a critical path as a centrifugal path with certain minimality conditions:

Definition. *A centrifugal path P, defined by vertices $v_0, \ldots, v_s$, is critical if*
   (i) $\lambda(P) = \lambda_T$,
   (ii) *For $i = 1, \ldots, s$, $|V(T_{v_i})| = w(v_{i-1})$, and $v_i \notin V(S)$ if $N^+(v_{i-1}) \not\subseteq V(S)$ (that is, v_i is a successor of v_{i-1} that is non-spinal (unless v_{i-1} is spinal with only one successor), with $|V(T_{v_i})|$ as small as possible for all such),*
   (iii) *subject to (i) and (ii), the weight sequence $\mathbf{w}(P)$, is minimal with respect to $<$,*
   (iv) *subject to (i), (ii) and (iii), the number of vertices of degree 1 adjacent to v_{s-1} is minimal.*

It follows that, if P is a critical path and Q is any centrifugal path, then $\mathbf{w}(P) \leqslant \mathbf{w}(Q)$; and if $\mathbf{w}(P) = \mathbf{w}(Q)$, then P and Q have the same length and the number of vertices of degree 1 adjacent to the penultimate vertex of P is less than or equal to the number of vertices adjacent to the penultimate vertex of Q.

4. Conclusion

So now we take an end edge of a critical path P as e_1 and some non-pendant edge, about which there is some choice, as e_2. This is possible if T is not spindly. Suppose that the pair of edges $\{e_1, e_2\}$ is replaced by a disjoint pair $\{f_1, f_2\}$. Let T' be defined by $V(T') = V(T)$ and $E(T') = E(T) - \{e_1, e_2\} + \{f_1, f_2\}$. We find that in T' either there are fewer critical paths than there are in T, or there is a path in T' that, if T' were isomorphic to T, would contradict the criticality of P. Consequently $\{e_1, e_2\}$ is a non-replaceable pair, as required.

5. Conjecture

It can be seen immediately that the 1-free trees are 2-free. Further, it can be shown that the 2-free trees are 3-free. Indeed, for the small trees that we have investigated, it appears that if a tree with n vertices is k-free then it is $(k+1)$-free (for $k < n-2$). It would be interesting to find out whether this result is true for all trees.

References

[1] C. R. J. Clapham and J. Sheehan, *Super-free Graphs*, Ars Combinatoria, to appear.
[2] C. St J. A. Nash-Williams, *The Reconstruction Problem*, Selected Topics in Graph Theory, Academic Press, 1978.
[3] J. Sheehan, *Fixing Subgraphs*, J. Combin. Theory (B) **12** (1972), 226–243.
[4] J. Sheehan and C. R. J. Clapham, *Edge-reconstruction and k-free graphs*, Utilitas (Proceedings of the British Combinatorial Conference), 1989, to appear.

J. Sheehan and C. R. J. Clapham
Department of Mathematical Sciences,
University of Aberdeen,
Aberdeen, UK

Fourth Czechoslovakian Symposium on
Combinatorics, Graphs and Complexity
J. Nešetřil and M. Fiedler (Editors)
1992 Elsevier Science Publishers B.V.

A Note on Reconstructing the Characteristic Polynomial of a Graph

SLOBODAN K. SIMIĆ

It is well known that the characteristic polynomial of any graph is determined up to an additive constant from its polynomial deck, i.e. the collection of the characteristic polynomials of the point deleted subgraphs. Here we prove that these constants are equal for any two connected graphs with the same polynomial decks, whenever the spectra of all subgraphs are bounded from below by -2.

1. Introduction

For a graph G, with the adjacency matrix A, its characteristic polynomial is defined by $\Phi(G; \lambda) = \det(\lambda I - A)$. If $G_1, G_2, \ldots, G_n$ are the point deleted subgraphs of G, let $\mathcal{P}(G) = \{\Phi(G_1; \lambda), \Phi(G_2; \lambda), \ldots, \Phi(G_n; \lambda)\}$ be the corresponding polynomial deck. By the well known formula

$$d/d\lambda\ \Phi(G; \lambda) = \sum_{i=1}^{n} \Phi(G_i; \lambda) \tag{1}$$

it follows that the characteristic polynomial of any graph G is determined by $\mathcal{P}(G)$ up to an unknown intergrating constant.

Question. *Given graphs G and H such that $\mathcal{P}(G) = \mathcal{P}(H)$, does it follow that $\Phi(G; \lambda) = \Phi(H; \lambda)$?*

This question is the spectral counterpart of the famous Ulam's reconstruction problem. It was first posed in [8], where some elementary observations were made. Further discussion on this topic (and similar ones) can be found in [10]. In this note we will answer the above question, provided G and H belong to a particular class of graphs which, among others, includes connected line graphs.

2. Preliminaries

The *line graph*, $L(G)$, is the graph whose points are the lines of G with two points being adjacent whenever the corresponding lines have a point in common. The *cocktail party graph*, $C(k)$, is the graph obtained from the complete graph on $2k$ points by deleting k independent lines. The *generalized line graph*, $L(G; a_1, \ldots, a_n)$, is constructed from a graph G with n points $v_1, \ldots, v_n$ and an n-tuple $(a_1, \ldots, a_n)$ of nonnegative integers by taking disjoint copies of $L(G)$ and $C(a_i)$, $i = 1, \ldots, n$, with additional lines joining a point in $L(G)$ with a point in $C(a_i)$ if the point in $L(G)$ corresponds to a line in G that has v_i as its endpoint. The graphs representable by root systems are described in [1]. For all other information on graph spectra see [3].

We now list some very nice results about graphs whose least eigenvalue is not less than -2.

Proposition 1 ([1]). *If G is a connected graph with $\lambda_{\min}(G) \geqslant -2$, then G is a generalized line graph, or a graph representable by the root system E_8.*

Proposition 2 ([7]). *If G is a graph with $\lambda_{\min}(G) < -2$, then there exists an induced subgraph H of G, such that $\lambda_{\min}(H) = -2$.*

Proposition 3 ([2]). *If G is a connected graph with $\lambda_{\min}(G) > -2$, then G is one of the following graphs:*
1° $L(T)$, where T is a tree;
2° $(L(T); 1, 0, \ldots, 0)$, where T is a tree;
3° $L(U)$, where U is unicyclic with an odd cycle;
4° one of the 573 graphs that are representable by E_8.

Remark 1. The exceptional graphs from 4° have at most eight points. A computer search for such graphs has been undertaken by F. C. Bussemaker; it resulted in 20 graphs on 6 points, 110 graphs on 7 points and 443 graphs on 8 points.

The next proposition is an immediate consequence of the former one and the explicit formulas for the multiplicity of -2 as an eigenvalue of line graphs [5], and of generalized line graphs [4].

Proposition 4. *If G is a connected generalized line graph having -2 as a simple eigenvalue, then G is one of the following graphs:*
1° $L(T; 2, 0, \ldots, 0)$, T is a tree;
2° $L(T, 1, 10, \ldots, 0)$, T is a tree;
3° $L(U)$, U is a bipartite unicyclic graph;
4° $L(U; 1, 0, \ldots, 0)$, U is a unicyclic graph;
5° $L(B)$, B is a nonbipartite bicyclic graph.

Proposition 5 ([9]). *If G satisfies $\lambda_{\min}(G) < -2$ and is minimal with respect to that property, then G has at most ten points.*

Remark 2. According to F. C. Bussemaker and A. Neumaier there are 1 812 minimal graphs with the least eigenvalue less than -2. Among these graphs there are 3, 8, 14, 67, 315 and 1 405 graphs on 5, 6, 7, 8, 9 and 10 points, respectively.

The next proposition is of crucial importance. Its partial proof can be found in [2]; especially, part 3° can be verified by examining the computer results of F. C. Bussemaker (see Remark 1).

Proposition 6. *If G is a connected graph on n points, then*
1° $\Phi(G;-2) = (-1)^n(n+1)$, *if* $G = L(T)$, *where T is a tree;*
2° $\Phi(G;-2) = (-1)^n 4$, *if* $G = L(T,1,0,\ldots,0)$, *where T is a tree, or if* $G = L(U)$, *where U is a nonbipartite unicyclic graph;*
3° $\Phi(G;-2) = (-1)^n(9-n)$, *if G belongs to E_8 and $6 \leqslant n \leqslant 8$.*

In what follows we provide the reader with some rather general observations regarding our question.

($\mathcal{O}_1$) The number of lines and of triangles of G are both determined by $\mathcal{P}(G)$.

($\mathcal{O}_2$) For any point of G, the number of lines and of triangles incident with it are both determined by $\mathcal{P}(G)$.

($\mathcal{O}_3$) The characteristic polynomial of G is reconstructible from $\mathcal{P}(G)$ whenever some polynomial in $\mathcal{P}(G)$ has a multiple root.

The proofs of the first two observations are straightforward. One needs only to know the fact that information on the number of lines and of triangles of some graph is contained in the coefficients of its characteristic polynomial (see [3], pp. 32). The third observation is a direct consequence of the interlacing theorem (see [8] as well).

Finally, by inspecting the characteristic polynomials of some small graphs we have:

($\mathcal{O}_4$) The characteristic polynomials of all connected graphs on up to seven points are reconstructible.

3. Main result

In this section we first state our main result and then give some basic of the proof.

Theorem (main result). *The characteristic polynomial of any connected graph whose point deleted subgraphs have their spectra bounded from below by -2, is reconstructible from its polynomial deck.*

Sketch of the proof. Suppose $\mathcal{P} = \{\Phi_1(\lambda), \Phi_2(\lambda), \ldots, \Phi_n(\lambda)\}$ is the polynomial deck of any graph that satisfies the assumption of the theorem. We have to prove that for any connected graph G such that $\mathcal{P}(G)$ equals $\mathcal{P}$, $\Phi(G;\lambda)$ is uniquely determined by (reconstructible from) $\mathcal{P}$.

Case 1: $\Phi_i(-2) \neq 0$ for all i. By the interlacing theorem, we now have that either $\lambda_{\min}(G) \geqslant -2$ or $\lambda_{\min}(G) < -2$. Since the latter possibility contradicts Proposition 2, it follows from Proposition 1 that G is either a generalized line graph, or a graph representable by the root system E_8. Moreover, the multiplicity of -2 as an eigenvalue of G cannot exceed 1.

Suppose first $n \geqslant 10$ (G has at least 10 points). Then G must be a generalized line graph; otherwise, if G is representable in the root system E_8, the same holds for its point deleted subgraphs, which contradicts Proposition 3 (any such graph has at most 8 points). Making use of Propositions 3 and 6, we get that G belongs to one of the following classes of graphs:

$\mathcal{A}_1 = \{L(T) \mid T$ is a tree$\};$

$\mathcal{A}_2 = \{L(T; 1, 0, \ldots, 0) \mid T \text{ is a tree}\}$;
$\mathcal{A}_3 = \{L(U) \mid U \text{ is a nonbipartite unicyclic graph}\}$;
$\mathcal{B}_1 = \{L(P, 1, 0, \ldots, 1) \mid P \text{ is a path whose end points are labeled by } 1\}$;
$\mathcal{B}_2 = \{C_n \mid \text{ with } n \text{ even}\}$;
$\mathcal{B}_3 = \{L(U, 1, 0, \ldots, 0) \mid$ where U is a unicyclic graph consisting of an odd cycle and possibly a hanging path whose endpoint (if any) is labeled by 1$\}$;
$\mathcal{B}_4 = \{L(B) \mid$ where B is a bicyclic graph consisting of two odd cycles (not necessarily of equal length) and a path of any length (possibly zero) between them$\}$.

Note that the graphs of $\mathcal{A}_1 - \mathcal{A}_3$ are those as in Proposition 3; they all have the least eigenvalue greater than -2. The graphs of $\mathcal{B}_1 - \mathcal{B}_4$ have -2 as a simple eigenvalue. They are derived from Proposition 4 by taking only those graphs whose subgraphs do not contain -2 in their spectra.

We next observe that the value of the characteristic polynomial at point -2 for each of the above graphs is fixed within a class (see Proposition 6). Therefore, to reconstruct the characteristic polynomial, we only need to decide, for given $\mathcal{P}$, to which of the above classes the corresponding graph G belongs. For the graphs from $\mathcal{B}_1 - \mathcal{B}_4$ we have the following advantage: they are all homeomorphic to one of certain six graphs of very simple structure, but not homeomorphic to any graph from $\mathcal{A}_1 - \mathcal{A}_3$. Thus, if we start from some collection $\mathcal{P}$, and recognize (by making use of observations $\mathcal{O}_1$ and $\mathcal{O}_2$ and, if necessary, forbidden subgraphs for generalized line graphs [4]) that the corresponding graph G belongs to $\mathcal{B}_1 - \mathcal{B}_4$, we have $\Phi(G; -2) = 0$. Otherwise, G must be from $\mathcal{A}_1 - \mathcal{A}_3$. Due to Proposition 6, we now have to decide whether G belongs to $\mathcal{A}_1$ or not. For this aim, we have a simple criterion: only to check if $\Phi(-2) = \pm 4$ for some i. If not, then $\Phi(G; -2) = (-1)^n(n+1)$; otherwise, $\Phi(G; -2) = (-1)^n 4$. So, if $n \geq 10$, we are done. If $n < 10$, we can choose between brute force or some similar line of reasoning based on observations $\mathcal{O}_1 - \mathcal{O}_4$ and the afore-mentioned propositions.

Case 2: $\Phi_i(-2) = 0$ for at least one i. Now if G is any (not necessarily connected) graph corresponding to $\mathcal{P}$, we have (by the interlacing theorem) that either $\lambda_{\min}(G) = -2$, or $\lambda_{\min}(G) < -2$. By Proposition 5, any such graph has at most 10 points. Consequently, our reconstruction statement is now true for all graphs with at least 11 points. The graphs with up to 10 points could be treated similarly as the small graphs in the previous case. A more detailed discussion regarding these exceptional graphs will be given in a forthcoming paper. □

Acknowledgement. The author wants to thank F. C. Bussemaker for being so kind as to enable him to use the results of his computer investigations in preparing the manuscript. Other thanks are due to D. M. Cvetković for stimulating discussions regarding the problem.

References

[1] P. J. Cameron, J. M. Goethals, J. J. Seidel, E. E. Shult, *Line graphs, root systems and elliptic geometry*, J. Algebra **43** (1976), 305–327.

[2] D. M. Cvetković, M. Doob, *On spectral characterizations and embeddings of graphs*, Linear Algebra Appl. **27** (1979), 17–26.

[3] D. M. Cvetković, M. Doob, H. Sachs, *Spectra of Graphs—Theory and Application*, Deutscher Verlag der Wissenschaften, Berlin Academic Press, New York (1980).

[4] D. M. Cvetković, M. Doob, S. K. Simić, *Generalized line graphs*, J. Graph Theory **5** (1981), 385–399.

[5] M. Doob, *An interrelation between line graphs, eigenvalues, and matroids*, J. Comb. Th. **15** (1973), 40–50.

[6] ———, *A surprising property of a least eigenvalue of a line graph*, Linear Algebra Appl. **46** (1982), 1–7.

[7] I. Gutman, D. M. Cvetković, *The reconstruction problem for the characteristic polynomials of graphs*, Publ. Elektrotehn. Fak. Ser. Fiz., vol. No 498–No 541, Univ. Beograd, 1975, pp. 45–48.

[8] V. Kumar, S. B. Rao, N. M. Singhi, *Graphs with eigenvalues at least-2*, Linear Algebra Appl. **46** (1982), 27–42.

[9] A. J. Schwenk, *Spectral reconstruction problems*, Ann. N. Y. Acad. Sci. **328** (1979), 183–189.

Slobodan K. Simić
Department of Mathematics,
Faculty of Electrical Engineering,
University of Belgrade,
110 00 Belgrade, Yugoslavia

Fourth Czechoslovakian Symposium on
Combinatorics, Graphs and Complexity
J. Nešetřil and M. Fiedler (Editors)
© 1992 Elsevier Science Publishers B.V. All rights reserved.

Exponential Constructions of Some Nonhamiltonian Minima

Z. SKUPIEŃ

Exponentially many n-vertex minimum nonhamiltonian (A) homogeneously traceable graphs and (B) bihomogeneously traceable oriented graphs are constructed. An analog of Sylvester's result on numerical semigroups is used.

1. Introduction

Assume that the adjective *maximum* (*minimum*) when applied to a (di)graph means that the *size* (i.e., the number of edges (arcs)) is as large [as small] as possible provided that the *order* (the number of vertices) is fixed.

We are going to construct exponentially many mutually nonisomorphic specialized nH (*nonhamiltonian*) n-vertex graphs or digraphs of minimum size.

Recall that following [6] a (di)graph G is called ht (*homogeneously traceable*) if the set of endvertices of H-paths (hamiltonian paths) of G coincides with the vertex set $V(G)$. A digraph D is called out-ht [in-ht] if each vertex is the first [last] vertex of an H-path of D. Then

$$\text{out-ht} \cap \text{in-ht} =: \text{bi-ht}$$

(*bihomogeneously traceable*, the name following [4]). In digraphs we differentiate between *cycles* and *dicycles* but the word *path* (if pertaining to digraphs) means dipath. A digraph is called *Hamiltonian* if it has an H-dicycle. An *oriented graph* is defined to be a digraph without 2-dicycles and without loops.

It is minimum htnH (ht and nH) graphs and minimum bi-htnH (bi-ht and nH) oriented graphs of which we are going to construct exponentially many. As a by-product we get as many cubic hamiltonian graphs each with a 1-factor nonextendible to any H-cycle. As many of arbitrary, maximal and minimum (cubic) hypohamiltonian graphs are constructed in [3], [5] and [8], respectively.

2. Nonhamiltonian minima

2.1 General construction

In each construction we are going to present, the output is a large graph which is built of disjoint copies, call them chips, of three small graphs each of which has four specified vertices a, b, c and d. Two of those graphs are of the same order, β, and are isomorphic but yield two distinct chips, which we call *flexible chips* and which differ only in the choice (or ordering) of the specified vertices. The third graph, which is either symmetric or rigid, and whose order we denote by α, yields a chip which we call *troublesome*. Each chip can be viewed as a "black box" with the boundary $abdca$ (oriented counterclockwise), cf. Figs. 1–3.

The output design (2.1):

Take r troublesome chips, take s flexible chips so as to have odd $r + s$, arrange all chips circularly so as to have all troublesome chips in one segment (control trouble by limiting r ($r = 1$ or $1 \leqslant r \leqslant \beta$) and by keeping all troublesome chips close to each other) and, for each pair of chips S and S^* such that S^* immediately follows S, add edges ca^* and db^* if the chips are undirected graphs; otherwise, add arcs ca^* and db^*.

Thus the resulting graph, whose order we denote by n, can be viewed as drawn on a cylindrical band. Moreover, r is small and

$$n = r\alpha + s\beta, \tag{2.2}$$

$$r + s \text{ is odd.} \tag{2.3}$$

The following cases are considered.

A. Undirected chips (called M-chips or skew-matchable chips) $\check{M}_6$, $\check{M}_6^!$ and M_4 (Fig. 1) each with a special 1-factor F; n ranges over all even integers $\geqslant 44$: $\alpha = 4$, $\beta = 6$, $\exists!(r, s) \geqslant (1, 4)$ and $r \leqslant 6$ (for $n = 42$, if r and s are positive then $s = 3$ and is too small). We assume that the union (still denoted F) of the F's over all chips involved is a fixed 1-factor of each output (2.1).

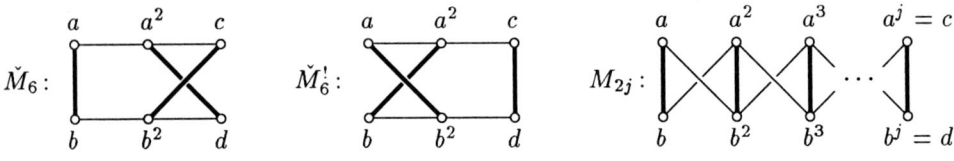

FIGURE 1. M-chips (bold edges in F's)

B. Directed chips (called slipways) S_4, S_4^-, S_6 (Fig. 2) and S_{2j+3}, $j = 2, 3, 4, 5$ (Fig. 3).

B_1. n ranges over all even integers $\geqslant 22$: $\alpha = 6$, $\beta = 4$, $\exists!(r, s)$ with $1 \leqslant r \leqslant 4$ (and $r + s \geqslant 3$) (for $n = 20$, if $(r, s) \geqslant (1, 0)$ in (2.2) then $(r, s) = (2, 2)$ and (2.3) is false).

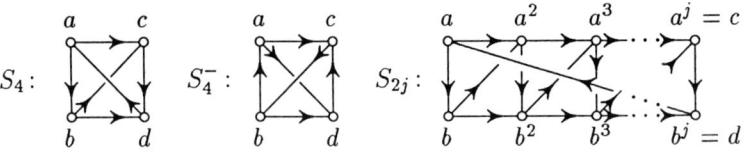

FIGURE 2. Hamiltonian slipways

B$_2$. n ranges over all odd integers ≥ 9: $\alpha = 2j + 3$ for $2 \leq j \leq 5$ ($n \equiv \alpha \pmod 8$), $\beta = 4$, $r = 1$, $s \geq 0$ but $s \neq 0$ if $\alpha = 7$ ($j = 2$; to preclude a 2-dicycle).

Notice that we impose some restrictions on the construction and on the range of n (e.g., $r > 0$) in order to simplify estimating the number of outputs.

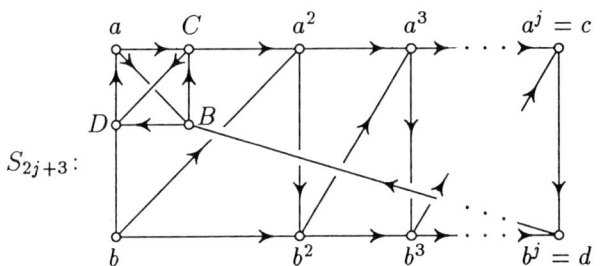

FIGURE 3. Nonhamiltonian slipways

2.2 The range of order n

Throughout this subsection, a and b stand for integers as do n, k, r, s throughout this paper. In order to justify the statements about the range of n in A and B above, we shall use the following analog of a result due to Sylvester [9]. Consider the following problem

$$n = ra + sb \quad \text{and} \quad (r, s) \geq (m_a, m_b) \qquad (2.4)$$

for some integers m_a and m_b.

Proposition 1. *If a and b are relatively prime positive integers, $a + b$ is odd, $\mu = \min\{a, b\}$ and $M = \max\{a, b\}$, then*

$$\min\{k \mid \forall n \geq k \;\exists (r, s): (2.4) \text{ and } (2.3) \text{ hold}\} =$$
$$= (2\mu - 1)(M - 1) + (\varepsilon + m_\mu)\mu + M m_M$$

where $\varepsilon \equiv (m_a + m_b) \bmod 2$, $\varepsilon \in \{0, 1\}$. Moreover, the uniqueness of (r, s) is implied by either of the following inequalities:

$$m_a \leq r < m_a + 2b, \qquad m_b \leq s < m_b + 2a.$$

Sylvester's result is that if (2.3) is dropped and $m_a = m_b = 0$, then the corresponding minimum is $(a-1)(b-1)$.

2.3 Related results

The output (2.1) reminds of a graph based on a composition of flip-flops (Chvátal's chips [2] of 1973) though the requirement (2.3) can be dropped and a Möbius band can be seen as involved therein, cf. [8] where a few kinds of flip-flops yield a variety of hypohamiltonian graphs (in particular cubic ones).

Another method of construction is used in [6] to prove the existential part of the result that a (nontrivial) bi-htnH oriented graph of order $n > 1$ (which, moreover, is of minimum size $2n$ or, equivalently, 2-diregular) exists iff $n \geqslant 7$. Namely, a single vertex of a fixed bi-htnH oriented graph D_i of order $i = 7, 8$ (or 9) is replaced by "boxes" S_{2j} of bigger and bigger orders ($j \to \infty$). On the other hand, the directed construction described above is used in [1] to produce a few new minimum bi-htnH oriented graphs of any order $n \geqslant 11$. The construction uses S_4, S_6^- and four chips obtained from S_{2j}^- ($j = 2, 3, 4, 5$) by replacing the vertex c in some way by S_4 (Figs. 2 and 3).

As for htnH graphs on n vertices, they are known to exist iff $n \leqslant 2$ or $n \geqslant 9$; moreover, their size $|E| \geqslant \lceil 5n/4 \rceil$ for $n \geqslant 9$ and then the minimum htnH graphs have the degree sequence

$$\{3\}^{n_3} \times \{2\}^{n_2} \quad \text{where} \quad n_3 = 2\lceil n/4 \rceil \text{ and } n_2 = n - n_3 \tag{2.5}$$

or, additionally, $\{4\} \times \{3\}^{\lfloor n/2 \rfloor} \times \{2\}^{\lfloor n/2 \rfloor}$ if $13 \leqslant n \equiv 1 \pmod 4$, see [7].

2.4 Properties of chips

Formally, a chip S is a pair,

$$S = (G; a, b, c, d), \tag{2.6}$$

where G, $G = G(S)$, is called the *graph* (or *digraph* if applicable) *of S* and a, b, c and d are four mutually distinct vertices (called *exterior vertices* of S or) of G which form the *vertex-quadruple* (a, b, c, d) of S. Depending on the existence or non-existence of certain routings in $G(S)$, a chip S can be a flip-flop or an M-chip if S is undirected or otherwise S can be a slipway.

Two chips S and S^* are called *isomorphic* if there is an isomorphism of $G(S)$ onto $G(S^*)$ which induces the transformation of the vertex-quadruple of S onto that of S^*. If i is the order of $G(S)$ [or $G(S^*)$] then i is called the *order* of S [of S^*], in symbols $S = S_i$ [$S^* = S_i^*$].

It appears (as we shall see) that the definitions of specific chips are D_2-invariant (in the undirected cases) or C_2-invariant (in the directed cases) where D_2 is the group of the rectangle (a dihedral group) and C_2 is the group of the horizontal arrow, $C_2 = \{\text{id}, \pi^-\}$ where π^- is the symmetry about the horizontal axis. If π is a symmetry then the *action* of π on S is assumed to consist in the reordering of the vertex-quadruple of S, e.g.,

$$\pi^-(S) := (G; b, a, d, c).$$

Therefore new useful chips can be obtained by the actions of some symmetries (which is used in [8] in the case of flip-flops).

Given vertices $x, y \in V(G)$, the pair xy is called *good* in G if G has an $x - y$ H-path. Similarly, the pair $\{xy, zt\}$ of pairs is called *good* in G if G has an $x - y$ path and a $z - t$ path which together span G and are disjoint.

Each (directed) chip S in B has the first three of following properties.
(i) Both ad and bc are good in G;
(ii) None of ac, bd or $\{ad, bc\}$ (also $\{ab, cd\}$ or $\{ac, bd\}$ in the undirected case only) is good in G;
(iii) (size minimum for B) Both the in- and the out-degree of each vertex of G is two with the exception that the in-degrees of both a and b as well as the out-degrees of both c and d are all one;
(iv) (size minimum for A) Each exterior vertex is of degree two and each remaining one of degree three;
(v) $\{ac, bd\}$ is good in G.

Given a subgraph F of G, F-good will mean that there exists a path-system which is required by the definition of "good" and includes F. Assume that replacing "good" with "F-good" in a condition (γ) gives the condition denoted by $(\gamma)_F$, $\gamma =$ i, ii. All chips introduced in A clearly satisfy (i)$_F$, (ii)$_F$ and (iv). Moreover, G is called F-Hamiltonian [F-traceable] or non-F-Hamiltonian [non-F-traceable] according to if G has or does not have an H-cycle [H-path] through F. Now the condition (ii) [resp. (ii)$_F$] together with (2.3) for $r + s \neq 1$ can be seen to ensure the nonhamiltonicity [non-F-hamiltonicity] of each output (2.1) (because any H-cycle intersects each edge-cut in an even number of elements). If $r + s = 1$, the non-(F)-hamiltonicity of the chip involved is required additionally. Notice that among M-chips in A only M_4 is F-hamiltonian. Thus we have

Proposition 2. *Each output (2.1) is non-F-Hamiltonian in case* A *and nonhamiltonian in case* B.

Let ι and τ stand for an initial and a terminal vertices in a spanning path-system of G with k components of the form: $\iota - y$ or $x - \tau H$-path ($y = c, d$; $x = a, b$; $k = 1$); $\iota - c \cup a - \tau$ or $\iota - d \cup b - \tau$ ($k = 2$); and $\iota - d \cup b - c \cup a - \tau$ (or $\iota - c \cup a - d \cup b - \tau$) ($k = 3$). Let $\eta(k, S)$ be the collection of all such path-systems with up to k components, $k = 1, 2, 3$.

A chip S is called *bihomogeneously k-coverable* if the range of ι and that of τ in $\eta(k, S)$ are both $V(G)$. A chip which is bihomogeneously k-coverable for $k = 3$ but not for $k = 2$ is called *bihomogeneously strictly 3-coverable*. So is each S_{2j+3} in Fig. 3 because $\tau = D$ requires three components: $\iota (= d)$, $b - c$, $aBCD$. Each chip S_{2j} in Fig. 2 is bihomogeneously 2-coverable, cf. Table where $a-d$ and $b-c H$-paths with $\iota \in \{a, b\}$ or $\tau \in \{c, d\}$ are omitted.

TABLE. 1- and 2-coverings (superscripts: $1 \leqslant i < j$ and $z \geqslant 2$)

S_{2j}	ι	Ends	τ	S_{2j+3}	ι	Ends	τ	ι	Ends	τ
	d, b^z	c	—		—	b	C, a^i	B	d, b	b, b^i
	—	b	a^i, a		C	c, a	d	D	c, a	a
	c, a^z	c, a	d		c, a^z	c, a	B			
	a	d, b	b^i, b		d, b^z	c, a	a	\multicolumn{3}{	c	}{See above for $\tau = D$}

Suppose G is undirected and has a 1-factor F. Then, let $\eta_F(k, S)$ be the subset of $\eta(k, S)$ whose each element includes F. Similarly, let $\tilde{\eta}_F(k, S)$ be the subset of

$\eta(k, S)$ whose each element Q intersects F in $F - e$ where e is an edge of F which is incident to an endvertex v of Q and $v = \iota$ or $v = \tau$. Now, S is called *homogeneously-$F^\sim$ k-coverable* if both the set of special endvertices v in $\widetilde{\eta}_F(k, S)$ and the union of ranges of ι and τ in $\eta_F(k, S)$ are $V(G)$. Furthermore, G is called *homogeneously $F^\sim$-traceable* if each vertex x is an endvertex of two H-paths: one through F and another through $F - e$ where e is the edge of F incident to x. Notice that all chips in A are homogeneously-$F^\sim$ 2-coverable (M_4 is 1-coverable).

A (directed) chip is called a *two-path* if it satisfies (v). Notice that (i) [resp. (i)$_F$] allows the extendibility of each 1- or 2-covering in $\eta(k, S)$ to an H-path [through all F's in remaining chips] in each output (2.1) in such a way that each of ι and τ remains an endvertex. Similar extendibility in case $k = 3$ when S is bihomogeneously strictly 3-coverable is possible if all remaining chips involved in (2.1) are two-paths. The chips in Fig. 2 can be seen to be both bihomogeneously 2-coverable and two-paths but S_{2j+3} in Fig. 3 is neither one nor the other. That is why $r = 1$ in case B_2 above.

Define an *M-chip* (or skew-matchable chip) to be an undirected chip S in (2.6) whose graph G has a fixed 1-factor F such that S is homogeneously-$F^\sim$ 2-coverable and conditions (i)$_F$ and (ii)$_F$ are satisfied. Similarly, a *slipway* is a chip which is directed and bihomogeneously 3-coverable and conditions (i) and (ii) are satisfied.

2.5 Main results

Theorem 3. *The number of n-vertex ⟨cubic graphs each with a 1-factor F which are homogeneously $F^\sim$-traceable but not F-Hamiltonian; minimum htnH graphs; minimum bi-htnH oriented graphs⟩, if non-zero (i.e., if ⟨even $n \geq 6$; $n \leq 2$ or $n \geq 9$; $n = 1$ or $n \geq 7$⟩) is bounded from below by an increasing exponential function of n.*

Proof. The chips in case A are minimum M-chips. Therefore if G, $G = G_F$, is an output (2.1) in case A then G is a cubic graph of order $n = 2k \geq 44$ with a 1-factor F and G is homogeneously $F^\sim$-traceable but not F-Hamiltonian. Notice that r copies of M_4 in (2.1) induce a bipartite segment of G. Assume that this segment in (2.1) is immediately preceded by $\check{M}_6$, $\check{M}_6^!$ and immediately followed by two copies of $\check{M}_6$, the remaining $s - 4$ positions being filled by $\check{M}_6$ and $\check{M}_6^!$ in an arbitrary way. Then the distribution of cycles C_5 in G helps to identify the whole F and all chips in their circular ordering. Vertices, however, are identified up to symmetry about the horizontal axis (or plane). Hence,

$$(\#G_F\text{'s}) = 2^{s-4} = 2^{(n-r\alpha)/\beta - 4} \geq 2^{n/6 - 8}$$

by (2.2).

For $j = 0, 1, 2, 3$, choose j out of k edges of F in G_F, subdivide each of them once and subdivide each remaining edge in F twice. The resulting graph can have any order $n := 4k - j \geq 85$, has the degree sequence (2.5) and can easily be seen to be htnH. The number of these graphs is $\geq 2^{n/12 - 8}$.

In case B each chip is really a minimum slipway and an oriented graph, and therefore each output is a minimum bi-htnH oriented graph. The distribution of triangles and ditriangles determines all chips and all vertices in (2.1). Hence the number of outputs is $2^s \geq 2^{n/4 - 6}$. □

In connection with Fig. 3, notice that replacing vertices a and d in S_{2j} by S_{2k} and S_{2m}, respectively, each vertex in either of some two ways, and adding arcs ca and db to the resulting chip can yield four new minimum bi-htnH oriented graphs of order $2(j+k+m-1) \geqslant 10$. Recursion in this construction does not work, however.

The only known n-vertex 2-diregular (and hence minimum) hypohamiltonian digraph is the cartesian product $\vec{C}_k \times \vec{C}_j$ where $n = jk$ in each case as well as $k = 2$ and $j = 2m - 1 \geqslant 3$ or (when it is an oriented graph) $k = 3$ and $j = 6m + 4 \geqslant 4$ or $k \geqslant 3$ and $j = mk - 1 \geqslant 3$ but $n \geqslant 12$; $k, m \in \mathbb{Z}$, see [10]. It is an open problem to find such minimum digraphs for remaining n's, $n > 6$, and to estimate the number of them.

Added in proof. For the proof of Proposition 1 (and its generalization), see Z. Skupień, A generalization of Sylvester's result on numerical semigroups, manuscript (1991).

Acknowledgement. Support of KBN of Poland in 1990, Grant P/05/003/90-2, is acknowledged.

References

[1] R. Balakrishnan and P. Paulraja, *Nonhamiltonian homogeneously traceable arc-minimal digraphs and their converse digraphs*, manuscript.

[2] V. Chvátal, *Flip-flops in hypohamiltonian graphs*, Canad. Math. Bull. **6** (1973), 33–41.

[3] J. B. Collier and E. F. Schmeichel, *New flip-flop constructions for hypohamiltonian graphs*, Discrete Math. **8** (1977), 265–271.

[4] S. Hahn and T. Zamfirescu, *Bihomogeneously traceable oriented graphs*, Rend. Sem. Mat. Univers. Politecn. Torino **9(2)** (1981), 137–145.

[5] P. Horák and J. Širáň, *On a construction of Thomassen*, Graphs Combinat. (1986), 347–350.

[6] Z. Skupień, *On homogeneously traceable nonhamiltonian digraphs and oriented graphs*, The Theory and Applications of Graphs (Proc. Kalamazoo 1980 Conf. (G. Chartrand, ed.), Wiley, 1981, pp. 517–527.

[7] ———, *Homogeneously traceable and Hamiltonian connected graphs*, Demonstratio Math. **7** (1984), 1051–1067.

[8] ———, *Exponentially many hypohamiltonian graphs*, Graphs, Hypergraphs and Matroids. III (Proc. Kalsk 1988 Conf., ed. by M. Borowiecki and Z. Skupień), Zielona Góra, 1989, pp. 123–132.

[9] J. J. Sylvester, *Problem 7382 (and "Solution by W. J. Curran Sharp")*, (Mathematics from) The Educational Times (with additional papers and solutions) **41** (1881), 21.

[10] C. Thomassen, *Hypohamiltonian graphs and digraphs*, Theory and Applications of Graphs (Proc. Kalamazoo 1976 Conf.), Lect. Notes Math. 642, Springer, 1978, pp. 557–571.

Z. Skupień
Instytut Matematyki AGH,
Mickiewicza 30, 30-059 Kraków,
Poland

Fourth Czechoslovakian Symposium on
Combinatorics, Graphs and Complexity
J. Nešetřil and M. Fiedler (Editors)
© 1992 Elsevier Science Publishers B.V. All rights reserved.

Hamiltonicity of Products of Hypergraphs

MARTIN SONNTAG

The investigation of products of hypergraphs from the point of view of traceability and hamiltonicity can be understood as a supplement to the corresponding considerations for products of undirected graphs. For certain classes of hypergraphs the pairs of hypergraphs $\mathcal{H}_1$, $\mathcal{H}_2$ are characterized such that their cartesian sum (normal product, lexicographic product) has a hamiltonian path or cycle. The specialization of these results to 2-uniform hypergraphs (i.e. to undirected simple graphs) yields not only known but also new assertions on hamiltonian properties of products of graphs.

1. Definitions

All hypergraphs considered in this article are supposed to be nonempty and finite without loops and multiple edges. Let $\mathbf{N}$ be the set of natural numbers, $\mathbf{N}^+ := \mathbf{N} - \{0\}$, $\mathcal{H} = (V, \mathcal{E})$ the hypergraph with the vertex set V and the edge set $\mathcal{E} \subseteq \mathcal{P}(V) - \{0\}$; moreover let $w = (v_0, e_1, v_1, e_2, \ldots, e_t, v_t)$ be an *edge sequence* of $\mathcal{H}$, i.e.

$$t \in \mathbf{N},\ \{v_0, v_1, \ldots, v_t)\} \subseteq V,\ \{e_1, e_2, \ldots, e_t)\} \subseteq \mathcal{E}$$

and

$$\forall k \in \{1, 2, \ldots, t\} : \{v_{k-1}, v_k\} \subseteq e_k \wedge v_{k-1} \neq v_k.$$

For $p \in \mathbf{N}^+$ the edge sequence w is called a *p-path* iff
(1) $\forall k, l \in \{0, 1, \ldots, t\} : k \neq l \Rightarrow v_k \neq v_l$ and
(2) $\forall k, l \in \{1, 2, \ldots, t\} : k \leqslant l \wedge e_k = e_l \Rightarrow l - k \leqslant p - 1 \wedge e_k = e_{k+1} = \ldots = e_l$.
A p-path id said to be a *path* iff $p = 1$.
The edge sequence w is a *cycle* iff $t \geqslant 3$, $v_0 = v_t$ and

$$\forall k, l \in \{1, 2, \ldots, t\} : k \neq l \Rightarrow v_k \neq v_l \wedge e_k \neq e_l.$$

As usual we call a p-path, a path or a cycle of $\mathcal{H}$ *hamiltonian* iff it contains all vertices of $\mathcal{H}$. The class of all hypergraphs possessing a hamiltonian p-path (path, cycle) is denoted by pT (T, H). For any hypergraph $\mathcal{H}$ we set

$$pT(\mathcal{H}) := \{w : w \text{ is a hamiltonian } p\text{-path of } \mathcal{H}\}.$$

Obviously, it holds ([2])
$$T = 1T \subset 2T \subset 3T \subset \cdots$$
In the following let $\mathcal{H}_1 = (V_1, \mathcal{E}_1)$ and $\mathcal{H}_2 = (V_2, \mathcal{E}_2)$ be hypergraphs with $|V_1| = r$ and $|V_2| = s$; furthermore we define $V := V_1 \times V_2$ to be the vertex set of our products of hypergraphs. $\mathcal{H}_1 + \mathcal{H}_2 = (V, \mathcal{E})$ is the *cartesian sum* of $\mathcal{H}_1$ and $\mathcal{H}_2$ iff
$$\mathcal{E} := \{\{x\} \times e \colon x \in V_1 \wedge e \in \mathcal{E}_2\} \cup \{e \times \{x\} \colon e \in \mathcal{E}_1 \wedge x \in V_2\}.$$
$\mathcal{H}_1 * \mathcal{H}_2 = (V, \mathcal{E})$ is the *normal product* of $\mathcal{H}_1$ and $\mathcal{H}_2$ iff
$$\mathcal{E} := \mathcal{E}(\mathcal{H}_1 + \mathcal{H}_2) \cup \{\{(x_1, y_1), \ldots, (x_k, y_k)\} \colon |\{x_1, \ldots, x_k\}| = |\{y_1, \ldots, y_k\}| = k \in \mathbb{N}$$
$$\wedge \exists e_1 \in \mathcal{E}_1 \, \exists e_2 \in \mathcal{E}_2 \colon \{x_1, \ldots, x_k\} \subseteq e_1 \wedge \{y_1, \ldots, y_k\} \subseteq e_2 \wedge (|e_1| = k \wedge |e_2| = k)\}.$$
$\mathcal{H}_1 \cdot \mathcal{H}_2 = (V, \mathcal{E})$ is the *lexicographic product* of $\mathcal{H}_1$ and $\mathcal{H}_2$ iff
$$\mathcal{E} := \{\{(x_1, y_1), \ldots, (x_k, y_k)\} \colon \{x_1, \ldots, x_k\} \in \mathcal{E}_1 \wedge |\{x_1, \ldots, x_k\}| = k \in \mathbb{N}$$
$$\wedge \{y_1, \ldots, y_k\} \subseteq V_2\} \cup \{\{x\} \times e \colon x \in V_1 \wedge e \in \mathcal{E}_2\}.$$

Obviously, all these products preserve q-uniformity, and in the case of 2-uniform hypergraphs (i.e. for undirected simple graphs) they coincide with the known graph theoretic products. Clearly, for q-uniform hypergraphs $\mathcal{H}_1, \mathcal{H}_2$ it holds:
$$\mathcal{E}(\mathcal{H}_1 * \mathcal{H}_2) \subseteq \mathcal{E}(\mathcal{H}_1 \cdot \mathcal{H}_2).$$

2. Results

In 1968 BEHZAD and MAHMOODIAN proved that bipartite graphs with odd number of vertices are the only traceable graphs possessing a non-hamiltonian cartesian sum (cf. [1]). For hypergraphs we obtain the following generalization of this assertion:

Theorem 1. ([2]) *Let $\mathcal{H}_1 \in T$, $\mathcal{H}_2 \in T$, $r \geqslant 2$ and $s \geqslant 2$. Then it holds:*
$$\mathcal{H}_1 + \mathcal{H}_2 \in H \quad \text{iff} \quad (r \cdot s \text{ even} \vee \exists i \in \{1, 2\} \colon \mathcal{H}_i \text{ is not a bipartite graph}).$$

Owing to $\mathcal{E}(\mathcal{H}_1 + \mathcal{H}_2) \subseteq \mathcal{E}(\mathcal{H}_1 * \mathcal{H}_2)$ every sufficient condition guaranteeing $\mathcal{H}_1 + \mathcal{H}_2 \in H$ implies $\mathcal{H}_1 * \mathcal{H}_2 \in H$. In order to characterize the pairs of hypergraphs the normal product of which is hamiltonian (for special classes of hypergraphs) we need two further notations. For $p \in \mathbb{N}^+$ let P_p denote the path with p vertices and
$$pT' := \{\mathcal{H} = (V, \mathcal{E}) \colon \mathcal{H} \in pT \wedge |\mathcal{E}| = \left\lfloor \frac{|V| - 2}{p} \right\rfloor + 1\},$$
i.e., pT' contains all $\mathcal{H} \in pT$ with minimal number of edges. Then for $p \in \mathbb{N}^+$, $\mathcal{H}_1 \in pT$, $\mathcal{H}_2 \in 2T$ with $r \geqslant 2$ and $s \geqslant 2$ each of the following conditions guarantees the hamiltonicity of $\mathcal{H}_1 * \mathcal{H}_2$:
(1) s even $\vee r = 2$;
(2) $r = 3 \wedge \sim (\mathcal{H}_1 \simeq P_3 \wedge \mathcal{H}_2 \in 2T' \vee \mathcal{H}_1 \in 2T' \wedge \mathcal{H}_2 \simeq P_3)$;
(3) $r \geqslant 4 \wedge \exists w \in 2T(\mathcal{H}_2) \, \exists i, i' \in \{0, 1, \ldots, s-1\} \exists e \in \mathcal{E}_2 \colon$
   $w = (v_0, e_1, v_1, e_2, \ldots, e_{s-1}, v_{s-1}) \wedge i < i' \wedge i, i'$ even $\wedge \{v_i, v_{i'}\} \subseteq e$;
(4) $r \geqslant 4 \wedge s \geqslant \left\lceil r / (\lfloor \frac{r-2}{p} \rfloor + 1) \right\rceil - 1$.

The proof of the sufficiency of some of these conditions requires to distinguish many cases. In [3] a sketch of this proof can be found.

Theorem 2. ([3]) Let $p \in \mathbb{N}^+$, $\mathcal{H}_1 \in pT'$, $\mathcal{H}_2 \in 2T$, $r \geqslant 2$ and $s \geqslant 2$. Then it holds:
$$\mathcal{H}_1 * \mathcal{H}_2 \in H \text{ iff } (1) \vee (2) \vee (3) \vee (4).$$

In the case of the lexicographic product we are able to characterize all pairs of hypergraphs $\mathcal{H}_1, \mathcal{H}_2$ possessing a hamiltonian path or a cycle in $\mathcal{H}_1 * \mathcal{H}_2$. But above all let $\mathcal{H} = (V, \mathcal{E})$ be a hypergraph and

$$K(\mathcal{H}) := \{w : w \text{ is an edge sequence in } \mathcal{H} \wedge V(w) = V\}.$$

For any edge sequence $w = (v_0, e_1, v_1, e_2, \ldots, e_{l_w}, v_{l_w})$ let

$$Z(w, \mathcal{H}) := \max_{v \in V} \left|\{k : k \in \{0, 1, \ldots, l_w\} \wedge v = v_k\}\right|,$$
$$Y(w, \mathcal{H}) := \min_{v \in V} \left|\{k : k \in \{0, 1, \ldots, l_w\} \wedge v = v_k\}\right|,$$
$$i(w) := v_0 \qquad \text{(the } \textit{initial vertex} \text{ of } w)$$

and
$$t(w) := v_{l_w} \qquad \text{(the } \textit{terminal vertex} \text{ of } w).$$

I.e., $Z(w, \mathcal{H})$ and $Y(w, \mathcal{H})$ is the maximal and the minimal number, respectively, of appearances of a vertex in the edge sequence w. Finally, by $P(\mathcal{H})$ we denote the *vertex path partition number* of $\mathcal{H}$, i.e., the minimal number of pairwise vertex and edge disjoint paths containing all vertices of $\mathcal{H}$.

Theorem 3. Let $\mathcal{H}_1, \mathcal{H}_2$ be hypergraphs with $s \geqslant 2$. Then it holds: $\mathcal{H}_1 \cdot \mathcal{H}_2 \in T$ iff
$$\exists w \in K(\mathcal{H}_1) : P(\mathcal{H}_2) \leqslant Y(w, \mathcal{H}_1) \leqslant Z(w, \mathcal{H}_1) \leqslant s.$$

Making use of the proof of this Theorem we can verify a corresponding assertion for the hamiltonicity:

Theorem 4. Let $\mathcal{H}_1, \mathcal{H}_2$ be hypergraphs with $r \geqslant 2$, $s \geqslant 2$. Then it holds: $\mathcal{H}_1 \cdot \mathcal{H}_2 \in H$ iff
$$\exists w \in K(\mathcal{H}_1) : P(\mathcal{H}_2) \leqslant Y(w, \mathcal{H}_1) \leqslant Z(w, \mathcal{H}_1) \leqslant s$$
$$\wedge \left(Z(w, \mathcal{H}_1) = s \Rightarrow \exists e \in \mathcal{E}_1 : \{i(w), t(w)\} \subseteq e\right).$$

For 2-uniform hypergraphs the last two Theorems provide shaper assertions than the known propositions for the lexicographic product of graphs published by TEICHERT in 1982 (cf. [1]).

References

[1] G. Schaar, M. Sonntag, H.-M. Teichert, *Hamiltonian properties of products of graphs and diagraphs*, Teubner-Texte Math. 108, Leipzig, 1988.
[2] M. Sonntag, *Hamiltonian properties of the cartesian sum of hypergraphs*, J. Inf. Process. Cybern. EIK **25** no. 3 (1989), 87–100.

[3] M. Sonntag, *Hamiltonicity of the normal product of hypergraphs*, J. Inf. Process. Cybern. EIK **7** no. 7 (1989), 415–433.

Martin Sonntag
Bergakademie Freiberg,
Fachbereich Mathematik,
9200 Freiberg/Sachs., Germany

Fourth Czechoslovakian Symposium on
Combinatorics, Graphs and Complexity
J. Nešetřil and M. Fiedler (Editors)
© 1992 Elsevier Science Publishers B.V. All rights reserved.

Non-Hamiltonian Simple 3-Polytopal Graphs with Edges of Only Two Types

MICHAL TKÁČ

1. Introduction

In this paper by a graph we mean a finite connected undirected graph with no loops or multiple edges. For any graph G let $v(G)$ denote the number of vertices and $h(G)$ the length of a maximum cycle. So, G is non-hamiltonian if and only if $h(G)$ is less than $v(G)$. A class of graphs is said to be *non-hamiltonian* if it contains at least one non-hamiltonian graph. The *shortness coefficient* $\varrho(\mathfrak{G})$ of an infinite class $\mathfrak{G}$ of graphs is defined by (see [5])

$$\varrho(\mathfrak{G}) = \liminf_{G \in \mathfrak{G}} \frac{h(G)}{v(G)}.$$

Let $G_3(p,q)$ denote the class of 3-connected trivalent planar graphs, i.e. simple 3-polytopal graphs, all of whose faces are p-gons and q-gons, $p, q \geqslant 3$, $p < q$.

In several papers, including [6] and [10] it has been shown that the shortness coefficient is less than one for many classes of graphs in $G_3(p,q)$.

Let $S(p,q)$ denote the class of simple 3-polytopal graphs in which all the edges are incident with two p-gons or a p-gon and a q-gon, $p \neq q$, $p, q \geqslant 3$.

Call an edge of a simple planar graph to be of the *type* (p,q) if the faces containing it are a p-gon and a q-gon, and assume we deal only with simplicial 3-polytopal graphs with exactly two types of edges. Then it is evident that such graphs can exist only if its edges are of the types (p,p) or (p,q), $p \neq q$, $p, q \geqslant 3$.

So, $S(p,q)$ is the class of simple 3-polytopal graphs the edges of which are of the type (p,p) or (p,q).

It is easy to see that $S(p,q)$ is a subclass of $G(p,q)$ if $p < q$ and of $G(q,p)$ if $p > q$.

In the papers [4] and [5] it has been shown that the class $S(p,q)$ is not infinite unless $6 \leqslant p \leqslant 10$ and $q = 3$, or $6 \leqslant p \leqslant 7$ and $q = 4$, or $p = 6$ and $q = 5$, or $p = 5$ and $q \geqslant 12$. According to Goodey, every member of $S(6,q)$ is hamiltonian, for $q = 3$ [2] and $q = 4$ [1]. The same has been shown by Jendroľ and Mihók for the class $S(5,12)$ [3]. In the paper [5] which has stimulated interest in the classes of graphs $S(p,q)$ and their shortness coefficients, Owens proved that $\varrho(S(5,q)) < 1$ for all $q \geqslant 28$ and he also asked whether there are some non-hamiltonian members in the classes $S(5,q)$ for $12 \leqslant q \leqslant 23$, or $q = 27$, and whether $\varrho(S(5,q)) < 1$, for $q = 24, 25, 26$.

It is easy to verify that the class $G(3,q)$ is equal to the class $S(q,3)$ for all $q \geqslant 4$. No graph of the class $G(3,q)$ can contain an edge which is of the type $(3,3)$, because the existence of such an edge which is contained in two triangles leads to a contradiction with the 3-connectivity of these graphs. This means that all results which have been proved for the classes $G(3,q)$ we can apply to the classes $S(q,3)$. So in [6] Owens has shown that $\varrho(S(q,3)) < 1$ for $q = 8, 9$ and 10.

These results are supplemented by the following theorem:

Theorem. (i) $\varrho(S(5,26)) \leqslant 209/210 < 1$.
  (ii) $\varrho(S(5,27)) \leqslant 439/440 < 1$.
  (iii) $\varrho(S(7,3)) \leqslant 415/416 < 1$.
  (iv) $\varrho(S(7,4)) \leqslant 1239/1240 < 1$.

The proofs of the parts (i) and (ii) of the Theorem will appear in [7]. The proofs of the parts (iii) and (iv) will be published elsewhere [8] and [9].

It seems to be still unknown whether there are any non-hamiltonian graphs in the classes $S(6,5)$ and $S(5,q)$ for $13 \leqslant q \leqslant 23$, and whether $\varrho(S(5,q)) < 1$ for $q = 24, 25$.

References

[1] P. R. Goodey, *Hamiltonian circuits in polytopes with even sided faces*, Israel J. Math. **22** (1975), 52–56.

[2] ———, *A class of hamiltonian polytopes*, J. Graph Theory **1** (1977), 181–185.

[3] S. Jendroľ and P. Mihók, *Note on a class of hamiltonian polytopes*, Discrete Math. **71** (1988), 233–241.

[4] S. Jendroľ and M. Tkáč, *On the simplicial 3-polytopes with only two types of edges*, Discrete Math. **48** (1984), 229–241.

[5] P. J. Owens, *Simple 3-polytopal graphs with edges of only two types and shortness coefficients*, Discrete Math. **59** (1986), 107–114.

[6] ———, *Non-hamiltonian simple 3-polytopes with only one type of face besides triangles*, Annals of Discr. Math. **20** (1984), 241–251.

[7] M. Tkáč, *Note on shortness coefficients of simple 3-polytopal graphs with edges of only two types*, Discr. Math., to appear.

[8] ———, *Note on shortness coefficients of simple 3-polytopal graphs with only one type of face besides triangles*, submitted.

[9] ———, *Simple 3-polytopal graphs with edges of only two types and shortness coefficients*, submitted.

[10] J. Zaks, *Non-hamiltonian simple 3-polytopes having just two types of faces*, Discr. Math. **29** (1980), 87–101.

Michal Tkáč
Department of Mathematics, Technical University,
Švermova 9, 040 00 Košice, Czechoslovakia

Fourth Czechoslovakian Symposium on
Combinatorics, Graphs and Complexity
J. Nešetřil and M. Fiedler (Editors)
© 1992 Elsevier Science Publishers B.V. All rights reserved.

On Spectra of Trees and Related Two-Graphs

S. V. TSARANOV

1. Trees and two graphs

Let T be a tree, $V(T)$ and $\varepsilon(T)$ be its vertex and edge sets. Define $\Omega = \Omega(T)$ as the set of all triples of edges of the tree T such that in each triple the following holds: *None of the three edges lies between the other two.*

The following property holds for the sets $\varepsilon(T)$ and Ω.

Every quadruple of edges contains an even number of triples from Ω.

This is easy to check by considering all possibilities for the quadruple. So, the object $\Delta(T) = (\varepsilon(T), \Omega(T))$ is a two-graph by the definition, cf. [Sei]. The above construction of the map $T \mapsto \Delta(T)$ from trees to two-graphs was proposed by Cameron to prove the theorem suggested in [SeTs].

Theorem 1 [Cam].
(1) The map $T \mapsto \Delta(T)$ is injective.
(2) The two-graph Δ is an image of a tree T, i.e., there exists a tree T such that $\Delta = \Delta(T)$, if and only if Δ contains no two-subgraphs "pentagon" or "hexagon".

Remark. The two-graph "pentagon" resp. "hexagon" arises from the ordinary pentagon resp. hexagon by choosing all those triples of vertices which are of the type indicated in the figure (5 triples for the pentagon and 12 triples for the hexagon).

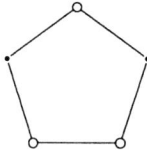 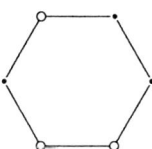

2. Switching classes

From the description of "pentagon" and "hexagon" it is easily seen which triples of vertices of an ordinary graph one can choose to obtain a two-graph. Let Γ be a graph having a vertex set $V(\Gamma)$ of order n and the edge set $\varepsilon(\Gamma)$. Let $\Omega(\Gamma)$ be the collection of those unordered triples of vertices which carry an odd number of edges in Γ. Then every quadruple of vertices contains an even number of triples from $\Omega(\Gamma)$, that is the pair $\Delta_\Gamma = (V(\Gamma), \Omega(\Gamma))$ is a two-graph. Let Γ' be a graph derived from Γ by a *switching* over a vertex v, that is, deleting the existing edges to v and adding all formerly nonexistent edges incident to v (in other words, by the addition mod 2 of a $K_{1,n-1}$ at v). With respect to this operation, all graphs are divided into *switching classes* of graphs. As switching doesn't change the set $\Omega(\Gamma)$ we have $\Delta_\Gamma = \Delta_{\Gamma'}$. Conversely, for a given two-graph $\Delta = (V, \Omega)$ one readily constructs the switching class corresponding to Δ. Thus we have, cf. [Sei], [BMS],

Theorem 2. *There is a one-to-one correspondence between the switching classes of graphs on n vertices and the two-graphs on n vertices.*

Let us return to the tree T and the two-graph $\Delta(T)$. How to interpret the corresponding switching class of graphs? To do this we choose an arbitrary orientation on the edges of T and define an ordinary graph Γ_T whose vertex set is the edge set $\varepsilon(T)$ of T. For every pair $x, y \in \varepsilon(T)$ we define adjacency in Γ_T as follows:

$\xrightarrow{x} \xleftarrow{y}$ are adjacent in Γ_T: put $e_{xy} = -1$;

$\xrightarrow{x} \xrightarrow{y}$ are nonadjacent in Γ_T: put $e_{xy} = 1$.

The matrix $E = (e_{xy})$ is the $(-1, 1)$ adjacency matrix of Γ_T. Switching of x in Γ_T amounts to changing the direction of x to the opposite, or, in terms of adjacency matrices, replacing E by DED where $D = \text{diag}(1, \ldots, 1, -1, 1, \ldots, 1)$ with -1 in the position of x. Notice that every graph from the same switching class of graphs is obtained by changing the direction of some edges of T. The spectrum of the $(-1, 1)$ matrix E is an invariant of the switching class so we call it the *spectrum* of the corresponding *two-graph*. Let $A = A(T)$ denote the ordinary adjacency matrix of T, I_k be the unit matrix of size k.

Theorem 3. *The matrices $3I_n + E$ and $2I_{n+1} - A$ have the same numbers of zero and negative eigenvalues.*

3. Representation of graphs.

In a way similar to [BCN] we call a $(p; q, r)$-*representation* of the graph Γ a map $\varphi: V(\Gamma) \to \mathbf{R}^k$, where $\mathbf{R}^k$ is the k-dimensional real space supplied with a scalar product $\langle \cdot, \cdot \rangle$, so that for vertices $x, y \in V(\Gamma)$

$$\langle \varphi(x), \varphi(y) \rangle = \begin{cases} p, & \text{if } x = y; \\ q, & \text{if } (x, y) \in \varepsilon(\Gamma); \\ r, & \text{if } (x, y) \notin \varepsilon(\Gamma). \end{cases} \quad (*)$$

The matrix $\Phi = (\langle\varphi(x), \varphi(y)\rangle)$ is the Gram matrix of pairwise scalar products of vectors from $\varphi(\Gamma)$. If the set $\varphi(\Gamma)$ generates $\mathbf{R}^k$ then the scalar product $\langle\cdot,\cdot\rangle$ is completely defined by the Gram matrix Φ. Thus for each graph on n vertices one can construct a representation of it in $\mathbf{R}^n$ such that $(*)$ holds.

We are ready now to prove theorem 3. Let the tree T have the vertex set $V(T) = \{0, 1, \ldots, n\}$ and all edges be directed towards 0. Let $\varphi \colon \Gamma \to \mathbf{R}^{n+1}$ be a $(2; -1, 0)$-representation of Γ with $\Phi = 2I_{n+1} - A$. We change the basis as follows:

$$f_t = e_0 + e_{i_1} + \ldots + e_{i_k}$$

where $0, i_1, \ldots, i_k = t$ is the shortest path from 0 to t.
Then

$$\langle f_0, f_t \rangle = \begin{cases} 2 & \text{if } s = t; \\ 1 & \text{if } (0, s) \subset (0, t) \text{ or } (0, s) \supset (0, t); \\ 0 & \text{otherwise.} \end{cases}$$

Hence we get a $(2;0,1)$-representation of the graph Γ_T defined at the end of the chapter 2, i.e.

$$\Phi' = (\langle f_s, f_t \rangle) = \begin{pmatrix} 2 & j^t \\ j & \frac{1}{2}(3I_n + E(\Gamma_T) + J_n) \end{pmatrix}$$

As the numbers of positive, zero and negative eigenvalues of a Gram matrix don't depend on a change of the basis, they are the same for Φ' and Φ. Changing the basis once again as follows:

$$g_0 = f_0,$$
$$g_s = \frac{1}{\sqrt{2}}(f_s - f_0),$$

we get the Gram matrix

$$\Phi'' = (\langle g_s, g_t \rangle) = \begin{pmatrix} 2 & 0^t \\ 0 & 3I_n + E(\Gamma_T) \end{pmatrix}$$

which proves theorem 3.

References

[BCN] A. E. Brouwer, A. M. Cohen, A. Neumaier, *Distance regular graphs*, Springer, 1989.

[BMS] F. C. Bussemaker, R. Mathon, J. J. Seidel, *Tables of two-graphs*, Report 79-WSK-05, Techn. Univ. Eindhoven, 1979; *Combinatorics and graph theory* (S. B. Rao, ed.), L. N. M. 885, Springer, 1981, pp. 70–112.

[Cam] P. J. Cameron, *Two-graphs and trees*, British Combinatorial Conference, Norwich, 1989, manuscript.

[Sei] J. J. Seidel, *A survey of two-graphs*, Proc. Intern. Coll. Teorie Combinatorie (Roma 1973), Accad. Nax. Lincei, Roma, 1976, pp. 481–511.

[SeTs] J. J. Seidel, S. V. Tsaranov, *Two-graphs, related groups, and root systems*, Bull. Soc. Math. Belgique, to appear.

S. V. Tsaranov
Inst. Systems Studies VNIISI,
Acad. Sci. USSR,
Moscow, USSR

Metrically Regular Square of Metrically Regular Bigraphs

VLADIMÍR VETCHÝ

1. Basic notations

Let X be a finite set, $\operatorname{card} X \geq 2$. For an arbitrary natural number D let $\mathbf{R} = \{R_0, R_1, \ldots, R_D\}$ be a system of binary relations on X. A pair $(X, \mathbf{R})$ will be called an *association scheme* with D classes if and only if it satisfies the axioms A1–A4:

A1. The system $\mathbf{R}$ is a partition of the set X^2 and R_0 is the diagonal relation, i.e. $R_0 = \{(x,x);\ x \in X\}$.

A2. For each $i \in \{0, 1, \ldots, D\}$, it holds $R_i^{-1} \in \mathbf{R}$.

A3. For each $i, j, k \in \{0, 1, \ldots, D\}$ it holds

$$(x,y) \in R_k \wedge (x_1, y_1) \in R_k \implies p_{ij}(x,y) = p_{ij}(x_1, y_1),$$

where

$$p_{ij}(x,y) = |\{z;\ (x,z) \in R_i \wedge (z,y) \in R_j\}|.$$

Then define

$$p_{ij}^k := p_{ij}(x,y), \text{ where } (x,y) \in R_k.$$

A4. For each $i, j, k \in \{0, 1, \ldots, D\}$ it holds $p_{ij}^k = p_{ji}^k$.

The set X will be called the *carrier* of the association scheme $(X, \mathbf{R})$. In particular, $p_{i0}^k = \delta_{ik}$, $p_{ij}^0 = v_i \delta_{ij}$, where δ_{ij} is the Kronecker-symbol and $v_i := p_{ii}^0$ and define $P_j := (p_{ij}^k)$, $0 \leq i, j, k \leq D$. (See [1].)

Given an undirected graph $G = (X, E)$ of diameter D we may now define $R_k = \{(x,y);\ d(x,y) = k\}$, where $d(x,y)$ is the distance from the vertex x to the vertex y in the standard graph metric. If $(X, \mathbf{R})$ gives rise to an association scheme, the graph G is called *metrically regular* (sometimes also called *distance regular*) and p_{ij}^k are said to be its *parameters*. In particular, a metrically regular graph with diameter $D = 2$ is called *strongly regular*.

Let $G = (X, Y)$ be an undirected graph without loops and multiple edges. The *second power* (or *square* of G) is the graph $G^2 = (X, E')$ with the same vertex set X and in which mutually different vertices are adjacent if and only if there is at least one path of length 1 or 2 in G between them.

The characteristic polynomial of the adjacency matrix A of a graph G is called the *characteristic polynomial* of G and the eigenvalues and the spectrum of A are called the *eigenvalues* and the *spectrum of G*. The greatest eigenvalue of G is called the *index* of G.

2. Conditions for metrically regular graphs of diameter D to have their square metrically regular

Because $G = (X, E)$ is metrically regular, the pair $(X, \mathbf{R})$ forms an association scheme with parameters p_{ij}^k, where

$$\mathbf{R} = \{R_0, R_1, \ldots, R_D\}, \qquad R_i = \{(x,y);\ x,y \in X, d(x,y) = i\}.$$

If $G^2 = (X, E')$ is metrically regular then the pair $(X, \mathbf{R}')$ forms an association scheme too, where $\mathbf{R}' = \{R_0', R_1', \ldots, R_{D'}'\}$ and $R_0' = R_0, \ldots, R_k' = R_{2k-1} \cup R_{2k}$. So, for its parameters $^2p_{i,j}^k$ it must hold

$$^2p_{ij}^k = \sum_{m,n=0}^{1} p_{2i-m,2j-n}^{2k-1} = \sum_{m,n=0}^{1} p_{2i-m,2j-n}^{2k}$$

$$1 \leqslant i, j, k \leqslant D'.$$

On the other hand, if A denotes the adjacency matrix of the metrically regular graph G and A_2 the adjacency matrix of G^2 it holds

$$A_2 = \frac{1}{p_{11}^2} A^2 + \frac{p_{11}^2 - p_{11}^1}{p_{11}^2} A - \frac{\lambda_1}{p_{11}^2} I.$$

So, if the eigenvalues of G are $\lambda_1 > \cdots > \lambda_k$ with respective multiplicities $m_1 = 1$, $m_2, \ldots, m_k$, the eigenvalues of G^2 are in the form

$$\mu_i = \frac{\lambda_i^2 + (p_{11}^2 - p_{11}^1)\lambda_i - \lambda_1}{p_{11}^2} \tag{1}$$

with multiplicities

$$m_i' = \sum_{j \in M_i} m_j, \qquad M_i = \{j;\ \mu_j = \mu_i\}.$$

It is proved for metrically regular bipartite graphs of diameter $D \leqslant 6$ that the spectrum of the metrically regular graph G^2 is in the form

$$S_p(G^2) = \left\{ \begin{matrix} \mu_1, & \mu_2 = \mu_{D+1}, & \mu_3 = \mu_D, & \mu_4 = \mu_{D-1}, & \ldots \\ 1, & m_2 + m_{D+1}, & m_3 + m_D, & m_4 + m_{D-1}, & \ldots \end{matrix} \right\}.$$

(It would be reasonable to conjecture that this holds for general D). Thus, from (1) we obtain for bipartite graphs $(p_{ij}^k = 0,\ i+j+k \equiv 1 \pmod{2})$:

$$-p_{11}^2 = \lambda_2 + \lambda_{D+1} = \lambda_3 + \lambda_D = \lambda_4 + \lambda_{D-1} = \cdots.$$

3. Bipartite graphs

$D = 3, 4$. The following results are proved in [2].

Theorem 1. Let G be a metrically regular graph with diameter $D = 3$ (4 distinct eigenvalues) and G^2 be strongly regular. Then it holds $\lambda_3 = -1$, $\lambda_2 > 0$.

Theorem 2. For every $k \in N$, $k \geqslant 2$ there is only one metrically regular bipartite graph $G = (X, E)$ with diameter $D = 3$, $|X| = 2k + 2$ so that G^2 is a strongly regular graph. The nonzero parameters of G are the following:

$p_{10}^1 = p_{23}^1 = p_{20}^2 = p_{13}^2 = p_{30}^3 = 1$ $\qquad v_0 = v_3 = 1$

$p_{12}^1 = p_{11}^2 = p_{22}^2 = k - 1$ $\qquad v_1 = v_2 = k$

$p_{12}^3 = k$ $\qquad S_p(G) = \left\{ \begin{array}{cccc} k, & 1, & -1, & -k \\ 1, & k, & k, & 1 \end{array} \right\}$

Construction of G: $G = (X = X_1 \cup X_2, E)$; $X_1 = \{v_1, \ldots, v_{k+1}\}$, $X_2 = \{u_1, \ldots, u_{k+1}\}$, $E = \{(v_i, u_j) \mid i, j = 1, 2, \ldots, k+1; i \neq j\}$

Theorem 3. There is only one table of parameters of an association scheme so that the corresponding metrically regular bipartite graph of diameter $D = 4$ (5 distinct eigenvalues) has a strongly regular square. The nonzero parameters of G are the following:

$p_{01}^1 = p_{20}^2 = p_{30}^3 = p_{40}^4 = p_{14}^3 = p_{24}^2 = p_{34}^1 = 1$ $\qquad v_0 = v_4 = 1$

$p_{11}^2 = p_{13}^2 = p_{33}^2 = 2$ $\qquad v_1 = v_3 = 4$

$p_{12}^1 = p_{23}^1 = p_{12}^3 = p_{23}^3 = 3$ $\qquad v_2 = 6$

$p_{22}^2 = p_{13}^4 = 4$

$p_{22}^4 = 6$ $\qquad S_p(G) = \left\{ \begin{array}{ccccc} 4, & 2, & 0, & -2, & -4 \\ 1, & 4, & 6, & 4, & 1 \end{array} \right\}$

The realization of this table is the 4-dimensional unit cube.

$D = 5$. The following theorem is proved in [3].

Theorem 4. There are only four tables of parameters of association schemes so that the corresponding metrically regular bipartite graphs of diameter $D = 5$ (6 distinct eigenvalues) have a metrically regular square. The nonzero parameters of G are the following:

$p_{10}^i = p_{45}^1 = p_{35}^2 = p_{25}^3 = p_{15}^4 = 1$ $\qquad k = p_{11}^2 = p_{44}^2 = p_{14}^3$

$p_{13}^2 = p_{24}^2 = p_{12}^3 = p_{34}^3 = k + 1$ $\qquad 2k = p_{12}^1 = p_{34}^1 = p_{13}^4 = p_{24}^4$

$p_{14}^5 = 2k + 1,$ $\qquad p_{23}^1 = p_{22}^4 = p_{33}^4 = 2k + 2,$ $\qquad v_0 = v_5 = 1$

$p_{22}^2 = p_{33}^2 = p_{23}^3 = 3k$ $\qquad 2k + 1 = v_1 = v_4$

$p_{23}^5 = 2(2k + 1)$ $\qquad 2(2k + 1) = v_2 = v_3$

$S_p(G) = \left\{ \begin{array}{cccccc} 2k+1, & k+1, & 1, & -1, & -k-1, & -2k-1 \\ 1, & 8 - \frac{12}{k+2}, & 6k - 5 + \frac{12}{k+2}, & m_3, & m_2, & 1 \end{array} \right\}, k \in \{1, 2, 4, 10\}$

In the case $k = 2$ the realization of this table is the 5-dimensional unit cube, for $k = 1$ the realization is shown in Fig. 1.

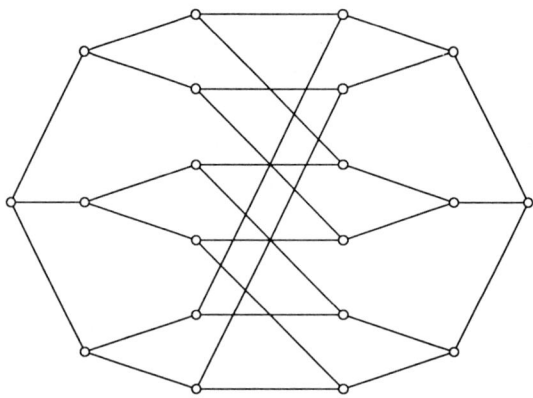

FIGURE 1

$D = 6$—The following theorem is proved in [4]

Theorem 5. *There is only one table of parameters of an association scheme with 6 classes so that the corresponding metrically regular bipartite graph of diameter $D = 6$ (7 distinct eigenvalues) has a metrically regular square. The realization of this table is the 6-dimensional unit cube.*

4. Conjecture

With respect to Theorems 1.–5. it would be reasonable to conjecture:

There is only one table of parameters of an association scheme with $2k$ classes ($k \geqslant 2$) so that the corresponding metrically regular bipartite graph of diameter $D = 2k$ has a metrically regular square. The realization of this table is the $2k$-dimensional unit cube.

References

[1] E. Bannai, T. Ito, *Algebraic Combinatorics I*, The Bejamin/Cummings Publishing Company, California, 1984.
[2] V. Vetchý, *Metrically regular square of metrically regular bigraphs I*, Arch. Math. Brno, to appear.
[3] _____, *Metrically regular square of metrically regular bigraphs II*, Arch. Math. Brno, to appear.
[4] _____, *Metrically regular square of metrically regular bipartite graphs of diameter $D = 6$*, to appear.

Vladimír Vetchý
Department of Mathematics,
VA Brno,
PS 13, 612 00 Brno, Yugoslavia

Fourth Czechoslovakian Symposium on
Combinatorics, Graphs and Complexity
J. Nešetřil and M. Fiedler (Editors)
© 1992 Elsevier Science Publishers B.V. All rights reserved.

Embedding of Graphs in the Complements of Their Squares

Mariusz Woźniak

The fact that every graph $G = (V, E)$ with $|V| = n$ and $|E| \leqslant n - 2$ can be embedded in its complement is well known. In the paper we characterize the graphs of order n and size at most $n-2$ which can be embedded in the complement of the square G^2 of G.

1. Introduction

We shall use standard graph theory notation. A finite, undirected graph G consists of a vertex set $V(G)$ and edge set $E(G)$. All graphs will be assumed to have neither loops nor multiple edges. For graphs G and H we denote by $G \cup H$ the vertex disjoint union of graphs G and H and kG stands for the disjoint union of k copies of graph G. A graph of order n and size m, i.e. an (n, m)-graph, is denoted by $G(n, m)$. The square G^2 of a graph $G = (V, E)$ is the graph defined as follows: $V(G^2) = V(G)$ and $E(G^2) = \{xy \colon \mathrm{dist}_G(x, y) \leqslant 2\}$.

An embedding of G in a graph H is an isomorphic mapping of G into H; in other words, there exists an embedding of G in H if H contains a subgraph which is an isomorphic copy of G. Denote by $\bar{G}$ the complement of a graph G of order n. If there is an embedding of G in $\bar{G}$ we say that there is a packing of two copies of G (into the complete graph K_n).

The following theorem was proved, independently, in [1], [3] and [4].

Theorem 1. *Let $G = (V, E)$ be a graph of order n. If $|E(G)| \leqslant n - 2$ then G can be embedded in its complement $\bar{G}$.*

This result has been improved in many ways. For example, if we assume that $G(n, n-2)$ is labeled then there exists a fixed-point-free embedding of G in $\bar{G}$ (cf. [5]). One can prove also (cf. [6]) that except for $K_3 \cup 2K_1$ and $K_4 \cup 4K_1$, there is a packing of three copies of a graph $G(n, n-2)$ into K_n. For other generalization and improvements see [2] or [7].

Our purpose is to prove the following

Theorem 2. *Let $G = (V, E)$ be a graph of order n. If $|E(G)| \leqslant n - 2$ then either G can be embedded in the complement of its square G^2 or G is isomorphic to one* of the following exceptional graphs:

$$C_5 \cup 2K_1, \; C_5 \cup K_2 \cup K_1, \; C_5 \cup 2K_2 \quad \text{or} \quad C_5 \cup K_1 \cup K_{1,n-7} \quad \text{for } n \geqslant 11.$$

The example of the star $K_{1,n-1}$ shows that Theorem 2 (as well as Theorem 1) cannot be improved by raising the size of G.

If G is an exceptional graph of Th. 2 then it is very easy to find an embedding of G in $\bar{G}$. Thus, since $G \subset \bar{G}^2$, Theorem 2 can be considered as an improvement of Theorem 1. In some cases this improvement is considerable. For instance if $G = K_{1,n-2} \cup K_1$ then $\bar{G}^2 = K_{n-1} \cup K_1$ and in this case all edges of K_n but one are covered by the edges belonging to $E(G)$ or to $E(\bar{G}^2)$.

The proof of Theorem 2 is given in section 3. In section 2 we consider some special cases.

2. Some lemmas

Throughout this paper we shall use the following terminology. In order to define an embedding of G into the complement of its square $\bar{G}^2$ it suffices to give a bijection $\alpha\colon V(G) \to V(G)$ such that $\alpha^*(E(G)) \cap E(G^2) = \emptyset$ where $\alpha^*(xy) = \alpha(x)\alpha(y)$. The edges of G^2 will be called *red* edges and the edges of $\alpha^*(E(G))$ will be called *black* edges. A vertex $x \in V(G)$ will be said to be *red* if it is incident only with red edges.

Lemma 3. *For every graph $G(n, n-2)$, $n \geqslant 3$, that is a union of a tree $T(n-1, n-2)$ and an isolated vertex u, there exist two embeddings α, β of G into the complement of G^2 such that if a vertex a is red with respect to α then a is not red with respect to β.*

Proof. The proof is by induction on n. It i easy to see that Lemma 3 holds in the case where $\mathrm{diam}(T) \leqslant 4$. Assume that $\mathrm{diam}(T) \geqslant 5$. Then there exist at least two vertices, a, b say, such that $G \setminus \{a\}$ and $G \setminus \{b\}$ have nontrivial components. Consider the graph $G' = G \setminus \{a, u\}$ and denote by $S_1, \ldots, S_l$ the nontrivial components of G'. Let $N_G(a) = \{x_1, \ldots, x_l\}$. Add one new vertex y_i to each S_i and apply the induction to each graph $S_i \cup \{y_i\}$. Let z_i be a red vertex in $S_i \cup \{y_i\}$. We can assume that $z_i \neq x_i$, $i = 1, \ldots, l$. Now, by identifying the vertex y_1 with z_2, y_2 with z_3, ..., y_l with z_1 we get an embedding α' of G' in the complement of G'^2 such that there is no black edge connecting any two vertices in $N_G(a)$. By putting $\alpha(a) = u$, $\alpha(u) = a$ and $\alpha(v) = \alpha'(v)$ for $v \in V(G')$ we obtain an embedding of G in the complement of G^2 with vertex a as a red vertex. Applying the above argument to the graph $G \setminus \{b\}$ we get another embedding with b as a red vertex. Then vertex a is not red. This finishes the proof of Lemma 3. □

The *proofs* of the following lemmas are not difficult and can be left to the reader.

Lemma 4. *For every graph $G(n, n-2)$, $n \geqslant 5$, $n \neq 7$, that is a union of a cycle C_{n-2} and two isolated vertices, there is a packing of G and G^2.*

Lemma 5. For every graph $G(n, n-2)$, $n \geqslant 8$, that is a union of at least two cycles and two isolated vertices there is a packing of G and G^2.

Lemma 6. Let G_1 be a union of k edge disjoint paths of length at least 2. We define graph G by adding to G_1 $k+1$ isolated vertices and one vertex of degree $2k$ adjacent to all ends of k paths. Then there is a packing of G and G^2.

3. Proof of Theorem 2

The proof is by induction on n. It is easy to see that a $(n, n-2)$-graph G is not connected and there are at least two components of G which are trees.

Suppose first that at least one component of G is a nontrivial tree T and let $G = T \cup R$. By Lemma 3 we can suppose that R is not trivial. Consider the graphs $T' = T \cup \{x\}$ and $R' = R \cup \{y\}$ obtained from T and R by adding two new vertices x, y ($x \neq y$, $x, y \notin V(G)$).

Suppose that R' is not an exceptional graph. By induction, there exist an embedding α of T' in the complement of its square and an embedding β of R' in the complement of its square. Denote by a and b the red vertices of T' and R' with respect to α and β, respectively. Identifying x with b and y with a we get a packing of G and G^2. The case where R' is an exceptional graph is left to the reader.

Suppose now that G has not a component which is a nontrivial tree. Then G has at least two isolated vertices. We shall distinguish several cases.

Case 1. $G(n, n-2)$ contains a triangle K_3. Denote by u one of the isolated vertices of G and put $V(K_3) = \{a, b, c\}$. Observe first that we can always choose one vertex of the triangle $K_3 \subset G$, a say, in such a way that the graph $G' = G \setminus \{a, u\}$ is not exceptional.

Let $k = \deg_G(a)$ and let $N_G(a) = \{x_1, \ldots, x_{k-2}, b, c\}$. Consider the graph G'' obtained from G' by adding the edges bx_i (if $bx_i \notin E(G')$), $i = 1, \ldots, k-2$. The graph G'' has $n-2$ vertices and at most $n-4$ edges. By induction, there is an embedding α' of G'' in the complement of its square. Observe that all edges connecting any two vertices of the set $N_G(a)$ are red. Since $G' \subset G''$, α' defines an embedding of G' in the complement of G''^2 without black edges between vertices belonging to $N_G(a)$. By putting $\alpha(a) = u$, $\alpha(u) = a$ and $\alpha(y) = \alpha'(y)$ for $y \in V(G')$ we get a packing of G and G^2.

Case 2. $G(n, n-2)$ has a vertex, a say, with $\deg_G(a) = 3$ and does not contain a triangle K_3.

Denote by u, v two isolated vertices of G and consider the graph $G' = G \setminus \{a, u, v\}$. Suppose that G' is not an exceptional graph. By induction, there exists a packing α' of G' and G'^2. If there are some black edges connecting vertices of $N_G(a)$, we redraw the black graph corresponding to α' using the isolated vertex v instead of one vertex of $N_G(a)$ in such a way that there are no more black edges between vertices of $N_G(a)$. This is possible since $|N_G(a)| = 3$ and G' does not contain a triangle. Now it is easy to obtain a packing of G and G^2 by permuting the vertices a and u. The case where G' is an exceptional graph is left to the reader.

Case 3. G does not contain a triangle K_3 and there is no vertex of degree 3 or 1 in G.

Denote by i the number of isolated vertices of G and by p the number of vertices of G with degree > 3. If $p = 0$ we apply Lemma 4 or Lemma 5 and if $p = 1$ we apply Lemma 6. So we may suppose $p \geqslant 2$.

Let $k = \min\{\deg_G(x): \deg_G(x) > 3\}$. We have $2n - 4 \geqslant pk + 2(n - p - i)$. Hence $2i \geqslant 4 + p(k - 2)$ and for $p \geqslant 2$ we get that $i \geqslant k$.

Let a be such a vertex of G that $\deg_G(a) = k$. Denote by $u_0, u_1, \ldots, u_{k-2}$ $k - 1$ isolated vertices of G and consider the graph $G' = G \setminus \{u_0, u_1, \ldots, u_{k-2}, a\}$. If G' is not an exceptional graph then there exists a packing α' of G' and G'^2. In order to obtain a packing α of G and G^2 we put $\alpha(a) = u_0$, $\alpha(u_0) = a$ and use $k - 2$ isolated vertices $u_1, u_2, \ldots, u_{k-2}$ to get out the black edges (with respect to α') from the set $N_G(a)$ which is possible since G' does not contain a triangle.

The case where G' is an exceptional graph is left to the reader.

Case 4. G does not contain a triangle K_3 and there is no vertex of degree 3 but at least one vertex of degree 1 in G.

Denote by G^* the graph obtained from G in the following way: We delete first all vertices of degree 1 in G, next all vertices of degree 1 in the obtained graph etc. Observe that G^* is an $(m, m - 2)$-graph. G^* satisfies the assumptions of Case 3. Let a be a vertex of G^* defined as in Case 3 with $\deg_{G^*}(a) = k$. Denote by $u_0, u_1, \ldots, u_{k-2}$ the isolated vertices of G^*. Let H be a tree induced (as a subgraph of G) by vertex a and all vertices x of G such that there is a path from x to a in G consisting of vertices deleted in the construction of G^*.

Let $G' = G \setminus \bigl(V(H) \cup \{u_0, u_1, \ldots, u_{k-2}\}\bigr)$. If G' is not an exceptional graph then there is a packing α' of G' and G'^2. As in Case 3 the vertices $u_1, \ldots, u_{k-2}$ are used to get out the black edges from $N_G(a) \setminus V(H)$. It is also easy to see that a packing β of $H' = H \cup \{u_0\}$ and H'^2 can be chosen in such a way that $\beta(a)$ is neither a nor its neighbor. Now it is easy to define a packing of G and G^2. □

Acknowledgment. This paper was partially supported by Polish Research Grant KBN P/05/003/90-2.

References

[1] B. Bollobás, and S. E. Eldridge, *Packings of Graphs and Applications to Computational Complexity*, Proceedings of the Fifth British Combinatorial Conference (Aberdeen, 1975), Congressus Numerantium XV, Utilitas Mathematica Publishing.

[2] B. Bollobás, *Extremal graph theory*, Academic Press, London, 1978.

[3] D. Burns and S. Schuster, *Every $(p, p - 2)$ Graph is Contained in its Complement*, J. Graph Theory **1** (1977), 277–279.

[4] N. Sauer, and J. Spencer, *Edge Disjoint Placement of Graphs*, Journal of Combinatorial Theory B **25** (1978), 295–302.

[5] S. Schuster, *Fixed-Point-Free Embeddings of Graphs in Their Complements*, Internat. J. Math.& Math. Sci. **1** (1978), 335–338.

[6] A. P. Wojda, and M. Woźniak, *Triple placement of graphs*, to appear.

[7] H. P. Yap, *Some Topics in Graph Theory*, London Mathematical Society, Lectures Notes Series 108, Cambridge University Press, Cambridge, 1986.

Mariusz Woźniak
Instytut Matematyki,
Akademia Górniczo-Hutnicza,
Al. Mickiewicza 30,
30-059 Kraków, Poland

Fourth Czechoslovakian Symposium on
Combinatorics, Graphs and Complexity
J. Nešetřil and M. Fiedler (Editors)
© 1992 Elsevier Science Publishers B.V. All rights reserved.

An $\frac{11}{6}$-Approximation Algorithm for the Steiner Problem on Graphs

A. Z. ZELIKOVSKY

1. Introduction

Given a graph $G = (V, E, d)$ with nonnegative edge lengths $d: E \to \mathbb{R}^+$ and a set $S \subseteq V$ of distinguished vertices, any tree within G is called a *Steiner tree* if it spans all members of S. The *Steiner problem* on graphs, which is NP-hard (see [3]), asks for a minimum length Steiner tree $T_{\min}$.

We shall use the following notations. The *metrical closure* of G is the complete graph $\bar{G} = (V, \bar{E}, \bar{d})$ which has edge lengths equal to shortest path distances in G. G_M is the subgraph of G induced by a vertex subset $M \subseteq V$. So $\bar{G}_S$ is the subgraph of $\bar{G}$ induced by S. Let us denote by $\delta_{\mathrm{mst}}(G)$ the total length of a minimum spanning tree (MST) of G. The famous *MST-algorithm* [2] approximates $T_{\min}$ by an MST of $\bar{G}_S$ with edges replaced by the corresponding shortest paths in G. The following inequality was proved:

$$\frac{\delta_{\mathrm{mst}}(\bar{G}_S)}{\delta(T_{\min})} \leq 2. \qquad (1)$$

This Steiner tree can be computed in time $O(|V|\log|V| + |E|)$ [4].

In this paper, we present an $\frac{11}{6}$-approximation algorithm which improves the MST-algorithm by taking into consideration some additional vertices outside S.

2. The algorithm

Some preliminary definitions: Given a metric closed graph $G = (V, E, d)$, to *contract* an edge $e = (v_1, v_2) \in E$ means to replace the ends of this edge by a vertex v. The distance between the resulting vertex v and any other vertex $v' \in V$ equals to $\min(d(v_1, v'), d(v_2, v'))$. The graph $G[e] = (V[e], E[e], d')$ is the metric closure of the contracted graph. For any triple $z = \{v_1, v_2, v_3\} \in V^3$ the graph $G[z]$ is obtained from G by contraction of edges of the triangle with vertices from z.

The algorithm goes as follows:

2.1. Algorithm.
(1) For every $z = \{s_1, s_2, s_3\} \in S^3$ find $d(z) = \min\limits_{v \in V} \sum\limits_{i=1}^{3} d(v, s_i)$ and $v(z) \in V$ minimizing this sum.

(2) $F = \bar{G}_S$; $W = \emptyset$; until $w > 0$ do:
   a) find $z^* \in S^3$ maximizing $w = \delta_{\text{mst}}(F) - \delta_{\text{mst}}(F[z]) - d(z)$;
   b) $F \leftarrow F[z^*]$, $W \leftarrow W \cup v(z^*)$.
(3) Find an approximate Steiner tree, T_H, using the MST-algorithm for the graph $G = (V, E, d)$ with the distinguished vertex set $S \cup W$.

In a few words, the presented algorithm (see step (2)) consequently finds the best MST-length reduction of the graph F, which initially coincides with $\bar{G}_S$, by adding to an MST of F three G-edges with a common end (such collection will be called a *star*) and removing the longest edges from each resulting cycle. After the star contraction we get the next graph F.

The implementation time of Algorithm 2.1 is $O(|V|^3 + |V||S|^3)$. The main result of this paper is the following

2.2. Theorem. $\sigma(T_H)/\sigma(T_{\min}) \leq \frac{11}{6}$.

In the rest of the paper we briefly describe the proof of Theorem 2.2.

3. The win of a greedy triangle sequence

Fix some minimum spanning tree T of $\bar{G}_S$ and numerate its edges in the increasing order $\{e_1, \ldots, e_{|S|-1}\}$: $d(e_i) \leq d(e_{i+1})$ ($i = 1, \ldots, |S| - 1$). Define an *index* $n: S^2 \to \{0, 1, \ldots, |S| - 1\}$ where $n(s_1, s_2)$ is the maximum index of tree edges belonging to the unique path between s_1 and $s_2 \in S$. $\text{ind}(u) = d(e_{n(s_1,s_2)})$. The following statements are obvious.

3.1. Lemma. *Any triangle in $\bar{G}_S$ contains a unique edge with the minimum index and two other edge indices equal each other.*

3.2. Corollary. *Any cycle in $\bar{G}_S$ has at least two edges with the maximum index.*

Now we shall introduce a system of independence in $\bar{G}_S$.

Let $H = \{h_i\}_1^r$ be the set of triangles of $\bar{G}_S$ and H^* be a subgraph with edges belonging to triangles of H. H is called *independent* if
(i) any two triangles have no common edges;
(ii) the cyclic rank of H^* equals r.

We shall associate a *star* in the graph $\bar{G}_S$ with a triple $\{s_1, s_2, s_3\} \subseteq S$. This star consists of a vertex $v \in V$ and three edges (v, s_i), $i = 1, 2, 3$. Denote $d(z) = \sum_{i=1}^{3} d(v, s_i)$ and $d(Z) = \sum_{z \in Z} d(z)$ for a set of stars Z. Further all definitions for stars are the same as for associated triangles.

Let I be the family of all independent star sets in $\bar{G}_S$.

Let $Z \in I$ and $\bar{G}_S[Z]$ be a graph obtained from $\bar{G}_S$ by the contraction of all Z members. The value

$$w(Z) = \delta_{\text{mst}}(\bar{G}_S) - \delta_{\text{mst}}(\bar{G}_S[Z]) - d(Z)$$

will be called the *win* of a star set Z.

Consider an arbitrary triangle of $\bar{G}_S$ with the ends s_1, s_2, s_3 and associate a star z with it. Assume that $n(s_1, s_2) \neq n(s_2, s_3)$. There are two cycles in the graph

$T \cup z$. By Lemma 3.1 the longest edges in these two cycles have indices $n(s_1, s_2)$ and $n(s_2, s_3)$, respectively. These two edges must be removed from $T \cup z$. Hence the win of z equals to

$$w(z) = d(e_{(s_1,s_2)}) + d(e_{(s_2,s_3)}) - d(z) \quad (n(s_1, s_2) \neq n(s_2, s_3)).$$

Let $A \cup B \in I$, $A \cap B = \emptyset$. The *conditional win* $w_A(B)$ is the win of the star set B in the graph $\bar{G}_S[A]$. Note that $w(A) + w_A(B) = w(A \cup B)$.

Now let us return to Algorithm 2.1. This algorithm gives an independent star sequence $H = \{h_i\}_1^r$. Let $H_k = \{h_i\}_1^k$. This star sequence is *greedy* relative to conditional wins, i.e. for all $k = 1, \ldots, r$

$$w_{H_{k-1}}(h_k) = \max\{w_{H_{k-1}}(z) \mid H_{k-1} \cup z \in I\}.$$

3.3. Lemma. *For a greedy star sequence H and an arbitrary independent star set Z the following inequality is true:*

$$2w(H) \geqslant w(Z). \tag{2}$$

The proof of this Lemma is based on the previous Lemma.

4. The star set induced by the minimum Steiner tree

4.1. Lemma. *There is an independent star set Z such that*

$$3[\delta_{\mathrm{mst}}(\bar{G}_S) - w(Z)] \leqslant 5\delta(T_{\min}). \tag{3}$$

Using zero length edges we transform a given instance of the Steiner problem into its metrical equivalent with a binary rooted tree as a minimum Steiner tree. Then we construct a spanning tree T associated with this Steiner tree. For a certain weighted edge graph of the tree T we find a lower bound of the maximum matching. There is an independent star set associated with this matching. This star set satisfies (3).

To prove Theorem 2.2 it is enough to write the sequence of the following inequalities for the greedy star sequence H:

$$6\delta(T_H) = 6\delta_{\mathrm{mst}}(\bar{G}_{S \cup W}) \leqslant 6\delta_{\mathrm{mst}}(\bar{G}_S \cup H) = 6\big(\delta_{\mathrm{mst}}(\bar{G}_S) - w(H)\big)$$

$$= 3\delta_{\mathrm{mst}}(\bar{G}_S) + 3[\delta_{\mathrm{mst}}(\bar{G}_S) - 2w(H)] \overset{(2)}{\leqslant} 3\delta_{\mathrm{mst}}(\bar{G}_S) + 3[\delta_{\mathrm{mst}}(\bar{G}_S) - w(Z)]$$

$$\overset{(3)}{\leqslant} 3\delta_{\mathrm{mst}}(\bar{G}_S) + 5\delta(T_{\min}) \overset{(1)}{\leqslant} 6\delta(T_{\min}) + 5\delta(T_{\min}) = 11\delta(T_{\min})$$

The Rayward-Smith's average distance heuristic gives a $\frac{4}{3}$-approximation for the special case of the Steiner problem for complete graphs with all edge lengths either 1 or 2 [1].

4.2. Remark. *Algorithm 2.1 is a $\frac{4}{3}$-approximation for the Steiner problem with edge lengths either 1 or 2.*

To find a rectilinear Steiner tree we can also use the presented algorithm.

4.3. Conjecture. *Algorithm 2.1 is a $\frac{11}{8}$-approximation for the Steiner problem in rectilinear metric.*

References

[1] M. Bern, P. Plassman, *The Steiner problem with edge lengths 1 and 2*, Inform. Process. Lett. **32** (1989), 171–176.

[2] C. El-Arbi, *Une heurisitque pour le probleme de l'arbre de Steiner*, RAIRO Operations Research **12** (1978), 207–212.

[3] R. M. Karp, *Reducibility among combinatorial problems*, Complexity of computer computation (R. E. Miller and J. W. Tatcher, eds.), Plenum Press, New York, 1972, pp. 85–103.

[4] K. Mehlhorn, *A faster approximation algorithm for the Steiner problem in graphs*, Inform. Process. Lett. **27** (1988), 125–128.

A. Z. Zelikovsky
Panfilova 59/2, kv. 2,
277008 Kishinev, USSR

Fourth Czechoslovakian Symposium on
Combinatorics, Graphs and Complexity
J. Nešetřil and M. Fiedler (Editors)
© 1992 Elsevier Science Publishers B.V. All rights reserved.

Distances Between Graphs

(Extended Abstract)

BOHDAN ZELINKA

1. Introduction

In this lecture we shall study distances between isomorphism classes, or shortly distances between graphs. An isomorphism class of graphs is the class of all graphs which are isomorphic to a given graph.

Various types of distances between such classes may be introduced; they yield a certain measure of dissimilarity of graphs. If we want to speak shortly, we may speak about a distance between graphs instead of a distance between isomorphism classes of graphs. In this case we must have in mind that two graphs whose distance is zero need not be identical, but are isomorphic.

Let G_1, G_2 be two graphs with the same numbers n of vertices. Then [Z1] the distance $\delta(G_1, G_2)$ is equal to n minus the maximum number of vertices of a graph which is isomorphic simultaneously to an induced subgraph of G_1 and to an induced subgraph of G_2.

Introducing this distance was motivated by a certain problem of V. G. Vizing [V1]. F. Kaden [K1] and F. Sobik [S2] have generalized this distance for graphs which need not have the same number of vertices; in this case, to the distance defined above the absolute value of the difference between the cardinalities of the vertex sets of G_1 and G_2 is added.

Similarly the tree distance δ_T can be defined [Z2]. Let T_1, T_2 be two trees with the same number n of vertices. The distance $\delta_T(T_1, T_2)$ is equal to n minus the maximum number of vertices of a tree which is isomorphic simultaneously to a subtree of T_1 and to a subtree of T_2.

The edge distance δ_E was introduced by V. Baláž, J. Koča, V. Kvasnička and M. Sekanina [B1]. Let $G_1 = (V_1, E_1)$, $G_2 = (V_2, E_2)$ be two graphs, let $G_{12} = (V_{12}, E_{12})$ be a graph whose number of edges is maximal among all which are isomorphic simultaneously to a subgraph of G_1 and to a subgraph of G_2. Then

$$\delta_E(G_1, G_2) = |E_1| + |E_2| - 2|E_{12}| + ||V_1| - |V_2||.$$

This distance has applications in organic chemistry and especially in medical chemistry.

The edge rotation distance δ_R was introduced by G. Chartrand, F. Saba and H.-B. Zou [C1]. Let x, y, z be three pairwise distinct vertices of a graph G; let x and y be adjacent, let x and z be non-adjacent in G. To perform the rotation of the edge xy to the position xz means to delete the edge xy from G and to add the edge xz to G. Now let G_1, G_2 be two graphs with the same number of vertices and the same number of edges. The edge rotation distance $\delta_R(G_1, G_2)$ is equal to the minimum number of edge rotations which are necessary for transforming the graph G_1 into a graph isomorphic to G_2.

The edge shift distance was defined by M. Johnson [J1]. An edge shift is a special case of an edge rotation. The rotation of the edge xy to the position xz is called an edge shift (along the edge xz), if y and z are adjacent. Let G_1, G_2 be two connected graphs with the same number of vertices and the same number of edges. The edge shift distance $\delta_S(G_1, G_2)$ is equal to the minimum number of edge shifts which are necessary for transforming the graph G_1 into a graph isomorphic to G_2.

The contraction distance δ_c was introduced in [Z3]. Let e be an edge a graph $G = (V, E)$, let u, v be its end vertices. To perform the contraction of the edge e means to replace u and v by one vertex which is adjacent to all vertices of $V - \{u, v\}$ which were adjacent to u or to v. We say that the graph H is a retract of the graph G if H is obtained from G by a finite number of edge contractions. Now let G_1, G_2 be two graphs with the same number n of vertices. The contraction distance $\delta_c(G_1, G_2)$ is equal to n minus the maximum number of vertices of a graph which is isomorphic simultaneously to a retract of G_1 and to a retract of G_2.

Some of the described distances may be transferred also to directed graphs.

2. Distance defined by means of subgraphs

Theorem 2.1. [Z1] *Let n be a positive integer, let k be a nonnegative integer. Let G_1, G_2 be two graphs, each having n vertices. Then the following two assertions are equivalent:*
  (i) *There exists a graph G with at most $n + k$ vertices which has two induced subgraphs G_1' and G_2' such that $G_1' \simeq G_1$, $G_2' \simeq G_2$.*
  (ii) *There exist isomorphic graphs G_1'', G_2'', each having at least $n - k$ vertices, such that G_1'' is an induced subgraph of G_1 and G_2'' is an induced subgraph of G_2.*

Theorem 2.2. [Z1] *Let $\mathcal{S}_n$ be the system of all isomorphism classes of undirected graphs with n vertices without loops and multiple edges. If $\mathfrak{G}_1 \in \mathcal{S}_n$, $\mathfrak{G}_2 \in \mathcal{S}_n$ and $n + k$ is the least number of vertices of a graph containing induced subgraphs from the classes $\mathfrak{G}_1$ and $\mathfrak{G}_2$, then we denote $\delta(\mathfrak{G}_1, \mathfrak{G}_2) = k$. The system $\mathcal{S}_n$ with the distance δ is a metric space.*

The metric δ is obviously the above defined distance δ.

Let $\mathcal{G}_n$ be the graph whose vertex set is $\mathcal{S}_n$ and in which two vertices are adjacent if and only if their distance δ is 1.

Theorem 2.3. *The distance of two arbitrary vertices in the graph $\mathcal{G}_n$ is equal to their distance δ.*

Theorem 2.4. *The diameter of $\mathcal{G}_n$ is $n-1$; this is the distance between the complete graph and its complement.*

3. Tree distance

Theorem 3.1. [Z2] *The distance δ_T is a metric on the set of all isomorphism classes of trees with n vertices.*

Theorem 3.2. [Z2] *Let T_1, T_2 be trees with n vertices, let k be an integer, $0 \leqslant k \leqslant n$. Then the following two assertions are equivalent:*
(i) *There exists a tree T with $n+k$ vertices which contains a subtree isomorphic to T_1 and a subtree isomorphic to T_2.*
(ii) *There exists a tree T_0 with $n-k$ vertices such that both T_1 and T_2 contain subtrees isomorphic to T_0.*

Now $\mathcal{G}_n$ will denote the graph whose vertex set is the set of all isomorphism classes of trees with n vertices and in which two vertices are adjacent if and only if their tree distance δ_T is equal to 1.

Theorem 3.3. [Z2] *The distance of arbitrary two vertices in $\mathcal{G}_n$ is equal to their tree distance.*

Theorem 3.4. [Z2] *The diameter of $\mathcal{G}_n$ is $n-3$. There exists exactly one pair of vertices of $\mathcal{G}_n$ whose distance is $n-3$.*

A tree is called a *caterpillar* if by deleting all its terminal vertices a path is obtained.

Theorem 3.5. [Z2] *The set of all isomorphism classes of caterpillars with n vertices induces a subgraph $\tilde{\mathcal{G}}_n$ of $\mathcal{G}_n$ with the property that the distance in $\tilde{\mathcal{G}}_n$ is the same as in $\mathcal{G}_n$. The diameter of $\tilde{\mathcal{G}}_n$ is $n-3$.*

4. Edge rotation distance

Theorem 4.1. [Z4] *Let G_1, G_2 be two graphs with the same number of vertices and the same number of edges. Then*
$$\delta(G_1, G_2) \leqslant \delta_R(G_1, G_2)$$
and the equality may occur.

Theorem 4.2. [Z4] *Let N be a positive integer. Then there exist graphs G_1, G_2 such that*
$$\delta_R(G_1, G_2) - \delta(G_1, G_2) = N.$$

Theorem 4.3. [Z4] *Let T_1, T_2 be two trees with the same number of vertices. Then*
$$\delta_R(T_1, T_2) \leqslant \delta_T(T_1, T_2)$$

Theorem 4.4. [Z4] *Let N be a positive integer. Then there exist trees T_1, T_2 such that*

$$\delta_R(T_1, T_2) - \delta_R(T_1, T_2) = N.$$

Theorem 4.5. [Z4] *Let S be the star with n vertices, let T be an arbitrary tree with n vertices. Let Δ be the maximum degree of a vertex of T. Then*

$$\delta_R(S, T) = n - 1 - \Delta.$$

Theorem 4.6. [Z4] *Let P be a path with n vertices, let T be an arbitrary tree with n vertices. Let $t(T)$ be the number of terminal vertices of T. Then*

$$\delta_R(P, T) = t(T) - 2.$$

Corollary. *Let S be the star with n vertices, let P be a path with n vertices. Then*

$$\delta_R(P, S) = n - 3.$$

5. Edge distance

Theorem 5.1. [Z5] *Let G_1, G_2 be two graphs with the same number of vertices, let their edge distance δ_E be equal to 1. Then a graph isomorphic to G_2 can be obtained from $G + 1$ by adding or deleting one edge.*

In this paragraph the graph $\mathcal{G}_n$ will be the graph whose vertex set is the set of all isomorphism classes of graphs with n vertices and in which two vertices are adjacent if and only if their edge distance δ_E is 1.

Theorem 5.2. [Z5] *The distance of two arbitrary vertices in $\mathcal{G}_n$ is equal to their edge distance δ_E.*

Theorem 5.3. [Z5] *The diameter of the graph $\mathcal{G}_n$ is $\frac{1}{2}n(n-1)$ and the unique pair of vertices of $\mathcal{G}_n$ having their distance equal to $\frac{1}{2}n(n-1)$ consists of the complete graph and its complement.*

A *self-complementary graph* is a graph which is isomorphic to its own complement. These graphs were studied independently by G. Ringel [R1] and H. Sachs [S1]. For the number n of vertices of a self-complementary graph $n \equiv 0 \pmod 4$ or $n \equiv 1 \pmod 4$ always holds. We can define an *almost self-complementary graph* as a graph which is isomorphic to a graph obtained from its complement by adding or deleting one edge.

Theorem 5.4. [Z5] *An almost self-complementary graph with n vertices exists if and only if $n \equiv 2 \pmod 4$ or $n \equiv 3 \pmod 4$.*

Theorem 5.5. [Z5] *If $n \equiv 0 \pmod 4$ or $n \equiv 1 \pmod 4$ then the radius of the graph $\mathcal{G}_n$ is equal to $\frac{1}{4}n(n-1)$ and each class containing a self-complementary graph is its central vertex. If $n \equiv 2 \pmod 4$ or $n \equiv 3 \pmod 4$, then the radius of $\mathcal{G}_n$ is equal to $\frac{1}{4}n(n-1) + \frac{1}{2}$ and each class containing an almost self-complementary graph is its central vertex.*

6. Edge shift distance

This distance will be studied here for trees with n vertices.

Theorem 6.1. [Z6] Let T_1, T_2 be two trees with n vertices, let Δ denote the maximum degree of a vertex. Then

$$\delta_S(T_1, T_2) \geq |\Delta(T_1) - \Delta(T_2)|.$$

Theorem 6.2. [Z6] Let T_1, T_2 be two trees with n vertices, let d denote the diameter. Then

$$\delta_S(T_1, T_2) \geq |d(T_1) - d(T_2)|.$$

Theorem 6.3. [Z6] Let S be the star with n vertices, let T be an arbitrary tree with n vertices. Then

$$\delta_S(S, T) = n - 1 - \Delta(T).$$

Theorem 6.4. [Z6] Let P be the path with n vertices, let T be an arbitrary tree with n vertices. Then

$$\delta_S(P, T) = n - 1 - d(T).$$

Theorem 6.5. [Z6] For any positive integer q there exists a positive integer n and two trees T_1, T_2 with n vertices such that

$$\delta_S(T_1, T_2) - \delta_R(T_1, T_2) = q.$$

Theorem 6.6. [Z6] Let T_1, T_2 be two trees with n vertices. Then

$$\delta_S(T_1, T_2) \leq 2n - 2 - \Delta(T_1) - \Delta(T_2).$$

Theorem 6.7. [Z6] For any positive integer q there exists a positive integer n and two trees T_1, T_2 with n vertices such that

$$\delta_S(T_1, T_2) - \delta_T(T_1, T_2) = q.$$

Theorem 6.8. [Z6] For any positive integer q there exists a positive integer n and two trees T_1, T_2 with n vertices such that

$$\delta_T(T_1, T_2) - \delta_S(T_1, T_2) = q.$$

Corollary. Let T_1, T_2 be two trees with n vertices for $n \geq 4$. Then

$$\delta_S(T_1, T_2) \leq 2n - 7.$$

7. Contraction distance

Theorem 7.1. [Z3] *For any integer $n \geq 6$ there exist graph G_1, G_2 with n vertices such that*

$$\delta_c(G_1, G_2) = \lfloor n/2 \rfloor - 2,$$
$$\delta(G_1, G_2) = 1.$$

Theorem 7.2. [Z3] *For any positive integer p there exist graphs G_1, G_2 such that*

$$\delta_c(G_1, G_2) = p,$$
$$\delta(G_1, G_2) = 2p.$$

Theorem 7.3. [Z3] *For two arbitrary trees T_1, T_2 with n vertices*

$$\delta_c(T_1, T_2) \leq \delta_T(T_1, T_2).$$

Theorem 7.4. [Z3] *Let T_1, T_2 be two trees with n vertices, let d denote the diameter. Then*

$$\delta_c(T_1, T_2) \geq d(T_1) - d(T_2).$$

Theorem 7.5. [Z3] *Let P be the path with n vertices, let T be a tree with n vertices. Then*

$$\delta_c(P, T) = n - 1 - d(T).$$

Theorem 7.6. [Z3] *Let T_1, T_2 be two trees with n vertices, let t denote the number of terminal vertices. Then*

$$\delta_c(T_1, T_2) \geq |t(T_1) - t(T_2)|.$$

Theorem 7.7. [Z3] *Let S be the star with n vertices, let T be a tree with n vertices. Then*

$$\delta_c(S, T) = n - 1 - t(T).$$

References

[B1] V. Baláž, J. Koča, V. Kvasnička, M. Sekanina, *A metric for graphs*, Časop. pěst. mat. **111** (1986), 431–433.

[C1] G. Chartrand, F. Saba, H.-B. Zou, *Edge rotations and distance between graphs*, Čas. pěst. mat. **110** (1985), 87–91.

[J1] M. Johnson, *An ordering of some metrics defined on the space of graphs*, Czech. Math. J. **37** (1987), 75–85.

[K1] F. Kaden, *Graph metrics and distance graphs.*, Graphs and Other Combinatorial Topics, Proc. Symp. Prague 1982 (M. Fiedler, ed.), Teubner, Leipzig, 1983.

[R1] G. Ringel, *Selbstkomplementäre Graphen*, Arch. Math. Basel **14** (1963), 354–358.

[S1] H. Sachs, *Über selbstkomplementäre Graphen*, Publ. Math. Debrecen **9** (1962), 270–288.

[S2] F. Sobik, *One some measures of distance between graphs*, Graphs and Other Combinatorial Topics, Proc. Symp. Prague 1982 (M. Fiedler, ed.), Teubner, Leipzig, 1983.

[V1] В. Г. Визинг, Некоторые нерешенные задачи в теории графов, Успехи мат. наук **23** (1968), 117–134.

[Z1] B. Zelinka, *On a certain distance between isomorphism classes of graphs*, Čas. pěst. mat. **100** (1975), 371–373.

[Z2] _____, *A distance between isomorphism classes of trees*, Czech. Math. J. **33** (1983), 126–130.

[Z3] _____, *Contraction distance between isomorphism classes of Graphs*, Čas. pěst. mat. **115** (1990), 211–216.

[Z4] _____, *Comparison of various distances between isomorphism classes of graphs*, Čas. pěst. mat. **110** (1985), 289–293.

[Z5] _____, *Edge-distance between isomorphism classes of graphs*, Čas. pěst. mat. **112** (1987), 233–237.

[Z6] _____, *Edge shift distance between trees*, Arch. Math. Brno, (submitted).

Bohdan Zelinka
Department of Mathematics,
Technical University, Liberec,
Czechoslovakia

Fourth Czechoslovakian Symposium on
Combinatorics, Graphs and Complexity
J. Nešetřil and M. Fiedler (Editors)
© 1992 Elsevier Science Publishers B.V. All rights reserved.

Domatic Number of a Graph and its Variants

(Extended Abstract)

BOHDAN ZELINKA

1. Introduction

In this lecture we shall treat some numerical invariants of graphs which are related to the concept of domination, namely the domatic number and its variants.

The word "domatic" used here has nothing in common with the same word used in crystallography. It was created from the words "dominating" and "chromatic" in the same way as the word "smog" was composed from the words "smoke" and "fog". This concept is a certain analogy of the chromatic number, but instead of independent sets, dominating sets are used in its definition.

A subset D of the vertex set $V(G)$ of an undirected graphs G is called *dominating* if for each $x \in V(G) - D$ there exists a vertex $y \in D$ adjacent to x. A *domatic partition* of G is a partition of $V(G)$, all of whose classes are dominating sets in G. The maximum number of classes of a domatic partition of G is called the *domatic number of* G and denoted by $d(G)$.

This concept was introduced by E. J. Cockayne and S. T. Hedetniemi in [C1]. The same authors together with R. M. Dawes [C2] have defined also an analogous concept, based on total dominating sets.

A subset D of $V(G)$ is called *total dominating*, if for each $x \in V(G)$ there exists $y \in D$ adjacent to x. The maximum number of classes of a partition of $V(G)$ into total dominating sets is called the *total domatic number of* G and denoted by $d_t(G)$.

In [C3] E. J. Cockayne and S. T. Hedetniemi have defined two variants of these concepts, namely the adomatic and the idomatic number; in [Z1] the total adomatic number was introduced. A dominating (or total dominating) set D in G is called *indivisible*, if it is not a union of two disjoint dominating (or total dominating respectively) sets of G. The minimum number of classes of a partition of $V(G)$ into indivisible dominating (or indivisible total dominating) sets is the *adomatic* (or *total adomatic*, respectively) *number of* G.

The adomatic number of G is denoted by $ad(G)$, the total adomatic number by $ad_t(G)$. The *idomatic number of* G, denoted by $id(G)$, is the maximum number of classes of a partition of $V(G)$ into sets which are simultaneously independent and dominating, in the case when such a partition exists; otherwise it is put $id(G) = 0$.

In [L1] R. Laskar and S. T. Hedetniemi have introduced the *connected domatic number* $d_c(G)$ of a graph G. It is the maximum number of classes of a partition of $V(G)$ into dominating sets which induce connected subgraphs of G.

In[B1] M. Borowiecki and M. Kuzak have defined a *k-dominating set* in a graph G, where k is a positive integer, this is a subset D of $V(G)$ with the property that for each $x \in V(G) - D$ there exists a vertex $y \in D$ such that the distance between x and y in G is at most k. The maximum number of classes of a partition of $V(G)$ into k-dominating sets is the *k-domatic number* $d_k(G)$ of C [Z2].

A *k-ply dominating set* in G is a subset D of $V(G)$ with the property that for each $x \in V(G) - D$ there exist k vertices $y_1, \ldots, y_k$ of D which are adjacent to x. The maximum number of classes of a partition of $V(G)$ into k-ply dominating sets is the *k-ply domatic number* $d^k(G)$ of G [Z3].

A subset D of $V(G)$ is called *complementarily dominating in G*, if for each vertex $x \in V(G) - D$ there exist vertices $y \in D$, $z \in D$ such that y is adjacent to x and z is not. The maximum number of classes of a partition of $V(G)$ into complementarily dominating sets is the *complementarily domatic number* $d_{cp}(G)$ of G [Z4].

Further variants can be obtained by taking edges instead of vertices. A subset D of the edge set $E(G)$ of G is called *dominating*, if for each edge $e \in E(G) - D$ there exists an edge $f \in D$ having a common end vertex with e. Such a set D is called *indivisible*, if it is not a union of two disjoint dominating sets (of edges). The maximum number of classes of a partition of $E(G)$ into dominating sets is the *edge domatic number* $ed(G)$ of G [Z5]. The minimum number of classes of a partition of $E(G)$ into indivisible dominating sets is the *edge adomatic number* $ead(G)$ of G [Z6].

For directed graphs we can introduce semidomatic numbers. Let G be a directed graph with the vertex set $V(G)$. A subset D of $V(G)$ is called *inside-semidominating* (or *outside-semidominating*) in G, if for each vertex $x \in V(G) - D$ there exists a vertex $y \in D$ such that the edge xy (or yx, respectively) is in G. The maximum number of classes of a partition of $V(G)$ into inside-semidominating (or outside-semidominating) sets is the *inside-semidomatic* (or *outside-semidomatic*) *number* of G and is denoted by $d^-(G)$ (or $d^+(G)$, respectively) [Z7]. A set which is simultaneously inside-semidominating and outside-semidominating is called *dominating*, by means of such sets the *domatic number* $d(G)$ of a directed graph G is defined.

Finally, the *antidomatic number* $\bar{d}(G)$ of an undirected graph G is the minimum number of classes of a partition of $V(G)$ into sets, none of which is dominating in G. [Z14].

We shall present some results concerning the above described concepts. Proofs can be found in the quoted references.

2. Domatic number and total domatic number

Theorem 2.1. [Z8] *Let G be a graph with the domatic number $d(G)$ and the total* domatic number $d_t(G)$. Then

$$\left\lfloor \frac{1}{2}d(G)\right\rfloor \leqslant d_t(G) \leqslant d(G).$$

Theorem 2.2. [Z9] *For every non-zero cardinal number a there exists a graph G in which each vertex has degree at least a and whose domatic number is 2.*

Theorem 2.3. [Z10] *For every non-zero cardinal number there exists a graph G in which each vertex has degree at least a and whose total domatic number is 1.*

(In the case of infinite graphs we take the supremum instead of the maximum.)

Theorem 2.4. [Z11] *Let k be a positive integer. Then the graphs of the cubes of dimensions $2^k - 1$ and 2^k have both the domatic number and the total domatic number equal to 2^k.*

3. Adomatic number and total adomatic number

Theorem 3.1. [Z12] *Let G be a disconnected graph without isolated vertices. Then $ad(G) = 2$.*

Theorem 3.2. [Z12] *Let G be a connected graph whose diameter is a least 3. Then $ad(G) = 2$.*

Theorem 3.3. [Z12] *Let a, n be integers such that $2 \leqslant a \leqslant n - 2$ or $2 \leqslant a = n$. Then there exists a connected graph G with n vertices such that $ad(G) = a$.*

Theorem 3.4. [Z1] *The total adomatic number of a complete graph with n vertices, $n \geqslant 2$, is equal to $\lceil n/3 \rceil$.*

4. Idomatic number

Theorem 4.1. [Z12] *Let c, d be positive integers, $2 \leqslant c \leqslant d$. Then there exists a graph G such that $id(G) = c$, $d(G) = d$.*

5. k-domatic number

Theorem 5.1. [Z2] *Let k, m be positive integers, $k \leqslant m$. Let G be a graph. Then $d_k(G) \leqslant d_m(G)$.*

Theorem 5.2. [Z2] *Let G be a connected graph with n vertices, let k be a positive integer. Then $d_k(G) \geqslant \min(n, k+1)$.*

Theorem 5.3. [Z2] *Let G be a path with n vertices, let k be a positive integer. Then $d_k(G) = \min(n, k+1)$.*

Theorem 5.4. [Z2] *Let k, n be two integers, $2 \leqslant k \leqslant n$ Then for each integer m such that $k + 1 \leqslant m \leqslant n$ there exists a tree T_m with n vertices such that $d_k(T_m) = m$.*

6. k-ply domatic number

Theorem 6.1. [Z3] *Let G be a graph, let k be a positive integer. Let $\delta(G)$ be the minimum degree of vertex of G. Then*

$$d^k(G) \leqslant \lfloor \delta(G)/k \rfloor + 1.$$

Theorem 6.2. [Z3] *Let K_n be the complete graph, let k be a positive integer. Then*

$$d^k(K_n) = \lfloor n/k \rfloor.$$

Theorem 6.3. [Z3] *Let $K_{m,n}$ be the complete bipartite graph, let k be a positive integer. Then*

$$\begin{aligned} d^k(K_{m,n}) &= 1 & \text{for } \min(m,n) < k, \\ d^k(K_{m,n}) &= 2 & \text{for } k \leqslant \min(m,n) < 2k, \\ d^k(K_{m,n}) &= \lfloor \min(m,n)/k \rfloor & \text{for } \min(m,n) \geqslant 2k. \end{aligned}$$

7. Connected domatic number

Theorem 7.1. [Z13] *For any y positive integer q there exists a graph G such that $d(G) - d_c(G) = q$.*

Theorem 7.2. [Z13] *Let $n \geqslant 3$ be an integer. For every integer k such that $1 \leqslant k \leqslant n-2$ or $k = n$ there exists a graph G with n vertices such that $d_c(G) = k$. For $k = n-1$ such a graph does not exist.*

Theorem 7.3. [Z13] *Let G be a connected graph, let n be the number of its vertices, let n_0 be the number of its saturated vertices. Then*

$$d_c(G) \leqslant \frac{1}{2}(n + n_0).$$

(A saturated vertex is a vertex adjacent to all others.)

8. Complementarily domatic number

Theorem 8.1. [Z4] *Let G be a disconnected graph. Then*

$$d_{cp}(G) = d(G).$$

Theorem 8.2. [Z4] *Let G be a connected graph with diameter at least 4. Then $d_{cp}(G) \geqslant 2$.*

Theorem 8.3. [Z4] *Let P_n be a path of length n. Then*
$$d_{cp}(P_1) = d_{cp}(P_2) = 1,$$
$$d_{cp}(P_n) = 2 \quad \text{for } n \geqslant 3.$$

Theorem 8.4. [Z4] *Let C_n be a circuit of length n. Then*
$$d_{cp}(C_3) = d_{cp}(C_5) = 1,$$
$$d_{cp}(C_4) = 2,$$
$$d_{cp}(C_n) = 2 \quad \text{for } n \geqslant 7,\ n \not\equiv 0 \pmod{3},$$
$$d_{cp}(C_n) = 3 \quad \text{for } n \geqslant 6,\ n \equiv 0 \pmod{3}.$$

9. Edge domatic number and edge adomatic number

Theorem 9.1. [Z5] *Let G be a graph. Then*
$$\delta(G) \leqslant \mathrm{ed}(G) \leqslant \delta_e(G) + 1,$$
where $\delta(G)$ is the minimum degree of a vertex of G and $\delta_e(G)$ is the minimum degree of an edge of G. These bounds cannot be improved.

(The degree of an edge is the number of edges which have a common end vertex with it.)

Theorem 9.2. [Z5] *Let T be a tree. Then $\mathrm{ed}(T) = \delta_e(T) + 1$.*

Theorem 9.3. [Z6] *Let G be a connected graph which is not a star. Then $\mathrm{ead}(G) \leqslant 3$.*

(The edge adomatic number of a star is equal to its number of edges.)

Theorem 9.4. [Z6] *Let G be a connected bipartite graph on vertex sets A, B with $|A| \geqslant 2$, $|B| \geqslant 2$. Then $\mathrm{ead}(G) = 2$.*

10. Semidomatic numbers

Theorem 10.1. [Z7] *Let d_1, d_2, n be positive integers such that $d_1 \leqslant \frac{1}{2}n$, $d_2 \leqslant \frac{1}{2}n$. Then there exists a tournament T with n vertices such that $d^-(T) = d_1$, $d^+(T) = d_2$*

Theorem 10.2. [Z7] *Let G be a directed graph. Then the following two assertions are equivalent:*
(i) *G contains a factor G_0 which is bipartite and has no sink.*
(ii) *$d^-(G) \geqslant 2$.*

Theorem 10.2'. [Z7] *Let G be a directed graph. Then the following two assertions are equivalent:*
(i) *G contains a factor G_0 which is bipartite and has no source.*
(ii) *$d^+(G) \geqslant 2$.*

Corollary. *If each cycle of a directed graph G has odd length, then*
$$d^-(G) = d^+(G) = 1.$$

11. Antidomatic number

Theorem 11.1. [Z14] *Let G be a disconnected graph. Then $\bar{d}(G) = 2$.*

Theorem 11.2. [Z14] *Let G be a connected graph without saturated vertices, let its diameter be at least 3. Then $\bar{d}(G) = 2$.*

Theorem 11.3. [Z14] *Let G be a connected graph without saturated vertices, let $\delta(G)$ be the minimum degree of a vertex in G. Then*

$$\bar{d}(G) \leqslant \delta(G) + 2.$$

Theorem 11.4. [Z14] *Let G be a graph without saturated vertices, let n be its number of vertices. The antidomatic number of G is equal to n if and only if n is even and G is obtained from the complete graph with n vertices by deleting the edges of a linear factor.*

Theorem 11.5. [Z14] *Let n, k be integers, $n \geqslant 2$. If n is odd, let $2 \leqslant k \leqslant n - 1$; if k is even, let $2 \leqslant k \leqslant n - 2$ or $k = n$. Then there exists a graph G with n vertices such that $\bar{d}(G) = k$.*

Theorem 11.6. [Z14] *Let C_n be a circuit of length n. Then $\bar{d}(C_4) = 4$, $\bar{d}(C_5) = 3$, $\bar{d}(C_n) = 2$ for $n \geqslant 6$.*

Theorem 11.7. [Z14] *Let G be a complete k-partite graph without saturated vertices. Then $\bar{d}(G) = 2k$.*

References

[B1] M. Borowiecki, M. Kuzak, *On the k-stable and k-dominating sets of graphs*, Graphs, Hypergraphs and Block Systems, Proc. Symp. Zielona Góra 1976 (M. Borowiecki, Z. Skupień, L. Szamkolowicz, eds.), Uniw. Zielona Góra, 1976.

[C1] E. J. Cockayne, S. T. Hedetniemi, *Towards a theory of domination of graphs*, Networks **7** (1977), 247–261.

[C2] E. J. Cockayne, *Domination of undirected graphs a survey*, Theory and Applications of Graphs, Proc. Michigan 1976 (Y. Alavi, D. R. Lick, eds.), Springer Verlag, Berlin-Heidelberg-New York, 1978.

[C3] E. J. Cockayne, R. M. Dawes, S. T. Hedetniemi, *Total domination in graphs*, Networks **10** (1980), 211–219.

[L2] R. Laskar, S. T. Hedetniemi, *Connected domination in graphs*, Techn. Report No. 414, Clemson Univ., Clemson South Carolina, 1983.

[Z1] B. Zelinka, *Adomatic and total adomatic numbers of graphs*, Math. Slovaca, (submitted).

[Z2] _____, *On k-domatic numbers of graphs*, Czech. Math. J. **33** (1983), 309–313.

[Z3] _____, *On k-ply domatic numbers of graphs*, Math. Slovaca **34** (1984), 313–318.

[Z4] _____, *Complementarily domatic number of a graph*, Math. Slovaca **38** (1988), 27–32.

[Z5] _____, *Edge-domatic number of a graph*, Czech. Math. J. **33** (1983), 107–110.

[Z6] _____, *Edge-adomatic numbers of graphs*, Čas. pěst. mat., (submitted).

[Z7] _____, *Semidomatic numbers of directed graphs*, Math. Slovaca **34** (1984), 371–374.

[Z8] _____, *Total domatic number of a graph*, Čas. pěst. mat., to appear.

[Z9] _____, *Domatic number and degrees of vertices of a graph*, Math. Slovaca **33** (1983), 145–147.

[Z10] _____, *Total domatic number and degrees of vertices of a graph*, Math. Slovaca **39** (1989), 7–11.

[Z11] _____, *Domatic numbers of cube graphs*, Math. Slovaca **32** (1982), 117–119.

[Z12] _____, *Adomatic and idomatic numbers of graphs*, Math. Slovaca **33** (1983), 99–103.

[Z13] _____, *Connected domatic number of a graph*, Math. Slovaca **36** (1986), 387–391.

[Z14] _____, *Antidomatic number of a graph*, Čas. pěst. mat., (submitted).

Bohdan Zelinka
Department of Mathematics,
Technical University, Liberec,
Czechoslovakia

The Space of Graphs and its Factorizations

ALEXANDER A. ZYKOV

Laziness is one of the most powerful motors of scientific progress. Observing that in the equalities "two flies plus three flies equal five flies" and "two elephants plus three elephants equal five elephants" the numerical result does not depend on the nature of the counted objects, and being lazy to repeat the names, we say and write simply "$2 + 3 = 5$". So the abstract notion of natural number appears and the arithmetical laws in general form are established.

Let us determine the set $\mathbf{M}(G)$ of all proper vertex colorings of a simple graph $G = (X, U)$, considering two colorings as distinct iff there is at least one pair x, y of vertices in G which take two different hues by the first coloring, but one and the same hue by the second. Thus the names of hues play no role, and each clique has a unique coloring.

For any incomplete graph G and each pair x, y of different nonadjacent vertices of G,

$$\mathbf{M}(G) = \mathbf{M}(G \cup \widetilde{xy}) \cup \mathbf{M}(G\langle xy\rangle) \tag{1}$$

and $\mathbf{M}(G \cup \widetilde{xy}) \cap \mathbf{M}(G\langle xy\rangle) = \emptyset$ where $G \cup \widetilde{xy} = (X, U \cup \{\widetilde{xy}\})$, $G\langle xy\rangle$ is obtained from G by identifying x with y (and identifying the superimposed edges). If at least one of the graphs $G \cup \widetilde{xy}$ and $G\langle xy\rangle$ is not complete, we can apply an equality of the type (1) to some other pair of distinct nonadjacent vertices, etc., till the union of $\mathbf{M}$'s of cliques only is obtained. For example,

$$\mathbf{M}\begin{pmatrix} b & c \\ | & | \\ a & d \end{pmatrix} = \mathbf{M}\begin{pmatrix} b & c \\ |\diagup| \\ a & d \end{pmatrix} \cup \mathbf{M}\begin{pmatrix} b & ac \\ \diagup \\ d & \end{pmatrix} =$$
$$= \mathbf{M}\begin{pmatrix} b & c \\ |\times| \\ a & d \end{pmatrix} \cup \mathbf{M}\begin{pmatrix} bd & c \\ \diagdown\diagup \\ a \end{pmatrix} \cup \mathbf{M}\begin{pmatrix} b & ac \\ \diagup \\ d \end{pmatrix} = \cdots$$

But we are lazy to write each time the attribute "$\mathbf{M}(\cdot)$" and to use the sign "$\cup$" which is absent on our typewriter and the simplest text editors, and therefore we write simply:

$$\begin{array}{c}b-c\\|\quad\quad|\\a\quad\quad d\end{array} = \begin{array}{c}b-c\\|\;/\;|\\a\quad\;d\end{array} + \begin{array}{c}b-ac\\/\\d\end{array} = \qquad\qquad\qquad(2)$$

$$= \begin{array}{c}b-c\\|\times|\\a\quad\;d\end{array} + \begin{array}{c}bd-c\\\backslash\;/\\a\end{array} + \begin{array}{c}b-ac\\/\\d\end{array} = \cdots,$$

considering each graph picture as some hieroglyph denoting the set of colorings of this graph. Moreover in order to avoid repetition of similar terms (as e.g. for the last 3-vertex summand in (2)), we apply the recursive equality (1) in the following way: add the new edge $\widetilde{xy}$ to the picture of G and add the summand $G\langle xy\rangle$ as a new picture. In our example the whole process is written as follows:

$$\begin{array}{c}b-c\\|\quad\quad|\\a\quad\quad d\end{array} = \begin{array}{c}b-c\\|\times_2|\\a\;\overset{\scriptscriptstyle 1}{-}\; d\\{\scriptscriptstyle 3}\end{array} + \begin{array}{c}b-ac\\\backslash\;/\\{}_4\;\;\;d\end{array} + \begin{array}{c}bd-c\\\backslash\;/\\d\end{array} + \begin{array}{c}b-c\\\backslash\;/\\a\end{array} + \begin{array}{c}ac\\|\\bd\end{array} \quad(3)$$

where the numbers 1, 2, 3, 4 at the edges indicate the order in which they are added.

The left side of (3) denotes the task, and the right side is a list of all colorings of the given graph without omissions and repetitions. Namely, the first summand corresponds to the coloring using four different colors, the second summand represents such a 3-coloring in which the vertices a and c have one and the same hue, the vertex b has another hue, and the vertex d has a third hue, etc.; the last summand shows the unique 2-coloring. If the addition of an edge and of the corresponding summand are considered as one step of the procedure, the total number of steps turns out to be exactly the same which is required for giving out the result to the customer.

As a positive consequence of our laziness, the thought arises whether "+" in (3) can be considered as some operation on graphs themselves. Being led by this idea, we introduce the *space of graphs* as the abelian group $\mathbf{S} = \{\lambda_1 G_1 + \ldots + \lambda_k G_k : \text{all } \lambda_i \in \mathbb{Z}; +\}$ of all formal linear combinations, with integral coefficients, of simple graphs with the natural addition operation. This group generated by all simple graphs is free, but the system of determining equations

$$G = G \cup \widetilde{xy} + G\langle xy\rangle \qquad\qquad(4)$$

for every incomplete G and each pair x, y of different nonadjacent vertices of G turns $\mathbf{S}$ into the factorspace (factor-group) $\mathbf{G}/(4)$ in which each element can be expressed uniquely (up to a permutation of the summands) as a linear combination of cliques.

Further factorization with respect to the graph isomorphism relation gives the factor-space $(\mathbf{G}/(4))/\simeq$ in which (3) assumes the form

$$\sqcap = K_4 + 3K_3 + K_2,$$

whence the numbers $r_i = r_i(\sqcap)$ of different i-colorings of the given graph are $r_4 = 1$, $r_3 = 3$, $r_2 = 1$, and $r_i = 0$ for all other i's. The space $(\mathbf{G}/(4))/\simeq$ is isomorphic to the additive group of the polynomial ring $\mathbb{Z}[\mathbf{x}]$ in one formal variable $\mathbf{x}$, where $K_n \longleftrightarrow \mathbf{x}^n$.

Another way to obtain some equivalent coloring polynomials consists in starting whith the abelian group $\mathbf{S_x}$ generated by all formal products $\mathbf{x}^m \cdot G$ with non-negative integers m. Evidently $\mathbf{S} = \mathbf{S_x}/(\mathbf{x} = 1)$. Factorization of the space $\mathbf{S_x}$ with respect to the determining equalities

$$G = G \cup \widetilde{xy} + \mathbf{x} \cdot G\langle xy\rangle \tag{5}$$

yields the factor-space $\mathbf{S_x}/(5)$ in which, for example

$$\begin{matrix}b & — & c\\ | & & |\\ a & & d\end{matrix} = |\mathord{\times}_2| + \mathbf{x} \cdot \begin{matrix}b & — & c\\ & \diagup & \\ a \mathrel{\mathrlap{\smash{-}}{\diagdown}} d & & \end{matrix} + \mathbf{x} \cdot \begin{matrix}b & — & ac\\ & \diagup & \\ & d & \end{matrix} + \mathbf{x} \cdot \begin{matrix}bd & — & c\\ \diagdown & & \\ & a & \end{matrix} + \mathbf{x} \cdot \begin{matrix}b & — & c\\ \diagdown & & \diagup\\ & ad & \end{matrix} + \mathbf{x}^2 \cdot \begin{matrix}& ac &\\ & | &\\ & bd &\end{matrix}. \tag{6}$$

Further factorization by the system of equalities $K_m = 1$ leads to the space in which

$$G = 1 + r_{n-1} \cdot \mathbf{x} + r_{n-2} \cdot \mathbf{x}^2 + \ldots + r_\gamma \cdot \mathbf{x}^\gamma,$$

where G is an n-vertex graph with chromatic number $\gamma = \gamma(G)$, and $r_i = r_i(G)$ is the number of its distinct i-colorings $(1 = r_n(G))$.

The summands in (6) which do not contain any clique with at least one numerated edge correspond bijectively to such colorings of G by which for any two distinct used hues there is a pair of adjacent vertices in G having these two hues. If we retain the notation K_m only for an m-clique without numerated edges, and denote by K'_m the m-clique with at least one such edge, the factorization of $\mathbf{S_x}/(5)$ by the system of equalities $K'_m = 0$ gives the factor-space in which

$$\begin{matrix}b & — & c\\ | & & |\\ a & & d\end{matrix} = \mathbf{x} \cdot \begin{matrix}bd & — & c\\ \diagdown & & \\ & a & \end{matrix} + \mathbf{x} \cdot \begin{matrix}b & — & c\\ \diagdown & & \diagup\\ & ad & \end{matrix} + \mathbf{x}^2 \cdot \begin{matrix}& ac &\\ & | &\\ & bd &\end{matrix},$$

i.e. the given graph has two achromatic 3-colorings and one achromatic 2-coloring. Further factorization by the system $K_m = 1$ leads to the space in which

$$\begin{matrix}b & — & c\\ | & & |\\ a & & d\end{matrix} = 2\mathbf{x} + \mathbf{x}^2 \text{ —the achromatic polynomial.}$$

If we factorize $\mathbf{S_x}$ by the relations

$$G = (G \setminus x) + \mathbf{x} \cdot O(G; x),$$
$$K_0 = \emptyset = 1,$$

where $x \in X$ and $O(G; x)$ is the set of vertices adjacent to x in G, we obtain the factor-space in which
$$G = \sum_{i \geqslant 0} f_i(G) \cdot \mathbf{x}^i;$$
here $f_i(G)$ is the number of i-cliques in G. Analogously, by the factorization of $\mathbf{S_x}$ with respect to the system
$$G = (G \setminus x) + \mathbf{x} \cdot \overline{O(\bar{G}; x)}$$
(here $\bar{H}$ is the complementary graph of H) the factor-space is obtained in which
$$G = \sum_{i \geqslant 0} e_i(G) \cdot \mathbf{x}^i,$$
where $e_i(G)$ is the number of i-heaps (independent i-vertex subsets) in G.

References

[1] А. А. Зыков, Основы теории графов, Наука, Москва, 1987.
[2] Alexander A. Zykov, *Fundamentals of Graph Theory*, BCS Associates, Moscow, Idaho USA, 1990.

Alexander A. Zykov
South Center of Acad. Sci. Ukrain. SSR,
Odessa 270 044, Thälmann Lane 6,
USSR

Problems Proposed at the Problem Session of the Prachatice Conference on Graph Theory

OLEG V. BORODIN (NOVOSIBIRSK)

Problem 1. The *entire chromatic number* $\chi_e(G)$ of a plane graph G is the minimal number of colors needed to simultaneously color the vertices, edges and faces of G so that every two neighbour elements receive different colors. In 1973 Kronk and Mitchem for any plane graph G conjectured that $\chi_e(G) \leq \Delta(G) + 4$, where $\Delta(G)$ is the maximal degree of G, and proved that $\chi_e(G) \leq 7$ if $\Delta(G) = 3$. They also observed that $\chi_e(K_4) = 7$.

Recently I confirmed this conjecture for all $\Delta(G) \geq 8$ and proved that moreover $\chi_e(G) \leq \Delta(G) + 2$ for all $\Delta(G) \geq 14$. The last bound whenever true is the best possible as shown by $K_{1,\Delta}$ (the central vertex plus Δ edges plus the infinite face are $\Delta + 2$ mutually neighbour elements).

The problem consists in finding the precise upper bound for $\chi_e(G)$ in the remaining cases: $\Delta(G) \in \{4, 5, \ldots, 13\}$.

Problem 2. Let $e_{i,j}$ be the number of edges joining vertices of degree i with those of degree j. For the plane graphs with the minimum degree 5, the following results were obtained: $e_{5,5} + e_{5,6} > 0$ (Wernicke, 1904), $4e_{5,5} + e_{5,6} \geq 60$ (Grünbaum and Shephard, 1981), and $2\frac{4}{7}e_{5,5} + e_{5,6} \geq 60$ (myself, submitted).

The second coefficient in the two last inequalities is clearly the best possible; the first one cannot be less than $2\frac{1}{7}$ as follows from a construction due to Fisk.

What is the minimal value of $a_{5,5}$ provided that

$$a_{5,5}e_{5,5} + e_{5,6} \geq 60?$$

Problem 3. Given a plane graph, let $f_{i,j,k}$ be the number of triangles whose boundaries contain vertices of degree i, j and k altogether. For the plane graphs with the minimal degree 5, the following results were obtained: $f_{5,5,5} + \frac{2}{3}f_{5,5,6} + \frac{3}{7}f_{5,5,7} + \frac{1}{4}f_{5,5,8} + \frac{1}{9}f_{5,5,9} + \frac{1}{3}f_{5,6,6} + \frac{2}{21}f_{5,6,7} \geq 20$ (Lebesgue, 1940), $\sum_{i+j+k \leq 18} f_{i,j,k} > 0$ (Kotzig, 1963), and $f_{5,5,5} + f_{5,5,6} + f_{5,5,7} + f_{5,6,6} > 0$ (myself, 1989).

No term of the last inequality can be removed without upsetting it as shown by constructions, and we also have $18f_{5,5,5}+9f_{5,5,6}+5f_{5,5,7}+4f_{5,6,6} \geqslant 144$ (submitted). All coefficients of this inequality, except probably the third, are the best possible. What about the third coefficient?

MIECZYSŁAW BOROWIECKI (ZIELONA GÓRA)

Let G be a finite simple graph. Suppose that with each vertex v of G we associate $k \geqslant 1$ distinct colours. The graph G is said to be *k-choosable* if, no matter what colours are associated, we can always make a choice consisting of one colour from each vertex, with distinct colours from adjacent vertices. The *choice number of* G, denoted by $\#(G)$, is equal to k if G is k-choosable but not $(k-1)$-choosable.

Prove or disprove
$$\#(G) \leqslant \eta(G),$$
where $\eta(G)$ is the Hadwiger number of G.

References

[1] P. Erdős, A. L. Rubin, H. Taylor, *Choosability in graphs*, Proc. West. Coast Conference on Combinatorics, Graph Theory, and Computing; Congressus Numerantium **26** (1979), 125–157.

[2] P. Vaderlind, *Choosability in graphs: Some results and open problems. A survey*, Proc. 7th Regional Scientific Session of Mathematicians, Kalsk, September 1988, Graphs, Hypergraphs and Matroids III, WSI Publ., Zielona Góra, 1989, pp. 157–163.

M. DOOB (WINNIPEG)

The spectral radius of a graph is the spectral radius of the 0-1 adjacency matrix. It is easy to show that if an edge is deleted from a graph, the spectral radius gets smaller.

Conjecture. If the edge is contained in only "large" cycles, then deleting the edge causes only a "small" change in the spectral radius. Can this conjecture be proved and made more precise?

D. FRONČEK (OPAVA), R. NEDELA (BANSKÁ BYSTRICA)

Problem of sequently realizable graphs. Let G be a graph without loops and multiple edges. By the neigbourhood of a vertex x in G we mean the subgraph of G induced by the set of all vertices adjacent to x. We say that G is a realization of some graph G_0 if the neigbourhood of any vertex x of G is isomorphic to G_0.

A graph G_0 is called k-sequently realizable (∞-sequently realizable) if there exists a sequence of graphs $G_1, G_2, \ldots, G_k$ (infinite sequence $G_1, G_2, \ldots, G_n, \ldots$) such that G_i is a realization of G_{i-1} for each $i = 1, 2, \ldots, k$ $(i \geqslant 1)$.

Zykov [1] proposed the problem to characterize these graphs for $k = 1$. Our problem is:

Which finite or infinite graphs are k-sequently realizable (for $k > 1$) or ∞-sequently realizable?

References

[1] A. A. Zykov, *Problem 30.*, Theory of graphs and its applicatioons. Proc. Symp. Smolenice 1963 (M. Fiedler, ed.), Prague, 1964, pp. 164–165.

GEŇA HAHN (MONTRÉAL)

1. A (simple) graph is bridged if in every cycle C of length at least 4 there are two vertices x, y whose distance in the graph is strictly less than their distance on the cycle $[d_G(x, y) < d_C(x, y)]$. The paths realizing the shorter distances are called bridges; this generalizes chordality.

Examples.

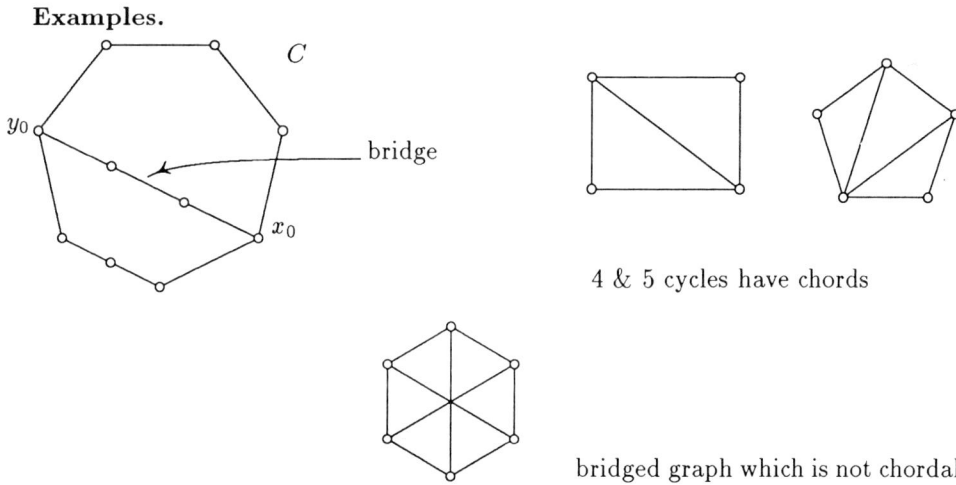

4 & 5 cycles have chords

bridged graph which is not chordal

Problem. Let G be an infinite bridged graph. Is it true that every finite set of vertices in G lies in a finite induced bridged subgraph of G?

Comments.
(i) true if diam $(G) = 2$ (Hahn, Laviolette, Sauer, Woodrow);
(ii) not known even if "induced" is dropped.

References. Hahn, Laviolette, Sauer, Woodrow, *On cop-min graphs*, submitted (in 1988) to Graphs and Combinatorics. Also papers by Anstree & Farber, Farber & Jamison, Farber.

2. There is a function $f: \mathbb{N} \to \mathbb{N}$ with the property that for any $k \in \mathbb{N}$, any complete graph on at least $f(k)$ vertices whose edges are coloured in such a way that each colour is used at most k times contains a Hamilton cycle with all edges of distinct colours.

References. Papers by Hahn; Alspach, Gerson, Hahn, Hell; Hahn, Thomassen.

Theorem. $f(k) \in O(k^3)$. *(Hahn, Thomassen)*

Private communication. Rödl: $f(k) \in O(K^2)$.

Conjecture (Hahn). $f(k) \in O(K)$.

3. (Calgary Problem). Is there a function $\varphi: \mathbb{N} \to \mathbb{N}$ such that every tournament whose vertices are coloured (in any way) with k colours contains a set S of at most $\varphi(k)$ vertices with the property that there is a monochromatic directed path from any vertex $x \notin S$ to some $y \in S$?

Comments.
(i) for $k = 2$, $\varphi(2) = 1$ (Sands, Sauer, Woodrow, JCT(B) 1986);
(ii) if $\varphi(3)$ exists then $\varphi(3) \geq 3$.

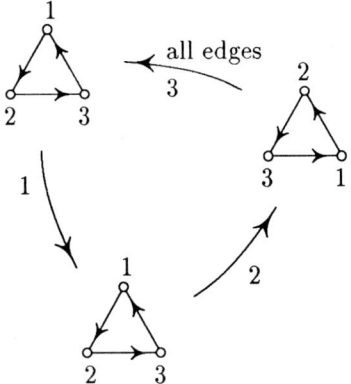

Reference. Sands, Sauer, Woodrow*) (JCT(B) 1986).

---

*) Theorem: Any digraph contains an independent set S which "absorbs by monochromatic paths" if $k = 2$. Works for finite or infinite!

P.S. If the tournament is infinite, is there a *finite* S?

M. Horňák, A. Nagy (Košice)

The k-neighbourhood $N_k(x, G)$ of a vertex x of a graph G is the subgraph of G induced by vertices at the distance k from x. A graph H is said to be $(1,2)$-realizable if there exists a graph such that for any vertex x of G the graphs $N_1(x, G)$, $N_2(x, G)$ and H are isomorphic.

The smallest examples of connected regular $(1,2)$-realizable graphs are C_5 and the cartesian product $K_3 \times K_3$. (The degree must be even). All known examples of regular $(1,2)$-realizable graphs are vertex-transitive. (There is an infinite class of them).

Does there exist a regular $(1,2)$-realizable graph which is not vertex-transitive?

C. A. J. Hurkens (Eindhoven)

Prove or disprove the following conjecture: Let k be a positive integer. Then every simple k-regular graph on $2k+1$ vertices contains, as induced subgraph, a selfcomplementary graph on 5 vertices.

Note 1. For a k-regular graph on $2k + 1$ vertices to exist k should be even.

Note 2. There exist only 2 selfcomplementary graphs on 5 vertices, namely the 5-cycle C_5, and the "bull's head" B:

C_5 \qquad bull's head B

Note 3. There are infinitely many examples of $2k$-regular graphs on $4k+1$ vertices not containing C_5.

Note 4. There are infinitely many examples of $2k$-regular graphs on $4k+1$ vertices not containing the bull's head B.

Note 5. It can be shown that in any minimal counterexample to the conjecture, each edge is contained in at least 2 triangles. As a corollary it follows that any counterexample contains at least 17 vertices.

François Jaeger (Grenoble)

Problem. We consider finite undirected graphs with multiple edges allowed. Write $k \mapsto (k_1, \ldots, k_r)$ for positive integers $k, k_1, \ldots, k_r$ if for every k-edge-connected graph $G = (V, E)$ there exists a partition $\{E_1, \ldots, E_r\}$ of E such that $G_i = (V, E_i)$ is k_i-edge-connected for $i = 1, \ldots, r$. Let $f(k_1, \ldots, k_r)$ be the minimum k such that $k \mapsto (k_1, \ldots, k_r)$. Determine the function f. First unsolved cases: $f(1, 2) \in \{5, 6\}$ and $f(2, 2) \in \{5, 6, 7, 8\}$.

Remarks.

1) $2 \sum_{i=1}^{r} k_i \geq f(k_1, \ldots, k_r)$ because $2r \mapsto \underbrace{(1, \ldots, 1)}_{r \text{ times}}$ by a result of Kundu. In fact $f(\underbrace{1, \ldots, 1}_{r \text{ times}}) = 2r$.

2) Trivially $f(k_1, \ldots, k_r) \geq \sum_{i=1}^{r} k_i$. Is equality possible?

A. V. Kostochka (Novosibirsk), A. F. Sidorenko (Moscow)

The notion of list-colouring was introduced by V. G. Vizing and P. Erdős, A. L. Rubin and H. Taylor.

A list Φ for the graph G with $V(G) = \{v_1, \ldots, v_n\}$ is a couple $\{\Phi(v_i)\}_{i=1}^{4}$ of lists for the vertices, where $\Phi(v_i)$ is a finite set of colours $\{\alpha_{j1}, \ldots, \alpha_{ji}\}$, which are admissible for v_i ($i = 1, \ldots, n$). Let's say that a colouring f of the vertices of G is a list-colouring for (G, Φ) if $f(v_i) \in \Phi(v_i)$ for any $1 \leq i \leq n$ and $f(v_i) \neq f(v_k)$ for any $(v_i, v_j) \in E(G)$. Denote by $F(G, \Phi)$ the number of list-colourings for (G, Φ) and
$$F_G(k) = \min\{F(G, \Phi) : |\Phi(v_1)| = |\Phi(v_2)| = \ldots = |\Phi(v_n)| = k\}.$$

The function $F_G(k)$ is an analogue of the chromatic polynomial $P_G(k)$ of the graph G, and for any chordal graph G it coincides with $P_G(k)$. But sometimes it does not.

Question. Is $F_G(k)$ a polynomial for any G?

Jan Kratochvíl, Jaroslav Nešetřil (Praha)

Intersection graphs of grid line segments and matrices. A 0-1 matrix A is called *good* if there are no six indices i, i_1, i_2, j, j_1, j_2 such that $a_{ij} = 0$, $a_{ij_1} = a_{ij_2} = a_{i_1j} = a_{i_2j} = 1$ and $i_1 < i < i_2$ and $j_1 < j < j_2$ (a "bad" pattern is illustrated in the figure).

$$\begin{pmatrix} & & 1 & & \\ & & \vdots & & \\ \cdots & 1 & \cdots & 0 & \cdots & 1 & \cdots \\ & & \vdots & & \\ & & 1 & & \end{pmatrix}$$

The matrix A is called *r-good* if it is good after a suitable permutation of rows (i.e. if the matrix PA is good for a suitable permutation matrix P), and it is called *rc-good* if it is good after suitable row and/or column permutations (i.e. if PAQ is good for suitable permutation matrices P, Q). We consider problems

R-GOOD
Instance: A $0-1$ matrix A.
Question: Is A r-good?

RC-GOOD
Instance: A $0-1$ matrix A.
Question: Is A rc-good?

and we pose the following question:

Problem. Is R-GOOD solvable in polynomial time?

Comments. The problems considered are tightly related to intersection graphs of grid line segments. These are (bipartite) graphs whose vertices can be represented by segments of the lines of a (rectangular) grid so that two segments cross iff the corresponding vertices are joined by an edge, and any two segments on the same line are disjoint. One can see that a graph $G = (X \cup Y, E)$ where $X = \{x_1, \ldots, x_n\}$, $Y = \{y_1, \ldots, y_m\}$ are its color classes, has a representation by horizontal segments $s(x_i)$ with coordinates h_i and vertical segments $s(y_i)$ with coordinates v_i so that $h_1 < h_2 < \ldots < h_n$ and $v_1 < \ldots < v_m$, if and only if the matrix $A = (a_{ij})$ defined by $a_{ij} = 0$ (resp. 1) if $x_i y_j \notin E$ (resp. $\in E$) is good. It follows that G is an intersection graph of grid line segments iff A is rc-good (this fact was observed independently by A. Z. Zelikovsky and A. Gorpinevich). The problem R-GOOD is thus a question whether a given bipartite graph is the intersection graph of segments with a prescribed ordering of the vertical segments. We hope that this restricted version might be polynomially solvable, while the general problem RC-GOOD is NP-complete [J. Kratochvíl: *A special planar satisfiability problem and some consequences of its NP-completeness*, submitted].

Zelikovsky and Gorpinevich suggested the following construction. Given a $0-1$ matrix A, define a graph G' whose vertices are pairs (i, j) such that $a_{ij} = 0$, and $(i, j)(i', j')$ is an edge iff $a_{ij'} = a_{i'j} = 1$. Zelikovsky proved that A is not rc-good, provided the chromatic number $\chi(G')$ is greater than four, and he conjectured [personal communication] that A is rc-good if $\chi(G') \leqslant 4$.

ZBIGNIEW LONC (WARSAW)

I would like to present a problem on shadows in Boolean lattices.

Let $\mathbf{B}_n$ be the Boolean lattice with n atoms and L_k be the k-th level in $\mathbf{B}_n$. Denote by $\mathcal{P}_t$ the class of partitions of L_k into t almost equal parts, i.e.

$$P_t = \{\{A_1, \ldots, A_t\}: A_1 \cup \cdots \cup A_t = L_k, A_i \cap A_j = \emptyset \text{ for } i \neq j$$
$$\text{and } ||A_j| - |A_i|| \leq 1 \text{ for each } i, j\}.$$

For $\mathcal{A} = \{A_1, \ldots, A_t\} \in \mathcal{P}_t$, let $s(\mathcal{A})$ be the maximum shadow of the sets $A_1, \ldots, A_t$, i.e.

$$s(\mathcal{A}) = \max_{i=1,\ldots,t} |\{b \in L_{k-1}: b \leq a \text{ for some } a \in A_i\}|.$$

Problem. Find $\min_{\mathcal{A} \in \mathcal{P}_t} s(\mathcal{A})$.

I suspect there is little hope of solving this problem completely. A few partial results for small k and for small $n - k$ are known. Any partial solutions of the problem like a nontrivial upper bound or an asymptotic behavior of $\min \mathcal{A} \in \mathcal{P}_t s(\mathcal{A})$ when k and t are fixed would be welcome.

J. PELIKÁN (PRAHA)

Open problem. Let's have an edge valuated graph $G = \{V, E\}$, where $V = \{v_1, v_2, \ldots, v_n\}$ is the set of vertices and E the set of edges.

The edge $(v_i, v_j) \in E$ is valuated by integer h_{ij}. The problem consists in finding a vertex valuation $k(v_i), v_i \in V$, which must satisfy the following conditions:
1) $k(v_i) \in \{h_{ij}; (v_i, v_j) \in E\}$,
2) $|M|$ is minimal, where

$$M = \{(r, s); \text{ there is } (v_i, v_j) \in E \text{ such that } r = k(v_i) \neq s = k(v_j)\}.$$

A. RUCIŃSKI (POZNAŃ)

Let $d_G = \dfrac{e_G}{v_G}$ and $m_G = \max_{H \subseteq G} d_H$, where e_G and v_G stand for the number of edges and vertices of a graph G, resp.

Problem. Given a graph G and an integer r, find

$$m_{cr}(G, r) = \inf\{m_F: F \to (G)_r^1\},$$

where $F \to (G)_r^1$ means that for every r-coloring of the **vertices** of F there is a monochromatic copy of G in F. At the moment it is only known that:
1° For all G and r,

$$\frac{1}{2} \leq \frac{m_{cr}(G, r)}{r \max_{H \subseteq G} d_H} \leq 1.$$

2° For all r and s,
$$m_{cr}(K_s, r) = d_{K_{(s-1)r+1}} = \frac{r}{2}(s-1),$$

3° $\quad \frac{4}{3} \leqslant m_{cr}(P_2, 2) \leqslant \frac{7}{5}.$

HORST SACHS (ILMENAU)

Problem on colouring ball packings. This is a problem of long standing often repeated at problem sessions.

An n-packing B is a finite collection of unit balls in n-dimensional space where any two balls of B are allowed to touch but not to overlap (i.e., the interiors of any two balls of B are disjoint).

Let χ_n denote the minimum number of colours that suffice for colouring the balls in any n-packing B such that any two balls of B which touch must have different colours. It is known that $\chi_2 = 4$ and $5 \leqslant \chi_3 \leqslant 10$. What is χ_n? In particular, what is χ_3?

J. SEDLÁČEK (PRAHA)

Let $U(v)$ be the set of all edges incident with a vertex v of G. A graph G is said to be set-magic if there is a set S and an injection f from $E(g)$ into the collection of all subsets of S so that

$$\bigcup_{e \in U(v)} f(e) = S$$

for every vertex v of G. In full, we say that G is set-magic with respect to S and to f.

It can be easily shown that G is set-magic if and only if G has at most one end-vertex.

Prove or disprove: If G is set-magic with respect to a set S and to an injection f then G is also set-magic with respect to a set S_0 with $|S_0| = |E(G)|$ and to an injection f_0 with

$$\{|f_0(e)|\,;\ e \in E(G)\} = \{1, 2, 3, \ldots, |E(G)|\}.$$

References

[1] J. Sedláček, *On set-magic graphs*, Graphs, Hypergraphs and Block systems. Proc. Zielona Góra 1976, pp. 247–253.

A. F. SIDORENKO (MOSCOW)

Problem. Is it true that for any graph G

$$R(G, K_3) \leqslant e(G) + v(G)?$$

Here R is the Ramsey number, K_3 is the triangle, $e(G)$ and $v(G)$ are the numbers of edges and vertices of G, respectively.

List of Participants

J.	Abrham	Toronto
A.	Ádám	Budapest
B.	Alspach	Burnaby
D.S.	Archdeacon	Burlington
R.	Armann	Merseburg
M.	Bača	Košice
V.	Bálint	Žilina
S.L.	Bezrukov	Moscow
R.	Bodendiek	Kiel
A.	Bondy	Paris
O.V.	Borodin	Novosibirsk
M.	Borowiecki	Zielona Góra
F.	Brenti	Ann Arbor
M.	Bučko	Košice
P.	Bugata	Košice
T.	Bugatová	Košice
J.	Bukor	Bratislava
G.	Burosch	Rostock
P.V.	Ceccherini	Roma
K.	Cechlárová	Košice
D.	Cieslik	Greifswald
Y.	Čorňáková	Košice
P.	Damaschke	Jena
J.H.	Dinitz	Burlington
M.	Doob	Winnipeg
T.	Dvořák	Praha
P.	Erdős	Budapest
P.L.	Erdős	Budapest
P.	Fiala	Praha
J.	Fiamčik	Prešov
M.	Fiedler	Praha
R.	Franek	Hamilton
D.	Fronček	Opava
F.	Göbel	Enschede
H.	Gropp	Heidelberg
F.	Guldan	Bratislava
P.	Gvozdjak	Bratislava
E.	Györi	Budapest
G.	Hahn	Orsay
F.	Harary	Les Cruces
B.	Hartnell	Halifax
A.S.	Hasratjan	Yerevan

I.	Havel	Praha
K.	Heinrich	Burnaby
P.	Hell	Burnaby
P.	Hliněný	Bílovec
C.T.	Hoang	Bonn
F.	Hoffmann	Berlin
J.	Holenda	Plzeň
P.	Horák	Bratislava
M.	Horňák	Košice
P.	Hrnčiar	Banská Bystrica
O.	Hudec	Košice
C.A.J.	Hurkens	Eindhoven
M.	Hužvár	Bratislava
P.	Híc	Trnava
T.	Ibaraki	Kyoto
J.	Ivančo	Košice
F.	Jaeger	Grenoble
S.	Jendroľ	Košice
J.	Jirásek	Košice
F.	Juhász	Budapest
V.	Jurák	Praha
M.	Juvan	Ljubljana
G.O.H.	Katona	Budapest
S.	Klavžar	Ljubljana
M.	Klešč	Košice
M.	Kochol	Bratislava
M.	Koebe	Greifswald
U.	Konieczna	Bydgoszcz
J.	Kópházi	Budapest
A.V.	Kostočka	Novosibirsk
L.	Koval	Bratislava
J.	Kratochvíl	Praha
M.	Kubale	Gdansk
A.	Kundrík	Košice
A.	Kurek	Poznań
J.	Kyppó	Jyváskylá
P.	Kyš	Bratislava
R.	Labahn	Rostock
J-M.	Laborde	Grenoble
M.	Laurent	Paris
C.F.	Laywine	St. Catharines
P.	Liebl	Praha
C.C.	Lindner	Auburn
M.	Loebl	Praha
Z.	Lonc	Warszawa

W.	Mader	Hannover
S.S.	Magliveras	Lincoln
H.M.	Mahmoud	Washington
A.	Malnič	Ljubljana
D.	Marušič	Ljubljana
L.S.	Mel'nikov	Novosibirsk
E.	Mendelsohn	Toronto
D.	Michalak	Zielona Góra
P.	Mihók	Košice
E.C.	Milner	Calgary
B.	Mohar	Ljubljana
M.	Mollard	Grenoble
J.W.	Moon	Edmonton
A.	Nagy	Košice
L.	Nebeský	Praha
R.	Nedela	Banská Bystrica
J.	Nešetřil	Praha
L.	Niepel	Bratislava
J.	Ninčák	Košice
V.	Nýdl	České Budějovice
F.	Olejník	Košice
P.	Ossona de Mendez	Paris
C.	Papadimitriou	Athens
S.	Pavlíková	Bratislava
C.	Payan	Grenoble
J.	Pelikán	Praha
M.	Petrovič	Kragujevac
J.	Plesník	Bratislava
S.	Poljak	Praha
G.	Pruesse	Toronto
V.	Puš	Praha
P.	Rajčáni	Bratislava
A.	Raspaud	Talence
C.	Rauzy	Marseille
G.	Rauzy	Marseille
A.	Récski	Budapest
I.	Rentner	Rostock
D.	Rogers	Aberdeen
A.	Rosa	Hamilton
A.	Ruciński	Poznań
M.	Ruszinko	Budapest
A.	Rycerz	Kraków
Z.	Ryjáček	Plzeň
H.	Sachs	Ilmenau
A.	Sali	Budapest

P.	Scheffler	Berlin
W.	Schöne	Chemnitz
J.	Sedláček	Praha
J.J.	Seidel	Eindhoven
N.	Seifter	Leoben
P.	Sekanina	Brno
K.	Seyffarth	Burnaby
J.	Sheehan	Aberdeen
M.	Sideri	Patras
A.F.	Sidorenko	Moscow
S.K.	Simić	Beograd
Z.	Skupień	Kraków
M.	Sonntag	Freiberg
D.	Sotteau	Orsay
E.	Stöhr	Berlin
M.M.	Sysło	Wrocław
J.	Širáň	Bratislava
M.	Škoviera	Bratislava
L.	Šoltés	Bratislava
M.	Tegze	Praha
M.	Tkáč	Košice
P.	Tomasta	Bratislava
M.	Trenkler	Košice
S.V.	Tsaranov	Moscow
M.	Ungar	Novi Sad
P.	Vacek	Vyškov
V.	Vetchý	Brno
J.	Vinárek	Praha
W.D.	Wallis	Carbondale
J.	Warnke	Rostock
K.	Weber	Warnemünde
M.	Woźniak	Kraków
A.Z.	Zelikovsky	Kishinev
B.	Zelinka	Liberec
Š.	Znám	Bratislava
A.A.	Zykov	Odessa
O.	Zýka	Praha
V.	Železník	Košice

Name Index

Abbott, 79
Abos, 272
Abrham, 1, 3, 278
Afrati, 67
Ajtai, 156, 159
Algor, 173
Alon, 77, 171, 173, 249, 250
Alspach, 210, 211, 378
Anderson, 43
Anstree, 378
Appel, 31, 35
Archdeacon, 5, 7, 11, 35, 134
Ashley, 129, 130, 134

Bača, 13, 16
Bachem, 284
Bacsó, 272
Baláž, 355, 360
Balakrishnan, 327
Bálint, 17
Bannai, 344
Baraev, 87, 90
Barbara, 253
Barbour, 146, 148, 149
Barnes, 154
Barnette, 11
Baum, 281, 284
Behzad, 35, 240, 330
Beineke, 71
Bender, 195, 202
Berge, 192, 202
Berman, 255, 257
Bermond, 278
Bern, 354
Biggs, 139
Blokhuis, 305, 306
Blum, 249
Bodendiek, 23, 24, 27–29, 35
Bollobás, 71, 76, 77, 156, 159, 194–196, 202, 348

Borodin, 31, 36, 375
Borowiecki, 39, 43, 285, 327, 364, 368, 376
Bosák, 1, 4
Bourbaki, 307
Boyd, 20
Brooks, 235, 236
Brouwer, 305–307, 339
Brown, 81, 83
Brückner, 134
Brylawski, 126
Bugata, 45, 49
Bui, 154
Bulitko, 45, 49
Burde, 126
Burns, 348
Burnside, 299
Burosch, 51, 56
Burtin, 156, 159
Buser, 151, 154
Bussemaker, 307, 316, 318, 339
Bydžovský, 85, 86, 90

Cameron, 300, 307, 318, 337, 339
Canfield, 195, 202
Caro, 188, 190
Catlin, 192, 202
Ceccherini, 51, 56
Chang, 72
Chang Chhao, 13
Chao Ko, 78, 79
Chartrand, 35, 240, 327, 356, 360
Chaudhuri, 154
Chen, 255, 256
Cheriton, 62
Chetwynd, 188, 190
Chistyakov, 213, 222
Christofides, 105
Christopher, 12
Chung, 71, 74

Chvátal, 163, 166, 272, 323, 327
Cieslik, 59, 62
Clapham, 309, 313
Clark, 151, 154
Cockayne, 363, 368
Cohen, 307, 339
Colbourn, 228
Cole, 86
Collier, 327
Comtet, 221
Connelly, 81, 83
Conway, 75, 303
Cook, 281, 283, 284
Cooper, 35, 79, 179
Cowell, 111
Coxeter, 20, 135, 302, 307
Crampin, 278
Crowell, 126
Cummings, 86
Cvetković, 318, 319
Dagan, 143
Dawes, 363, 368
DiPaola, 90
Dirac, 74
Djoković, 54, 56
Doelder, 306
Dolev, 272
Doob, 319, 376
Dow, 72
Drake, 72
Du, 62
Dupont, 272
Dvořák, 63
Dyer, 155, 157, 159
Edmonds, 192
El-Arbi, 255, 256, 354
Eldridge, 348
Ellis, 291
Entringer, 151, 154
Erdős, 33, 36, 69–73, 75, 76, 79, 156, 159, 191, 193, 202, 279, 376, 380
Faber, 71
Fajtlowicz, 76

Fan, 259
Fan Chung, 69
Faradzhev, 87, 90
Faragó, 43, 229, 233
Farber, 378
Faudree, 77
Favaron, 43, 187, 190
Feller, 222
Fellows, 248
Fenton, 111
Fiedler, 361, 377
Fisk, 375
Flajolet, 222
Fleischner, 94, 100
Fonlupt, 285
Fort, 228
Foster, 240
Foulds, 159
Fox, 126
Frank, 272
Frankl, 76, 79, 301, 307
Freund, 250
Frieze, 155, 157, 159, 195, 202
Fronček, 81, 376
Fučík, 110, 111
Fujisawa, 291
Füredi, 71, 72, 301, 307
Gaddum, 167, 170
Gallai, 94, 100
Gao, 272
Garcia-Molina, 253
Garey, 62, 67, 154, 159, 166
Gerson, 378
Gilbert, 62, 256
Giles, 282, 285
Gillies, 111
Glagoljev, 155, 156, 159
Goethals, 307, 318
Golomb, 1, 4
Golumbic, 143, 272
Goodey, 333, 334
Goodman, 231, 233
Gorpinevich, 381

Graaf, 306
Graham, 51, 53–56, 62, 67, 69, 159
Gram, 299, 300, 339
Gronau, 304, 307
Gropp, 85, 90
Gross, 11
Grossman, 202
Grötschel, 284
Grünbaum, 7, 11, 34, 36, 113, 116, 129, 134, 294, 295, 375
Guichard, 43
Gutman, 319
Guy, 75
Gvozdjak, 29
Gyárfás, 74, 77
Győri, 191, 202

Haemers, 304, 305, 307
Häggkvist, 77
Hahn, 327, 377, 378
Hajnal, 76–78
Haken, 31, 35
Hakimi, 105
Halberstam, 74
Hales, 43
Halin, 42
Hamilton, 278
Hanan, 62
Hanani, 228
Hansen, 105
Hanson, 79
Hao Wang, 45
Harary, 43, 51, 57, 83, 229, 233, 236
Hartsfield, 10, 11
Havel, 52, 57, 67
Hayasaka, 278
Hedetniemi, 231, 233, 363, 364, 368
Hedlund, 228
Hedrlín, 244
Heffter, 10, 11
Hell, 49, 378
Hermes, 135
Hesse, 86, 90
Heydemann, 43

Higman, 297, 303, 304
Hilton, 188, 190
Hindman, 71
Hirsch, 250
Hobbs, 202
Hoffman, 154
Horňák, 93, 94, 101, 379
Horák, 327
Hsu, 72
Hudec, 103, 105
Hurkens A.J.C., 107
Hurkens C.A.J., 107, 379
Hwang, 62, 255, 256

Imase, 255, 257
Itai, 259
Ito, 179, 344
Ivančo, 113

Jablonski, 155, 159
Jaeger, 94, 101, 117, 380
Jamison, 378
Jendroľ, 7, 11, 129, 135, 333, 334
Jiangang, 43
Jiugiang, 43
Johnson D.S., 62, 154, 166, 250, 272
Johnson P.D., 279
Johnson M., 356, 360
Jonish, 126
Jucovič, 16, 35, 36, 135
Jurák, 137

Kaden, 355, 361
Kahn, 72
Kahr, 45
Kane, 43, 229, 233
Kannan, 272
Kaposi, 111
Kariv, 105
Karoński, 146, 149, 192, 202
Karp, 354
Karplus, 272
Kashiwabara, 291
Katerinis, 44
Kauffman, 126

Kaufmann, 272
Kernighan, 272
Kimura, 179
Kirkman, 135
Knuth, 222
Ko-Wei Lih, 13, 16
Koča, 355, 360
Koebe, 141, 143, 272
Kolchin, 222
Komjáth, 78
Komlós, 156, 159
Konieczna, 145, 149, 157, 159
König, 305, 307
Korshunov, 155, 156, 159
Korte, 272, 284
Kostochka, 36, 151, 155, 156, 158, 159, 380
Kotzig, 1, 4, 31, 33, 34, 37, 116, 278, 279, 375
Kou, 255, 257
Koutsoupias, 249, 250
Král, 111
Krantz, 205, 206
Kratochvíl, 161, 166, 248–250, 301, 307, 380, 381
Krentel, 250
Kronk, 32, 37, 375
Kruskal, 62
Kuh, 273
Kumar, 319
Kundrík, 167
Kundu, 380
Kurek, 171
Kuzak, 364, 368
Kvasnička, 355, 360
Kwasnik, 44

Labbé, 105
Lai, 202
LaPaugh, 272
Larson, 72, 73
Laskar, 364, 368
Laufer, 278
Laviolette, 378

Lawler, 72
Lebesgue, 33, 37, 375
Lee, 167, 169
Leighton, 154
Lenz, 226, 228
Leon, 179
Lesniak, 240
Lick, 236
Lie, 175
Liebeck, 211
Lin, 175, 179
Lindner, 228
Lindström, 279
Lint, 306, 307
Lipson, 122, 124, 125
Livingston, 67
Li Nien, 13
Loebl, 181, 185
Lonc, 187, 190, 381
Longyear, 179
Lovász, 71, 72, 209, 211, 307
Love, 62
Luczak, 191

MacGregor, 154
Mahmoodian, 330
Mahrhold, 156, 157, 159
Mallows, 299, 307
Malnič, 205
Mani, 129, 134, 135
Marek-Sadowska, 272
Markowsky, 255, 257
Marušič, 10, 11, 209, 211
Mathon, 90, 307, 339
Matsuyama, 255, 257
McCanna, 7, 11, 135
McDiarmid, 171, 173
McKee, 11
Medyanik, 94, 101
Megiddo, 105, 249, 250
Mehlhorn, 354
Meir, 20, 213, 222
Mel'nikov, 32, 151
Melkman, 305, 307

Melter, 51, 57
Mendelsohn, 223, 228
Metelka J., 86, 90
Metelka V., 86, 90
Meyer, 43
Michalak, 40, 43, 44, 229
Middendorf, 272
Mihók, 235, 333, 334
Milas, 179
Miller, 354
Millett, 126
Milner, 79
Minieka, 105
Mitchell, 231, 233
Mitchem, 32, 37, 167, 169, 236, 375
Mohanty, 43, 229, 233
Mohar, 154, 205
Moon, 20, 213, 222
Moore, 45, 255
Morris, 62
Morávek, 52, 57, 67
Moser, 20, 135, 307
Mulder, 237, 240
Möhring, 291
Müller, 241, 244

Nagy, 49, 379
Nash-Williams, 192, 309, 313
Nebeský, 237, 240
Nedela, 376
Negami, 7, 11
Nešetřil, 74, 75, 161, 380
Neumaier, 304, 305, 307, 316, 339
Nordhaus, 167, 170
Novák, 86
Nowakowski, 55, 57
Nu, 72
Nýdl, 241, 244

Ore, 31, 37
Orlin, 282, 285
Otter, 222
Oudaise, 130, 135
Owens, 334

Palmer, 229, 233
Pao Chhi-Shou, 13
Papadimitriou, 67, 245, 250, 251
Papageorgiou, 67
Parsons, 205, 206, 210, 211
Paulraja, 327
Payan, 192, 203
Peeters, 105
Pelikán, 382
Pengelley, 10, 11
Persky, 272
Phelps, 90
Pinkus, 305–307
Pinter, 143
Pisier, 74, 75
Plassman, 354
Plesník, 255, 257
Plummer, 31, 34, 37
Poljak, 181, 185, 246, 249, 250
Pollak, 62, 256
Prather, 111
Pratt, 246, 250
Pudlák, 246, 250
Pultr, 244

Rado, 78, 79
Rao, 307, 319, 339
Raspaud, 259
Rayward-Smith, 255, 257, 353
Recski, 261, 271, 272
Reed, 171, 173
Rényi, 191, 202, 222
Richards, 255, 256
Richter, 7, 11, 134
Ringel, 10, 11, 31, 32, 36, 37, 358, 361
Rival, 55, 57
Robertson, 27, 30
Robinson, 201, 203
Rodeh, 259
Rödl, 75, 76, 78
Rogers, 275, 278, 279
Ronghua, 44
Rosa, 1, 4, 90, 228

Rose, 230, 233
Roth, 74
Rothschild, 192, 202
Rousseau, 77
Roy, 94, 100
Rubin, 36, 376, 380
Ruciński, 146, 149, 191, 192, 193, 200, 202, 382
Ruzsa, 74
Rycerz, 281, 285

Saba, 356, 360
Sabidussi, 52, 57
Sachs, 319, 358, 361, 383
Saito, 278
Saks, 249
Sands, 378
Sapozhenko, 155–157, 159, 160
Sauer, 348, 378
Savický, 250
Saxl, 211
Schaar, 331
Schäffer, 250
Scheffler, 287, 291
Schelp, 77
Schmeichel, 327
Schöne, 293, 295
Schrijver, 281, 285
Schumacher, 29, 35
Schuster, 348
Schweikert, 272
Schwenk, 319
Sebö, 272
Sedláček, 13, 16, 383
Seidel, 166, 297, 306, 307, 318, 339, 340
Sekanina, 355, 360
Servatius, 12
Sevast'yanov, 213, 222
Seymour, 27, 30
Shalaby, 223, 228
Shank, 11
Sheehan, 309, 313
Shelah, 78

Shen Hao, 223, 228
Shephard, 7, 11, 36, 113, 116, 129, 134, 375
Shrikhande, 73
Shult, 307, 318
Sideri, 251
Sidon, 73
Sidorenko, 380, 384
Siegel, 272
Simić, 315, 319
Simonovits, 70
Sims, 303, 304
Singhi, 73, 319
Sipser, 154
Skowrońska, 44
Skupień, 285, 321, 327, 368
Sloane, 299, 307
Smith, 62, 255
Sobik, 355, 361
Songlin, 43
Sonntag, 329, 331
Sós, 71, 77, 307
Sotteau, 43
Spence, 303, 307
Spencer, 79, 156, 159, 250, 348
Sreedharan, 295
Stahl, 10, 12
Staten, 76
Stein, 145, 149
Steinitz, 5, 12
Stepanov, 222
Stern, 226, 228
Stewart, 16
Steyaert, 222
Stockmeyer, 154
Stone, 70, 71
Stout, 67
Straus, 43, 229, 233
Stromberg, 222
Stromquist, 134
Strong, 272
Strzyzewski, 271, 272
Sudborough, 291

Sylvester, 321, 323, 327
Szamkolowicz, 368
Szemerédi, 78, 156, 159
Szymanski, 273
Širáň, 24, 27, 29, 327
Škoviera, 155, 160

Takahashi, 255, 257
Tamir, 105
Tarjan, 62
Tatcher, 354
Taylor D.E., 297, 307
Taylor H., 36, 376, 380
Taylor W., 78
Teichert, 331
Thisse, 105
Thistlethwaite, 126
Thomason, 250
Thomassen, 327, 378
Tkáč, 333, 334
Todd, 250
Toft, 37
Toman, 155, 156, 160
Trahtenbrot, 45
Trnková, 244
Trotter L., 281, 284
Trotter W., 71, 74
Truszczyński, 187, 190
Tsaranov, 307, 337, 340
Tucker, 11
Turán, 70, 73
Turgeon, 278
Turner, 291
Turzik, 246, 250
Tuza, 74, 250

Ullman, 272

Vaderlind, 376
Vavasis, 249, 250
Vetchý, 341, 344
Vince, 192, 193, 200, 203
Vitray, 11
Vizing, 33, 37, 57, 74, 355, 361, 380
Vogel, 126

de Vries, 86

Wagner, 23, 24, 27–29, 35, 67
Wallis, 175, 179
Wang, 255, 257
Waxman, 255, 257
Weber, 145, 149, 155–159
Weiss, 213, 222
Welsh, 126
Welzl, 249
Wernicke, 37, 375
Weselowsky, 62
White, 10, 12, 86, 127, 139, 236
Whitney, 124, 127
Whitty, 107, 111
Widmayer, 255–257
Wilbrink, 305, 306
Wilson, 11, 228
Winkler, 51, 53–57
Winter, 255, 257
Witt, 302
Wojda, 348
Woldar, 11
Woodall, 44, 229, 233
Woodrow, 378
Wormald, 195, 201, 203
Woźniak, 345, 348

Yang Hui, 13
Yannakakis, 249–251, 291
Yap, 349
Yoshimura, 273

Zacharias, 86, 91
Zaks, 113, 116, 334
Zamfirescu, 327
Zara, 306
Zelikovsky, 255, 257, 351, 381
Zelinka, 81, 83, 355, 361, 363, 368
Zhuravljov, 155, 156, 160
Zieschang, 126
Zou, 356, 360
Zýka, 161
Zykov, 45, 49, 371, 374, 377

Subject Index

F-factor, 181–183
F-hamiltonian, 325
NP-complete, 162, 164
k-profile, 175
k-tree, 229–231
r-regular pseudograph, 195
abelian group, 372, 373
achromatic number, 167
achromatic polynomial, 373
adomatic number, 363, 367
Alon-Bollobás conjecture, 77
Alon-Bollobás problem, 77
amalgamation, 8, 23, 28
antidomatic number, 364, 368
arboricity, 171
assigned coloring, 32
association scheme, 341, 342, 344
asymmetric digraph, 49
Banach space, 59, 305
bananas surface, 23
binary flowgraph, 107
binary graph, 108
binary tree, 167
binding number, 39, 40, 42, 43, 229
bipartite graph, 78
bipartition, 6, 64
bipartition class, 271
block design, 73
Borsuk-Ulam theorem, 248
bracket polynomial, 125
Brooks' theorem, 235, 236
Brouwer fixpoint, 246, 247
Brouwer's theorem, 246
Bulitko's theorem, 45
cartesian dimension, 53
cartesian product, 52, 239, 327, 379
cartesian sum, 329
caterpillar, 267, 357
Cauchy's theorem, 215

Cayley graph, 10
cell-decomposition, 93
centrally symmetric polyhedra, 134
characteristic polynomial, 318, 342
Chevalley's theorem, 247
chordal graph, 142, 377, 380
chromatic graph, 77
chromatic hypergraph, 72
chromatic index, 74, 188
chromatic number, 35, 70, 71, 77, 78, 167, 205, 265, 363, 373, 375, 381
chromatic polynomial, 380
circle graph, 142
coloring, 32, 33, 72, 86, 164, 167, 265, 371–373, 380, 382, 383
coloring theory, 31
complementary graph, 87, 374
complexity, 103, 156, 161, 181, 245, 256, 269, 290
complexity class, 246
convex polyhedron, 129, 134
convex polytope, 13, 15, 16
coterie, 251, 252
Coxeter group, 301, 302
Coxeter-Dynkin graph, 302
cube, 69, 343, 344, 365
cubic graph, 34, 74, 85, 86, 90, 151, 153, 326
cubic hamiltonian graph, 321
cubical dimension, 63
cubical graph, 63, 64, 66, 155, 158
cycle cover, 259
cycle matroid, 124
cyclic block design, 276
cyclic coloring, 31, 34, 35
cyclic degree, 34
design, 90, 137, 223, 226
difference set, 73, 275–277
distance matrix, 104, 246

distance of graphs, 52
domatic number, 363, 364, 366
domatic partition, 363
edge adomatic number, 364
edge chromatic number, 33
edge coloring, 55, 63
edge distance, 355, 358
edge domatic number, 364
edge rotation distance, 356
edge shift distance, 356
edge-partitions of graphs, 187
edge-reconstructible, 309
Edmonds' theorem, 192
Edmonds-Gallai theorem, 183, 184
eigenvalues, 342
embedding, 5, 6, 8–11, 24–26, 51, 52, 54, 114, 124, 287, 288, 291, 345–347
Erdős-Stone-Simonovits theorem, 76
Euler graph, 299
eulerian graph, 1
extended 4-profile, 175
extremal binding number, 42
extremal problem on graphs, 191

family-packing, 182
Fermat prime, 211
Four Color Problem, 31

Gate-Matrix-Layout-Problem, 287
Goeritz matrix, 125
good family packing, 181
graceful graph, 1
graceful labelling, 276
graceful numbering, 1
graceful valuation, 1
graph isomorphism, 6, 8, 45–47, 52, 77, 81, 89, 125, 129, 131, 161, 165, 188, 210, 211, 235, 241, 243, 310, 322, 345, 355–358, 372, 379
graph metric, 341

Hadamard matrix, 175, 177
Hallian graph, 230, 231
Hamilton cycle, 159, 209–211, 246, 378

Hamilton path, 209, 246
hamiltonian, 63, 201, 209–211, 333
hamiltonian cartesian sum, 330
hamiltonian cycle, 72, 163, 329
hamiltonian path, 162, 321, 329, 331
hamiltonian properties, 210
hamiltonian slipway, 322
hamiltonicity, 210, 211, 325, 329
Hamming distance, 146
harmonious k-coloring, 167
harmonious chromatic number, 167, 169
harmonious partition, 168
Hilbert basis, 282–284
hypercube, 63
hypohamiltonian graph, 321, 323, 327

intersection graph, 141, 142, 205, 206, 264, 267, 268, 381
interval graph, 142, 265, 267, 287
interval supergraph, 287, 288
irreducible embedding, 51, 53
isometric embedding, 52, 53, 55
isometric subgraph, 51, 52
isoperimetric number, 151, 154

Kauffman polynomial, 117, 119, 122, 124–126
Klein's bottle, 7
Kotzig's theorem, 33, 34, 113
Kuratowski graphs, 27, 28, 125
Kuratowski theorem, 27
Kuratowski-type theorem, 24

language, 245
lexicographic product, 329–331
line graph, 211, 299, 315–318
link, 9, 108, 117, 119–121, 125
Lovász's Local Lemma, 248

Möbius strip, 93–97, 99
magic edge labelling, 13
magic labelling, 13–16
Manhattan model, 263, 265
matching, 40, 156, 158, 181, 183–185, 187, 188, 190, 195, 202, 353

matching number, 187
matrix of distances, 103
matroid, 117, 124, 125, 182–184, 192
matroid partition problem, 184
Menger's theorem, 108
metric, 145, 205, 351, 354, 357
metric closure, 351
metric space, 51, 205, 356
metrical closure, 351
metrical equivalent, 353
metrically regular bipartite graph, 342–344
metrically regular graph, 341, 342
mini-max criterion, 103
minimal spanning tree, 59, 60
minimum cost tree, 255
minimum spanning tree, 255, 351, 352
Möbius band, 323

Nash-Williams arboricity theorem, 192
nonhamiltonian, 321, 333, 334
nonhamiltonicity, 325
normal product, 329, 330
nowhere-zero 4-flow, 259
NP, 141
NP-complete, 181, 269, 270, 381
NP-completeness, 246
NP-hard, 61, 103–105, 151, 255, 287, 351

packing, 17, 181, 183–185, 192, 223–226, 345–348, 383
packing algorithm, 18
partition, 104, 152, 153, 167, 175, 182, 187, 201, 230, 232, 239, 251, 275, 341, 364, 380
partition number, 331
permutation graph, 142
Petersen graph, 87, 209–211, 238, 303
planar graph, 31–35, 55, 103, 124, 125, 129, 169
planar matroid, 124, 125
plane graph, 32

plane tree, 42
Platonic polyhedra, 13
Poisson distribution, 213
polyhedron, 5, 8, 129, 133, 282
polytopal graph, 294, 333
polytope, 32, 293, 294
products of hypergraphs, 329
profile, 175, 176
projective polyhedron, 7
projective-planar polyhedra, 5
pseudosurface, 23

Ramsey number, 384
Ramsey's theorem, 75
random cubical graphs, 156
random graph, 77
random induced subgraphs, 155
random matching, 195
random spanning subgraphs, 155, 156
random trees, 213
realizability, 81, 86, 377, 379
realization of a starlike tree, 81
reconstruction, 309, 317, 318
regular complete multipartite graph, 11
regular graph, 1, 85, 121, 151, 164, 194, 195, 201, 202, 303, 379
regular matroid, 125
regular plane graph, 117
regular starlike tree, 81, 83
regular subgraph, 76, 259
regular two-graph, 297, 303–306
Reidemeister moves, 118–120, 123, 124
Reidemeister's theorem, 118
retract of graph, 356
rounding property, 281–283

Seidel switching, 161
self-complementary graph, 358, 379
self-dual, 5–10
selfdual, 129, 133, 134
selfdual convex polyhedron, 129, 130, 134
semidomatic number, 364

sequently realizable graph, 376
signed matroid, 124
simultaneous colorings, 33
Six Color Conjecture, 32
Smith's theorem, 246
space of graphs, 372
spanning subgraphs, 145
spanning tree, 153, 191, 192, 353
spectrum, 306, 338, 342
Sperner's lemma, 246
spherical polyhedron, 5, 6
spherical self-dual polyhedra, 7
spider graph, 141–143
spider representation, 141, 142
spinal distance, 311
square of graph, 341, 344, 345, 347
star arboricity, 171
star-triangle transformation, 122, 124, 125
starlike tree, 81, 82
Steiner minimal tree, 59, 61
Steiner point, 59, 60
Steiner problem, 255, 353
Steiner system, 85, 86, 89
Steiner tree, 141, 261, 263, 351–353
Steiner triple system, 303
strongly regular graph, 343
surface, 5–8, 93, 121

switching, 161–165, 297–299, 301, 302, 306, 338
switching polynomial, 162, 163
symmetric difference, 161
symmetric digraph, 46
symmetric graph, 247
symmetry group, 129, 133, 134
Taylor's formula, 217
three-chromatic graph, 71
three-chromatic., 72
Tits' theorem, 302
Trahtenbrot-Zykov problem, 45
trapezoid graph, 142
tree distance, 355, 357
Tsaranov group, 301, 302
Tucker's lemma, 248
Tutte polynomial, 125
two-chromatic graph, 70
two-graph, 300
Ulam's reconstruction problem, 315
vertex location problem, 103, 104
Vizing's theorem, 33
VLSI routing, 261
VLSI-design, 287
Weyl group, 302
Whitney's 2-isomorphism theorem, 125
Wilson's theorem, 224